Canada's Changing North

Canada's Changing North

Edited and with an Introduction by
WILLIAM C. WONDERS

Revised Edition

McGill-Queen's University Press
Montreal & Kingston • London • Ithaca

© McGill-Queen's University Press
ISBN 0-7735-2590-4 (cloth)
ISBN 0-7735-2640-4 (paper)

Legal deposit third quarter 2003
Bibliothèque nationale du Québec

Printed in Canada on acid-free paper.

First edition published by McClelland and Stewart, 1971.

McGill-Queen's University Press acknowledges the support of the
Canada Council for the Arts for our publishing program. We also
acknowledge the financial support of the Government of Canada
through the Book Publishing Industry Development Program (BPIDP)
for our publishing activities.

National Library of Canada Cataloguing in Publication

Canada's changing North / edited and with an introd. by William
C. Wonders. – Rev. ed.

ISBN 0-7735-2590-4 (bound). – ISBN 0-7735-2640-4 (pbk.)

1. Canada, Northern. I. Wonders, William C., 1924–

FC3956.C327 2003   971.9   C2003–900322–1
F1090.5.C36 2003

Typeset in 10/12 Sabon and Avenir Book. Book design, cartography,
and typesetting by zijn digital.

Contents

Introduction   xi

DEFINING THE NORTH

1   The Arctic Basin and the Arctic: Some Definitions   3
    John E. Sater

2   An Attempt to Regionalize the Canadian North   8
    Louis-Edmond Hamelin

PHYSICAL NATURE

3   Physiographic Regions of Northern Canada   15
    Indian and Northern Affairs Canada

4   Arctic Landforms   18
    J. Ross Mackay

5   How the Mackenzie River Was Made: Translation of a Slavey
    Legend   21
    John Tetso

6   Wind Chill in Northern Canada   24
    M.K. Thomas and D.W. Boyd

7   The Ecology of Snow   34
    W.O. Pruitt, Jr

8   Permafrost Map of Canada   43
    R.J.E. Brown

9   Reconnaissance Vegetation Studies on Western Victoria Island,
    Canadian Arctic Archipelago   50
    S.A. Edlund

10   Climate and Zonal Divisions of the Boreal Forest Formation in
     Eastern Canada   61
     F.K. Hare

11   Organic Terrain and Geomorphology   76
     Norman W. Radforth

     ABORIGINAL PEOPLES
12   The Fragments of Eskimo Prehistory   85
     William E. Taylor, Jr

13   The Inuit Sea Goddess   94
     Darlene Coward Wright

14   The Northern Athapaskans: A Regional Overview   96
     C. Roderick Wilson

15   The Northern Algonquians: A Regional Overview   102
     Jennifer S.H. Brown and C. Roderick Wilson

     HISTORICAL PERSPECTIVES
16   The Identification of Vinland   109
     Alan Cooke

17   Early Geographical Concepts of the Northwest Passage   113
     Theodore E. Layng

18   Voyageurs' Highway: The Geography and Logistics of the
     Canadian Fur Trade   120
     Eric W. Morse

19   Fur Trading Posts in the Mackenzie Region up to 1850   130
     John K. Stager

20   The Sponsors of Canadian Arctic Explorations, 1844–1859: The
     Franklin Search and Rae's Surveys   138
     John E. Caswell

21   Albert Peter Low   153
     F.J. Alcock

22   The North in Canadian History   158
     W.L. Morton

     CHANGE AMONG THE ABORIGINAL PEOPLES/FIRST NATIONS
23   Changing Settlement Patterns among the Mackenzie Eskimos
     [Inuit] of the Canadian North Western Arctic   163
     M.R. Hargrave

24   Changing Patterns of Indian Trapping in the Canadian
     Subarctic   174
     James W. Van Stone

25a Frequency of Traditional Food Use by Three Yukon First Nations Living in Four Communities 190
Eleanor E. Wein and Milton M.R. Freeman

b Native Subsistence Fisheries: A Synthesis of Harvest Studies in Canada 192
Fikret Berkes

26 This Land Is Our Life 193
Jim Bourque

27 Bewildered Hunters in the Twentieth Century 195
Abe Okpik

28 The Great White Hope 200
Ed Struzik

29 Northern Aboriginal Toponymy 207
Randolph Freeman

ECONOMIC RESOURCES
30 Resource Development 213
Robert M. Bone

31 Forty Years of Northern Non-Renewable Natural Resource Development 219
W.W. Nassichuk

32 The Montferré Mining Region Labrador-Ungava 242
Graham Humphrys

33 Gold Hurry!: Hemlo's Golden Giant 251
John Wroe

34 Diamonds under Ice 258
Jamieson Findlay

35 The James Bay Power Project 266
Peter Gorrie

36 The Economic Impact of Northern National Parks (Reserves) and Historic Sites 276
Dick Stanley and Luc Perron

37 Indigenous Tourism Development in the Arctic 281
Claudia Notzke

38 The Porcupine Caribou Herd: Action Alert 283
Canadian Arctic Resources Committee

TRANSPORTATION AND COMMUNICATION
39 Transportation North of 60°N 287
Indian and Northern Affairs Canada

40   Communications North of 60°N   298
Indian and Northern Affairs Canada

NORTHERN SETTLEMENTS
41   A Winter at Fort Norman   307
Maurice R. Cloughley

42   Fermont: A Design for Subarctic Living   310
William O'Mahony

43   Rankin Inlet: From Mining Town to Commercial Centre   314
Stacey J. Neale

44   Arctic Housing Update   318
Gabriella Goliger

POLITICAL CHANGE
45   Integration of Territory into the Administrative Map of
Quebec   325
John Ciaccia

46   Aboriginal Land Claims   328
Parks Canada

47   Tree Line and Politics in Canada's Northwest Territories   332
William C. Wonders

48   Nunavut: Canada's New Arctic Territory   344
William C. Wonders

49   The Quest for Provincial Status in Yukon Territory   348
Steven Smyth

NORTHERN CHALLENGES
50   Davis Inlet: A Community in Crisis   361
Stephen Jewczyk

51   A Critical Look at Sustainable Development in the Canadian
North   364
Frank Duerden

52   The Northern Native Labour Force: A Disadvantaged Work
Force   380
Robert M. Bone and Milford B. Green

53   Regional Perspectives on Twentieth-Century Environmental
Change: Introduction and Examples from Northern
Canada   387
John D. Jacobs and Trevor J. Bell

54   Insights of a Hunter on Recent Climatic Variations in
Nunavut   396
Peter Ernerk

55   Introduction to Mackenzie Basin Impact Study: Summary of
      Results   399
      S.J. Cohen et al.

56   The Ecosystem Approach: Implications for the North   402
      Robert F. Keith

57   The Polar Continental Shelf Project   410
      E.F. Roots

58   Requiem for a Fossil Forest   417
      Ed Struzik

59   Sovereignty, Security, and Surveillance in the Arctic   424
      Canadian Arctic Resources Committee

60   Arctic Sovereignty: Loss by Dereliction?   427
      Donald M. McRae

61   Commentary: The Canadian Polar Commission   441
      Whit Fraser

62   The Arctic Council: Will It Be Relevant?   445
      Chester Reimer

Introduction

Much has changed in the Canadian North since this book was published three decades ago. Indeed, in many ways, it now might be better titled "Canada's Changed North." Yesterday's "old North" of remote isolation, whose inhabitants for the most part led a traditional lifestyle only slightly modified from that of their predecessors, has disappeared. Though some aspects of the old pattern still survive, particularly in the Arctic, they involve fewer numbers of northerners, often only seasonally or on a part-time basis. The changes already observable thirty years ago and noted in the "Introduction" to the original edition, have now resulted in a North which finds itself increasingly involved with many of the same issues (economic, social, political) that concern the rest of Canada. At the same time that it is more and more integrated into the larger framework, it is determined to develop its own approach to such issues. Devolution of political power from the older centres, both federal and provincial, is facilitating this process. Control of the resource base to finance it, however, still remains largely outside the North, and the political fragmentation of the region continues to restrict an integrated regional approach.

Canadians have become more aware of their North in recent years, partly because of at least some introduction to the topic in school and post-secondary curricula, partly because of news media coverage of controversy between "developers" and "conservationists," and partly because of Aboriginal land claims. The former longitudinal perspective of Canada implied in its official motto "A Mare Usque Ad Mare" ("From Sea to Sea") has not yet been modified to "A Mare Usque Ad Mare Usque Ad Mare", but increasingly a latitudinal perspective that includes our third oceanic coast is being recognized. Awareness, however, should be accompanied by knowledge to ensure an informed and sympathetic public attitude towards this final national frontier for the future mutual benefit of northerners and the majority of Canadians living far to the south.

The goals of this present revision remain the same as those of the original edition. It seeks to provide accurate, reliable information on the North to contribute to the knowledge bank of the reader. As before, it also emphasizes the changing nature of Canada's North. The North has not only changed, but continues to change and will continue to do so in the future. In southern Canada it is common for many "old timers" revisiting the scenes of their youth to be amazed at the transformations, with former fields and pastures replaced by freeways and sprawling city suburbs, and with former single-family residential districts replaced by towering apartment blocks and commercial structures. Though there is nothing on the same scale in today's North, the degree of change is comparable, with modern housing replacing colourful but substandard cabins and shacks, and Ski-Doos replacing dog teams, while television, computers, and the Internet are now almost as universal as in southern Canada.

The book, as before, is intended primarily for students and for the general reading public. Accordingly, an attempt has been made to steer a middle course in making the selections between the narrowly specialized articles of the professional scholar and the extremely generalized accounts of the popular writer. Suitable writings on some worthwhile topics have not been found. The diversity of disciplines with an interest in the North adds to the complexity of the problem. It may well be that the book seeks an impossible goal, but it is hoped that it will prove interesting as well as informative.

It is important first of all to define the North since few other geographic regions are subject to such widely differing interpretations. The long-established practice by governments of arbitrarily selecting a particular degree of latitude to define the North still continues despite its artificiality. At the national level, 60°N has long marked the southern boundary of Canada's northern territories, which were until recently (and are still partially) controlled by the federal government. Within the provincial lands to the south, other latitudes have frequently been selected by provincial governments in an attempt to define their "norths" – e.g., 55°N in Alberta (Royal Commission on Northern Alberta, 1958); 53°N in Manitoba for tourism purposes but generally, with a few exceptions, extending as far south as the northern boundary of township 21 (Manitoba Northern Affairs Act, RSM 1988); 55°N in Ontario (Royal Commission on the Northern Environment, 1978);

and 55°N in Quebec (James Bay and Northern Quebec Agreement, 1975). While such boundaries facilitate government administration, they often ignore and complicate transboundary realities. Likewise, many other single criteria may satisfy the needs of one particular group or discipline but not those of others.

John E. Sater summarizes several of the various divisions of the North that have been used based on single criteria. He then sets out the fundamental duality within the enormous area of the North long recognized by geographers and other scientists, with its division into Arctic and Subarctic components. Approximately one third of all of Canada is included within the Arctic and another third within the Subarctic. Even within these fundamental subdivisions there are many regional variations. Until recently it was rare to find both physical and cultural conditions reflected within regional subdivision systems and even more so to have such systems able to be adjustable to changing realities with time. The system devised by Louis-Edmond Hamelin, however, incorporates both these elements and has been widely accepted by geographers and others, including the Government of the Northwest Territories.

Many of the cultural changes in the North are the result of technological changes. These latter may have modified the significance of the physical nature of the area, but they have only modified it and done so at considerable cost. Any understanding of the North must begin with an appreciation of the natural environment, which continues to be a fundamental (some would say "dominant") influence on the area. Elsewhere in Canada one sometimes heard in the recent past that "man had conquered nature," erroneous though this has proved to be, but in the North nature continues to have a formidable presence, refusing to be "conquered" and at best permitting humans' presence in a demanding co-existence. The book's second part sets out some of the most important physical characteristics of the area, which includes more diversity than is often assumed. Research scientists continue to expand our northern knowledge in this regard. This knowledge is enriched by increased appreciation of the traditional legends of the Aboriginal peoples about that same environment, as illustrated by the Slavey legend of the origin of the Mackenzie River.

The Government of Canada defines "Aboriginal peoples" as "the descendants of the original inhabitants of North America. The Canadian Constitution recognizes three groups of Aboriginal people – Indians, Métis people and Inuit. These are three separate peoples with unique heritages, languages, cultural practices and spiritual beliefs" (Department of Indian Affairs and Northern Development, Ottawa, 1997). The third part includes articles on the two most numerous of these groups, the Inuit of the Arctic and the Indians of the Subarctic. In recent years "Inuit" has replaced "Eskimo" as the preferred term, though the latter continues to be used widely by anthropologists and even by some of the Aboriginal people themselves. Similarly, the term "First Nations" has come into common usage still more recently to replace the term "Indian," as in "First Nations people" for the Indian people in Canada, and some now substitute the term for "band" in the names of their communities. The federal government points out, however, that "although the term First Nation is widely used, no legal definition of it exists" (ibid).

Part four provides a survey of some of the historical perspectives on the Canadian North, beginning with the arrival of the first Europeans. The Northwest Passage has been a persistent theme in the area from the earliest explorations down to the present. Much of the early exploration of the North and its first European settlement were also linked with the fur trade. Though many individuals played important roles in expanding our knowledge of the geography of the North, Albert Peter Low was particularly noteworthy. As a late explorer he provided us with accurate knowledge of the vast Labrador area and laid the base for the great post–World War II iron ore developments in that area. He also reflected the increasing concern with the Arctic in heading up the Canadian government's expedition aboard the "Neptune" in 1903. Finally, W.L. Morton suggests that our northern experience has given Canada a distinctive identity in which we should take greater pride.

Arrival of the Europeans, or Whites as they are often identified in the North, resulted in major changes for the Aboriginal peoples. Traditional ways of life were dramatically altered, first by the fur trade and later by the industrial demands of the "outside" world, by government policies, etc., as is pointed out in part five. Some of the repercussions were catastrophic, such as the impact of commercial whaling, which led to the ultimate extinction of the original Inuit population in the Western Arctic. (The present Aboriginal population there, the Inuvialuit, are more recent immigrants from Arctic Alaska, with a high degree of White admixture.) A reliance on fur harvest for the fickle world of fashion and the campaigns of outside "animal rights" lobbies left many northern Aboriginal communities dependent upon government welfare for survival. Yet they depend on the harvest of wildlife for much of their basic food needs. Little wonder that many northern Aboriginal people have been bewildered by the modern world in which they find themselves. Nevertheless, a younger, better educated generation among them is now better qualified to deal with that world. With growing political power through self-government, they are increasingly self-confident and demanding full partnership in the "new North."

Symbolically, the renaming of some communities in the Northwest Territories (e.g., Wha Ti replacing Lac La Martre), in Nunavut (e.g., Iqualuit replacing Frobisher Bay), and in northern Quebec (e.g., Chisabi replacing Fort George and Kuujjuaq replacing Fort Chimo) reflects this attitude.

The North is endowed with major economic resources among which minerals recently have been dominant, though these can be developed only if the value is sufficiently great that such problems as great distances, low temperatures, etc. can be overcome or if the supply is unusually vast. Part six provides examples of these. Furs and gold have been longtime representatives of the former and diamonds the most recent; iron ore, oil, and natural gas have been included among the latter. The Montferré region of Quebec-Labrador illustrates the massive scale of costly development involved in transforming such a potential mineral resource into actual production. Despite huge capital investment, however, mining communities are particularly susceptible to world competition and to decisions made in far off head offices, as well as to inevitable ore exhaustion. Schefferville at the north end of the Labrador Iron Trough was incorporated as a thoroughly modern resource town in 1955. By 1981 it had a population of 4,500 people, but a company decision a year later shut down the mine and town, and its permanent population is now only 280 residents, most of whom are Aboriginals.

Hydroelectric power is another important resource in parts of the North. In recent years massive hydro projects have been put in place in the Subarctic to meet the demand for power from southern regions. These have had severe impacts on northern Aboriginal peoples, resulting from massive dams and reservoirs and diversion of major rivers, with little or no prior consultation with the Aboriginal peoples involved – e.g., in the upper Mackenzie basin in the Northwest, on the Nelson and Churchill Rivers in northern Manitoba, and particularly in the James Bay area of Quebec. The latter project drew national and international attention. In addition to leading to new political institutions for the area (e.g., creation of the Baie-James Municipality for the Subarctic area immediately involved, and the establishment of the Nunavik Region for the Arctic area immediately to the north), the firm opposition of the Cree First Nation was a major deterrent to the full implementation of the project. Other Aboriginal peoples have learned the lesson. When the Churchill Falls was developed for power in Labrador (1966–74), the Aboriginal peoples were not involved. Further development of the downstream hydro power in Newfoundland-Labrador now being planned is dependent upon the full participation of the Aboriginal people there.

Tourism is a relatively new resource in the region. The tourism potential of the North is seen by most Aboriginal peoples as compatible with their traditional values, and increasing numbers are involved in its development. Creation of northern parks and wildlife reserves at times, however, encounters opposition from the mining industry, which fears its exclusion from such areas, and even some Aboriginal people question the relative benefit they might derive from these two competing land uses.

Transportation and communication are absolutely essential, yet among the most difficult problems in the North. Vast distances, small population numbers, and a long severe winter season severely restrict these critical services and contribute to their high costs. Part seven summarizes the nature of these services in the Northwest Territories prior to the creation of Nunavut, but the basic facts remain the same. Of necessity, air transportation is particularly important in the North despite its cost.

Most northern settlements were small and characterized by a relatively slow pace of life and log cabin dwellings well into the mid–twentieth century, as illustrated by the description of an earlier Fort Norman (now Tulita) in part eight. Since then, many have been transformed (in composition if not in size), first to provide modern accommodation for workers from "outside," but now for resident northerners. Newer houses are now little different from those in southern Canada, except perhaps for greater insulation, larger fuel-oil storage tanks, and sanitary waste collection facilities.

Many of the most significant changes in Canada's North in recent decades have involved political change. Some of these have arisen from policy decisions of a provincial government such as that in Quebec, where hydro power development, a wish for the province to establish a stronger presence in its northern territory, and increasing self-confidence among its Aboriginal population and their demand for recognition have combined to produce changed political relationships. Perhaps the greatest political change arose from the 1973 decision of the Supreme Court of Canada's acknowledging the existence of Aboriginal title to Aboriginal traditional lands and the subsequent inclusion of "Aboriginal rights" in the new Canadian Constitution. Since then the federal government has been engaged in settling comprehensive claims throughout the territories and attempting to deal with other specific claims across the nation, while many provincial governments now also find themselves deeply involved. In the Northwest Territories such previously "academic" matters as the definition of "tree line" took on practical political importance as seen in part nine.

The federal government has also been following a process of devolution of authority from Ottawa to the territories in recent years.

Northerners now have a much larger direct say within the region and have even forced the creation of a new territory, Nunavut. Though many programs are now administered entirely by territorial legislatures or on a shared basis with the federal government, control over the essential land resource base still has not been transferred, and Yukon's dream of provincial status still remains a dream.

Many challenges remain for Canada's North, reflecting the impact of change. Several of these are outlined in the final section. The plight of Davis Inlet in Labrador demonstrates that even with the best of intentions, government decisions sometimes can create more problems than they solve. "Sustainable development" is a widespread goal in the North, but is it really achievable? Despite the availability of a much improved education system in recent decades, a large number of Aboriginal students drop out early, and the northern Aboriginal labour force was still considered disadvantaged in the 1980s. Rates of substance abuse and of suicide continue to be well above those "outside" among youthful northern Aboriginals disillusioned by the very limited employment opportunities even if they have achieved a higher level of formal education. Employment by territorial governments assists, but it does not solve the problem. (In 1989 the Tungavik Federation of Nunavut estimated that half the population received social assistance some time during the year, and in some cases the proportion exceeded 90 per cent of the community.)

Even the physical character of the North is undergoing change, as has been demonstrated by recent and current climatic change or variation. This raises such fundamental questions as its impact on permafrost, on vegetation, on animal populations such as muskoxen and caribou, on sea ice conditions, on higher sea levels, etc. Scientific research into the essential nature of the North, such as the outstanding work of the Polar Continental Shelf Project in the High Arctic, is increasingly essential in this regard. The question of Canadian sovereignty in the Arctic continues so far as the United States is concerned, which is cause for unease among some. Nevertheless, the end of the Cold War has made for a new era of international cooperation in which the Canadian North is playing a leading role, which promises still more changes for the future.

I wish to acknowledge the reprint permission of the original publishers and/or authors of the journals and books from which these selections have been made. For economy I have eliminated many of the footnotes in the original text except for explanatory notes and direct quotations. For the same reason, original maps and diagrams have been reduced to an absolute minimum. It has not been possible to include photographs that in some cases illustrated the original articles. In a few cases there have been changes since the original articles were written, but they have not reduced the articles' importance in the over-all picture. Very brief factual notes have been added where it was felt absolutely necessary.

The proposal and early encouragement for a new edition of *Canada's Changing North* originated with Professor Jeremy Mouat of Athabasca University and has been followed by most useful suggestions through-out from his colleague Professor Johanne Kristjanson of the same uni-versity. Margaret Swanson, university librarian at the University of Victoria, has kindly provided full access to that university's library resources in my retirement years, and several return visits to the rich collections of the University of Alberta libraries, particularly those of the Canadian Circumpolar Institute, have made it possible to search a wide spectrum of northern journals in revising this work. Joanne Noel, research librarian in the Departmental Library of Indian and Northern Affairs Canada, has been very helpful in steering reprint permissions through the labyrinth of federal government bodies involved. Louise Maffett, executive director of the Royal Canadian Geographical Soci-ety, has kindly assisted in seeking out some elusive material for publi-cation.

Finally, the appearance of this new edition has been made possible through the enthusiasm and support of Philip J. Cercone, executive director and editor-in-chief of McGill-Queen's University Press, while his assistant, Margaret Levey, and the coordinating editor, Joan McGil-vray, have dealt with the numerous practical matters involved in an efficient and prompt manner. I thank them for their valued assistance.

William C. Wonders
Victoria, British Columbia
8 October 2002

DEFINING THE NORTH

1 The Arctic Basin and the Arctic: Some Definitions

JOHN E. SATER

Considerable misunderstanding still exists about the portion of the earth that we call the Arctic. If some of this misunderstanding stems from an imprecise terminology, the reason is that the well-known terms from the past do not suffice to express the subtleties that are now known to characterize the region. Certain terms that are commonly used are unhesitatingly translated by many into a few, rather limited – and sometimes wrong – ideas. Among these terms now rendered inexplicit by an increase in knowledge is the word "arctic" itself.

Within a particular discipline, its use as an adjective allows the delineation of a precise boundary concerning that subject, but to define it as a noun, or as an adjective for more general application, by conglomerating the various boundaries of the many disciplines would only be an exercise in confusion. Consequently the reader of these pages will find it more meaningful if he views the term "arctic" as a group of concepts and attributes that is not yet closed to amendment – as a word in the process of being redefined, and certainly not as a territory with precise boundaries.

Wherever its boundaries, the Arctic is partially obscured by its distance from use, but more by its inhospitable environment that has limited the numbers of those who could experience it. Only within the past few decades have there been concerted efforts to learn exactly the processes and elements that make the region what it is. These efforts have had a depth and success not previously achieved primarily because

SOURCE: John E. Sater, *The Arctic Basin* (rev. ed., Washington: Arctic Institute of North America, 1969), 1–4. Reprinted by permission of the author and publisher.

modern technology and geopolitical interests have offered a technical capacity and stimulus that had been lacking before.

As with many other areas of inquiry, the Arctic is a subject in which the more that is learned, the more, it is realized, there is to learn. When first encountered by Western man, a single simple term sufficed to describe it. The Greek word for bear – *arktos* – was an ample designation for the region in a period when men were conscious of the constellations, and the Great Bear, or Big Dipper, was a well-known sign in the northern sky. The early visits were probably limited to distant views of glaciers and snow-covered mountains or cautious probings of the southern edges of ice-filled seas, but their accomplishment became part of our history and the memory remained. Gradually more people penetrated more deeply into the region, some even settling there, but their contacts were still few and served to illuminate only small areas or locales. The rise of astronomy and cartography eventually led to the defining of an Arctic Circle: the line above which the sun would not rise on the winter solstice nor set on the summer solstice. The definition is valid under most circumstances, and at the North Pole the sun does not appear for nearly three months at mid-winter. However, because of the manner in which the atmosphere refracts light, the sun may be seen even when it is slightly below the horizon, and on some occasions, overcast excluded, will not be visible even when it is above the horizon.

Usually three expressions are intertwined – arctic, polar, and high latitude. They have come to be used interchangeably in conversations about the northern portion of our globe, but always with a lack of precision. Literally, the Arctic is the area north of 66°30'N, but the inappropriateness of the terms of definition weakens its utility. Furthermore, the area is not best known for the variation of the hours of daylight and darkness but for climatic conditions that are little related to the precision of an astronomic definition. "Polar" is less exact, therefore more mutable, but it may be related to many locations. The geographic, geomagnetic, dip, and cold poles and the pole of inaccessibility are all facets of the north polar region but each has its counterpart in the Southern Hemisphere. "High latitude" is convenient, because of its inexactness, but presumably supplements middle and low latitude while retaining the inflexibility of all mathematical divisions of the earth's surface. Assuming that it denotes the area north (or south) of 60°N (or S), it does not add or include sufficient distinction to warrant differentiating it from arctic. All these expressions convey the same approximate area and help to convey meaning to an idea, but each is unsuited or inadequate to serve as an all-inclusive definition, or designation ...

Oceanographers and marine biologists are concerned with the seas and the life therein, and their definition of arctic is therefore limited and applicable only to the seas. They define arctic water as having a temperature at or near 0°C and a salinity of approximately 30‰. The surface water of the Arctic Ocean exhibits these characteristics as do portions of the adjacent seas. These quantities may be established with accuracy and their limits drawn on a map. The Arctic Basin itself may be drawn on a map, for it is a structural feature of the earth's crust that is filled with a portion of the world's ocean ...

On the land, the occurrence of permafrost gives the geologists, and others, a significant and definite means of establishing an Arctic limit. This phenomenon is characterized by the presence of perennially frozen ground a decimeter to a meter or so below the surface, which may extend to depths of hundreds of meters. In more northerly locations, this phenomenon occurs beneath virtually all the land surface, while farther south it is interrupted by areas of unfrozen ground until a southern limit is reached beyond which it is not found. Permafrost is of interest to botanists because the roots of plants that grow in the ground above it cannot penetrate into it, nor can water. As a result, the root systems of plants are restricted in size and the ground may become very moist in summer. It is of interest and concern to engineers because it has tremendous significance to anything constructed above or in it.

As a result of the behavior of the earth's magnetic field, the aurora borealis is seen as a colourful and unique attribute of the northern regions. The field also creates problems in the use of magnetic compasses in the region and may severely disrupt radio communications. Physicists therefore may delineate an Arctic region in which there are severe magnetic disturbances, auroral displays, and significant and prolonged radio blackouts.

The most useful definitions of arctic, in terms of their meaningfulness to man, are those based on climatic factors. The weakness of these definitions is their number and the fact that no one of them has been accepted by all concerned. The 50°F [10°C] isotherm, and the 43°F [6°C] isotherm, the mean winter temperature in relation to summer means, and other parameters have been suggested but not agreed to. Further, the condition most widely associated with the term arctic – cold – is most applicable in terms of its duration rather than its intensity. Thus while the Arctic is cold longer than in the lower-latitude regions, the areas of extreme cold in the Northern Hemisphere are located in the Subarctic.

Tree line is probably the most useful definition of arctic for it is related to the climate and adds a clear visual element that all can see. As a result of the climatic, and soil, conditions there is a northern limit of

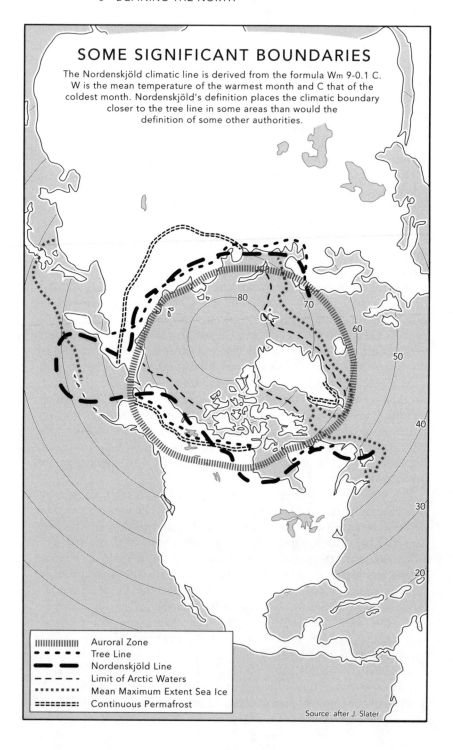

SOME SIGNIFICANT BOUNDARIES

The Nordenskjöld climatic line is derived from the formula Wm 9-0.1 C. W is the mean temperature of the warmest month and C that of the coldest month. Nordenskjöld's definition places the climatic boundary closer to the tree line in some areas than would the definition of some other authorities.

‖‖‖‖‖‖‖‖‖‖	Auroral Zone
▪ ▪ ▪ ▪ ▪ ▪	Tree Line
▬ ▬ ▬ ▬	Nordenskjöld Line
‑ ‑ ‑ ‑ ‑	Limit of Arctic Waters
▪▪▪▪▪▪▪▪▪▪	Mean Maximum Extent Sea Ice
════════	Continuous Permafrost

Source: after J. Slater

trees, and treeless mountains and tundra are the lands that are usually thought of as "the Arctic." Unfortunately, there are disadvantages, too, in using tree line as the basis of a definition. Not only is it liable to change in time, but also it must be subdivided in a manner similar to the oceanographer's high-, mean-, and low-tide lines. Thus there is a northern limit of continuous forest, a northern limit of erect trees, and a northern limit of species.

A concept as complex and extreme as the Arctic necessitates establishing another region wherein the various circumstances are mitigated and grade into those of the temperate regions. Thus has come into use the term and concept Subarctic: a vast forested area of climatic transition. But definitions require limits or boundaries and boundaries separate those things that are distinct. The profusion of conditions that are not coincident with each other but which have Arctic aspects virtually precludes the meaningful definition of an "Arctic." However, it is useful to list the factors that in sum make the Arctic the distinct region that it is for it is these factors that give meaning to the concept and the word.

The Arctic:
is located in the higher latitudes of the Northern Hemisphere;
receives less solar radiation and, hence, surface warming because of the oblique angle of incident solar radiation;
is cold longer than temperate regions;
is distinguished by the absence of trees;
is characterized by the occurrence of permafrost under its land surface;
consists mainly of an ice-covered ocean;
experiences a wide range in the duration of daylight and darkness;
receives no more precipitation than most deserts but has many lakes and rivers which may freeze solid;
is very sparsely settled;
lies at considerable distance from the centres of population;
lies between major population centres of the present day;
is not now of economic importance because its resources have only just become economically exploitable.

2 An Attempt to Regionalize the Canadian North

LOUIS-EDMOND HAMELIN

The regionalization of the Canadian North is a question which has been approached before. The nordic area is as varied as temperate countries; therefore boreal Canada, occupying half a continent, cannot be considered as a homogeneous region.

Types of Regionalization

The partition of the North can be envisaged from several different angles. Firstly it can be seen longitudinally (that is vertically), in this way recalling that the exploited North is merely a continuation of the south. From west to east, these projections are made around fundamental axes: (1) the Pacific coast; (2) Central Alberta-Columbia-Yukon-Alaska; (3) Alberta-Mackenzie basin-west Arctic; (4) Winnipeg, Manitoba-Hudson Bay; (5) Saint Lawrence Valley-Quebec peninsular Labrador-east Arctic.

On the other hand a detailed economic study would specify that, north of the very irregular section linked by road or rail, there are hundreds of isolated hamlets forming as many miniscule independent regions which have no communication, and which are virtually uninhabited.

In a country like Canada – strongly characterized by its wide seasonal temperature range – a complete regionalization would necessitate distinguishing between a summer north and a winter north; in the cold

SOURCE: North 11, no.4 (July-Aug. 1964): 16–19. Reprinted by permission of the author and publisher.

season Schefferville has some characteristics of the Far North, where-as in July its climate is "temperate." In the whole boreal world, not only is the winter north very different from the summer north, but the latter varies far more latitudinally than the former.

A fourth type of regionalization, the one we shall be considering here, would be concerned mainly with zones; that is, it would aim to divide the North latitudinally. In this way a near-north would evidently be distinguished from a distant-north. In fact, the division of the North into zones is not as simple as this might lead one to believe.

Indices of Latitudinal Regionalization

This type of regionalization uses the average values of the data for both the cold and the hot season. For instance, the Schefferville region, with its very cold winters and its merely cool summers, has some pockets of permafrost. These patches of frozen soil, associated with both the snow and the vegetation, become one of the characteristics of the "mean annual North" of Schefferville.

Other writers have dealt with the problem of latitudinal regionalization by adopting one specific criterion and considering it to be representative of the whole. Koppen used the 50°F (10°C) isotherm for the hottest month. Nordenskjold chose the 5°C isotherm as the southern limit of an "Inner Arctic" (Haut-Arctique). Still, from the point of view of temperature, the Subarctic zone is generally limited in the south by at least the 50°F (10°C) isotherm for a period of three or four months. Other writers have based their northern limits on frontiers – principally of vegetation; in these cases the tundra defines the Arctic. Research workers emphasizing human phenomena take into consideration the type of agriculture, the population distribution, and the means of transport. Even the telephone land-lines have been used to determine the southern frontier of the Canadian North. In the USSR, the rate of increase in the cost of exploitation as one moved north served to delineate the regions of Siberia. By using a single criterion, or by emphasizing one principal criterion, these different writers are unable, at least in theory, to define the North as perfectly as they would if they took into consideration several different elements.

To help us understand as well as possible the complex notion of the North, we have worked out a well-balanced global index.[1] We have retained ten standards: (1) latitude; (2) permafrost; (3) number of days above 42°F (5°5C); (4) negative thermal index from 65°F (18°3C); (5)

---

[1] The idea occurred to us following student seminars on Arctic frontiers given by Pierre Houde in 1959, and Hughes Morrisette in December, 1960. We made direct use of the manuscript of our assistant, Miss Cynthia Wilson, "Southern Limits of the Canadian North," (1962) 6 p., 28 c., bibl.

length of freeze up; (6) vegetation; (7) communications; (8) population (native and white); (9) exploitation of resources; (10) cost of goods. The ten categories chosen are intended to represent both physical and human phenomena; however, we have not included any psychological elements. The heterogeneous nature of these criteria means that the total score of values is not dependent on one single aspect, and we take into consideration duration, nature, frequency or mere quantities. For each section we have allowed a maximum value of 100 and a minimum of zero; for instance, latitude scores 100 at the pole and zero at the 47th parallel. For Schefferville, the values of each criteria are respectively: 20, 50, 50, 60, 50, 40, 30, 60, 50, 30, which makes a total of 440. Plotted on a map, these different number-indices have enabled us to make a quantitative differentiation of the Canadian North. We can see by how much James Bay is less nordic than the gulf of Amundsen; similar comparisons of the east with the west are equally possible (for instance, Schefferville with Edmonton). Finally, the distances between the "isonordic lines" enables us to establish a gradient for the North.

Latitudinal Regionalization in Canada

Comprehensive calculations of the global index enable us to distinguish four latitudinal norths in Canada. Each is situated beyond a southern base, which we will consider first.

These southern nuclei in Canada are related to the exploitation of inexhaustible resources – especially non-marginal agriculture – to a stable population, and to a multiple communications network throughout the year. This southern base extends over less than 10 per cent of Canada,[2] but takes in 91 per cent of the country's population. The Canadian Shield, the Cordillera, and the Appalachians partly account for the discontinuity and limited area of these patches of settlement. Yet, this Canada, reputedly temperate, is not completely free of nordic characteristics; indeed winter can cause the annual nordic index to be as high as 100.

Along this southern belt, there stretches from east to west a Near North, (or Lower North, "Pseudo" North, or Little North). This stepping stone to the North has an average index of 250; it encompasses the majority of Newfoundland, passes through Sept-Îles and includes Lake St John, Abitibi, the iron bridge north of Lake Superior, the Peace River, and it ends at Kitimat. This Near North – a land of raw materials, forestry, and mining (but with little agriculture) – has a permanent population of more than a million. It is the economic suburb of the

---

[2] These and the subsequent calculations were very kindly made by Benoit Dumont, an agronomist for the provincial government.

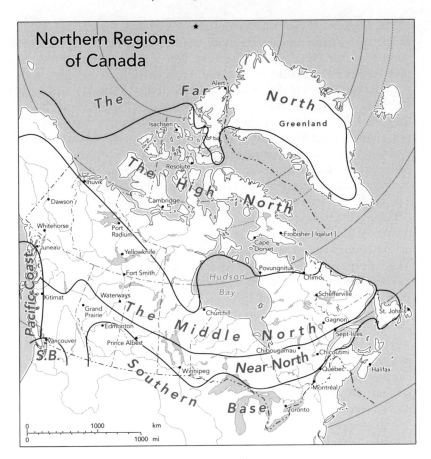

Northern Regions of Canada

The Far North

Greenland

The High North

The Middle North

Near North

Southern Base

S.B.

Pacific Coast

Alert
Isachsen
Resolute
Inuvik
Dawson
Cambridge
Whitehorse
Port Radium
Juneau
Yellowknife
Fort Smith
Kitimat
Waterways
Grand Prairie
Edmonton
Vancouver
Prince Albert
Winnipeg
Churchill
Hudson Bay
Frobisher [Iqaluit]
Cape Dorset
Povungnituk
Chimo
Schefferville
Gagnon
Sept-Isles
Chibougamau
Chicoutimi
Québec
Montréal
Toronto
St. John
Halifax

0   1000   km
0   1000 mi

southern bases with which it has many communications both from north to south and from east to west.

Above this extends the largest nordic zone in Canada – the Middle North (or Mid-North, Hemi-North, or peri-Arctic); this belt stretches from Labrador to the Yukon, taking in Hudson Bay. Climatically, it is mainly a Subarctic region. Economically, it is the domain of the pioneers' "incursions," with no communications latitudinally. The mean global index of this North varies between 400 and 500. The lines of communications, arranged vertically, are worthy of comment: rail from Schefferville and Churchill, road from Alaska, and the river route of the Mackenzie. This region, of 100,000 inhabitants, marks the end of the continuous extension northwards of the southern bases.[3]

3 These bases plus their extensions into the Near North and the Middle North make up the inhabited area of Canada (useful Canada to A. Siegfried; southern Canada to others).

The High North has a mean global index of 750. It is the beginning of the true North, and differs more from the Middle North than the latter from the Near North. The High North extends from the channels between the mainland and the Arctic islands; it includes both peninsulas (Ungava and Keewatin), islands (Baffin, Victoria), and some stretches of water (Hudson Strait, Foxe "basin," Parry Straits). Including the Eskimos [Inuit], only 20,000 people live in this fourth section of Canada's northern territory. The winter-summer range is very wide. A desert with oases of settlement, this country is economically very deficient. Government expenditure (defence, research, administration, equipment) is far greater than the amount of private capital invested. The High North seems like the "Government's North."

Finally, the Far North, with a nordic index of 900, is a country permanently frozen, both on the sea (the pack ice in the Canadian sector of the Arctic Ocean) and on the land (glaciers, including Ellesmere). This empty region nevertheless covers 10 per cent of Canada.

These four nordic regions are divided by a transition zone, rather than by a single line, and each includes some extraneous elements according to local changes in the altitude, the resources and the people. Together the four belts make up the Canadian North.

If this global index were applied to other countries in the boreal world, the majority of Greenland and Spitzbergen would belong to the Far North, whereas the periphery of Greenland, as well as a section of Iceland, would be in the High North. The southern two-thirds of Alaska would be in the Middle North.

This present index, the basis of a latitudinal regionalization of the Canadian North, is the outcome of only one piece of research; the author feels it would be necessary to "sophisticate" the mathematical side of it. However imperfect it may be, it does help us understand the differential nordic intensity of Canada. Other indices, based on a different combination of criteria, are equally possible.

PHYSICAL NATURE

3 Physiographic Regions of Northern Canada

INDIAN AND NORTHERN AFFAIRS CANADA

Northern Canada may be divided into five major geologically determined physiographic regions (see figure on following page). The most extensive is the Canadian Shield, which comprises much of the NWT mainland, and the southeastern Arctic Islands. The terrain is rocky and hilly, with numerous lakes. Structurally, the Shield consists of a series of broad arches and basins, which once formed more dramatic relief but are now eroded to a nearly level surface. Ancient igneous and metamorphic rocks predominate throughout the Canadian Shield, although sandstone plains occur in Central Keewatin [Kivalliq]. Younger sedimentary rocks are preserved within broad shallow basins, such as Hudson Bay and Foxe Basin.

The Inuitian Region includes most of the Queen Elizabeth Islands. Here the sedimentary rocks are gently to steeply folded. Faulting and erosion have separated the islands into a number of groups. The broad channel forming the Northwest Passage, known in its various parts as M'Clure Strait, Viscount Melville Sound, Barrow Strait, and Lancaster Sound, lies along such a major structural line.

The islands of the Sverdrup Basin are underlain by thousands of metres of sedimentary rock younger than those of the Canadian Shield. These layers give rise to low-lying terrain, usually less than 600 m in elevation, but rising to over 750 m on western Melville Island. The

SOURCE: Indian and Northern Affairs Canada, *Canada's North, The Reference Manual* (rev. ed. Ottawa: DIAND, 1990) Section 2.4.1. Reproduced with the permission of the Minister of Public Works and Government Services Canada, 2001.

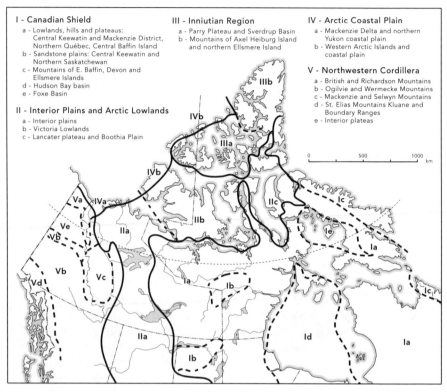

I - Canadian Shield
a - Lowlands, hills and plateaus:
   Central Keewatin and Mackenzie District,
   Northern Québec, Central Baffin Island
b - Sandstone plains: Central Keewatin and
   Northern Saskatchewan
c - Mountains of E. Baffin, Devon and
   Ellsmere Islands
d - Hudson Bay basin
e - Foxe Basin

II - Interior Plains and Arctic Lowlands
a - Interior plains
b - Victoria Lowlands
c - Lancater plateau and Boothia Plain

III - Inniutian Region
a - Parry Plateau and Sverdrup Basin
b - Mountains of Axel Heiburg Island
   and northern Ellsmere Island

IV - Arctic Coastal Plain
a - Mackenzie Delta and northern
   Yukon coastal plain
b - Western Arctic Islands and
   coastal plain

V - Northwestern Cordillera
a - British and Richardson Mountains
b - Ogilvie and Wermecke Mountains
c - Mackenzie and Selwyn Mountains
d - St. Elias Mountains Kluane and
   Boundary Ranges
e - Interior plateas

SOURCE: Geological Survey of Canada, May 1254A, modified after Bostock, 1970

northern part of the Inuitian Region is mountainous, rising to over 2,000 m on Axel Heiberg and Ellesmere islands.

Flanking the Canadian Shield to the north and west are extensive areas of near-horizontal sedimentary rocks, mostly carbonates, which form the Interior Plains and Arctic Lowlands. These underlie undulating lowland plateau areas which comprise much of Victoria and Prince of Wales islands and extend southward to form the Mackenzie Lowlands to the west of Great Slave and Great Bear Lakes.

The Arctic Coastal Plain is a low-lying zone adjacent to the Arctic Ocean and extends southward from Meighen Island through Prince Patrick and Banks islands to the Mackenzie Delta and the northern Yukon coast. It is made of unconsolidated Tertiary and Quaternary sand, gravel, and silt.[1] Quaternary sediments in this region often have

---

[1] In the geologic time scale the Tertiary period began about 65 million years ago and ended between 1.5 and 2 million years before the present. The Quarternary period followed the Tertiary period and includes the present.

high ice content, particularly in the Mackenzie Delta and the Yukon Coastal Plain.

The Northwestern Cordillera includes all of the Yukon (except the coastal plain) and the Mackenzie Mountains of NWT. This region of mountains and intermontane plateaus is developed on complex geological structures.

In northern Yukon, the British Mountains rise to over 1,500 m. They join the northern tip of the north-south trending Richardson Mountains which extend southward to link with the ranges of the Mackenzie Mountains. Much of northern and central Yukon consists of a series of intermontane plateaus. Elevations in the Yukon and Porcupine plateaus rise to 1,800 m, and peaks in the Ogilvie and Wernecke mountains exceed 2,200 m. In the extreme southwest Yukon, coastal mountain ranges of the ice-capped St. Elias Mountains culminate in Mount Logan (6,050 m), the highest point in Canada.

4 Arctic Landforms

J. ROSS MACKAY

The landforms of the Canadian Arctic are of great variety, ranging from the white, glacier-clad mountain peaks of islands in the eastern Arctic, to the monotonously flat green alluvial river-built lowlands of the mainland in the western Arctic. However, the distinctiveness of the Arctic topography cannot be found in the glacier peaks nor the alluvial plains, because similar peaks and plains occur far to the south in temperate and tropical lands. The uniqueness of the Arctic landforms is superficial, and confined mainly to the upper ten to fifty feet [3–15 m] of the ground.

When the large continental glacier covered the greater part of Canada, some thousands of years ago, most parts of the Arctic islands and the Arctic mainland were buried beneath a sea of ice. Even today, relic glaciers cover sizeable portions of Axel Heiberg, Baffin, Devon, and Ellesmere Islands. The effects of past glaciation are everywhere apparent, especially to the air traveller.

Along coastal areas, the observer may see prominent gravelly beaches extending inland, often for distances of many miles, to heights which may exceed five hundred feet [152 m]. As many people know, there has been an emergence of the land, relative to the sea, since the glaciers melted away, and the high beaches record this vertical change. The beaches, which parallel the coast, are similar to beaches along present coasts, except for their greater altitude above sea level. The ages of some of the beaches have been determined from radiocarbon dating of

SOURCE: I.N. Smith (ed), *The Unbelievable Land* (Ottawa: Department of Northern Affairs and Natural Resources, Queen's Printer, n.d.), 60–2. By permission of the author and publisher.

marine shells and driftwood, so that the general pattern of uplift of the land can now be sketched in for the past ten thousand years. In inland areas far away from the coast, the air traveller may see other evidence of glaciation, such as long flutings miles in length or sinuous sand and gravel ridges. These features are also present in more southerly parts of Canada, but are more frequently hidden from view by vegetation growth. The large flutings, which would resemble marks left by giant fingers drawn over the ground, record the direction of glacier movement. The sand and gravel ridges, which are called eskers, are the casts of former glacial streams. Some eskers are over a hundred miles [160 km] long and many of them are plotted on maps of the National Topographic Series.

Most of the land area in the Canadian Arctic is underlain by permafrost or perennially frozen ground. Permafrost is a condition of the ground defined upon the basis of temperature; that is, permanently frozen ground that never thaws even in the hottest time of the summer. In temperature, it is like the inside of a home freezer which never defrosts. The depth to the top of the permafrost may range from a few inches [cm] to several feet [m], but the bottom of the permafrost may be many hundreds of feet [m] down.

In areas underlain by permafrost, the surface of the ground is frequently broken up into eye-catching geometric patterns of circles, ovals, polygons, and stripes to form what is known as patterned ground. A very distinctive type of patterned ground is the tundra polygon which is widespread in the Arctic and may be easily seen either on the ground or from an airplane. The tundra polygons resemble enormous mud cracks, such as those of a dried-up muddy pool, but with diameters of from fifty to a hundred feet [15–30 m]. The tundra polygons may be nearly as regularly shaped as the squares on a checkerboard, but most are irregular, somewhat like the markings on turtle shells. The boundary between two adjacent polygons is a ditch. Beneath the ditch there is an ice wedge of whitish bubbly ice which tapers downwards, like the blade of an axe driven into the ground. Some ice wedges are probably at least several thousand years old. Eskimos [Inuit] frequently dig ice cellars at the junction of two or more large ice wedges. On a smaller scale, the ground observer may see stones arranged in circles or garlands a few feet across, like stone necklaces; or the ground may have stripes trending downhill. Although there has been much study on these forms, their origins are not fully understood. Most experts would agree, however, that they result from processes involving frost action.

Of particular interest to people in the western Arctic are the conical ice-cored hills called pingos, an Eskimo [Inuit] word for hill. The pingos are most numerous near the Mackenzie Delta, where there are nearly 1,500 of them. The pingos may reach a height of 150 feet [45 m] and so are prominent features in the landscape. They are found typically in shallow or drained lakes and are believed to have grown as the result of the penetration of permafrost into a thawed lake basin. Each pingo has an ice core of clear ice. If the ice core should melt, a depression with a doughnut shaped ring enclosing a lake is left behind. In areas of finely grained material, such as silts, large horizontal ice sheets may occur. These ice sheets are usually formed of dirty banded ice, quite different from the crystal clear pingo ice. The tops of the ice sheets frequently lie only several feet below the ground surface; their thicknesses may range from over five to over thirty feet[1.5–9 m]. Although the mechanism of ice sheet growth is still the subject of laboratory and field investigations, they have probably grown in place by the freezing of water sucked up from below by a "wick" action. The ice sheets are best observed along eroded coasts of seas, lakes, and rivers, especially along the Yukon coast and the Mackenzie Delta area. Some settlements and airstrips are built over such ice sheets.

In conclusion, the landform features of the Arctic are distinctive only where climate can exert its influence. The major relief features of hills and valleys, plains and plateaus, are not unique. It is the skin-deep or superficial features such as the bare raised gravelly beaches, the long parallel glacial flutings, the presence of permafrost, the widespread tundra polygons, the ice-cored pingos, and the horizontal ice sheets which impart character to the Arctic landscape.

5 How the Mackenzie River Was Made
Translation of a Slavey Legend

JOHN TETSO

Way back in the days of bow and arrow, we had spring, summer, fall, and winter, just like we have now. One year – we do not know what year it was – something went wrong. It was spring now, and the pretty leaves were on the trees. The birds of every kind came from the south, built their nests, and laid their eggs in them. The cow moose had their calves born and hovered over them protectingly. Certain fishes that spawn in the spring, went up creeks to do the work of reproduction.

The people put away their snowshoes, moved their camps to sunny hillsides, and took it easy, soaking up the sun. Their hard, cold winter days are over, and long, lazy days are here. Little did they know what nature had in store for them.

One morning, they awoke to find a little snow on the ground and cold. The north wind was blowing hard and more snow was coming down. Soon, everything was covered with a heavy blanket of snow. Lakes froze over, and some waterfowl perished, being caught unprepared. There were rivers, but they were small ones. These rivers too, froze and the water ceased to run in them.

At first people thought it strange; it made some of them laugh, some thought it might be just a late spring snowfall. Soon, they were forced to put their snowshoes on, as more snow kept coming down and by now was knee-deep.

SOURCE: *North* 14, no. 3 (May–June 1967): 13. Reprinted by permission of the author and publisher.

On and on it snowed, the wind also kept whistling through the woods, piling huge drifts many feet deep. After many days, the wind died down, and the snow ceased to fall. A very bitter cold gripped the land under its blanket of white. The days grew shorter, just like when it's winter, and long, long cold nights.

Across the vast land, starvation stalked both man and beast. Some villages were lucky, others were not. Once in a while, someone finds a frozen calf, given up by the mother, but a little calf in a large village was not enough to keep everyone alive, and soon, some villages had no smoke coming from them.

The great Master knew the suffering of the poor people, so He decided on a plan to help them, and at the same time made a big river for us, to drink from, to fish in, and travel on. As the people in those days eat mostly meat, our Master sent them a great big ball of dried meat, which dropped somewhere in the east from here. When it dropped, the huge ball started to roll westward. Two young men in white garments (angels) were also sent. These men had long poles with a spear at one end of their poles, and as the great ball rolled, they attacked it with their poles, peeling great chunks of the dried meat.

When the man on the left peeled some off, he attacked with such vigour, it made the huge ball roll more to the right and it did the same with the man on the right, winding as our river does today. Instead of decreasing in size, as some of it was being peeled off, the great ball grew bigger and bigger as it rolled across the land and into the sea, thus making the river wider towards the mouth.

A great, wide path was left where the huge ball rolled, leaving no trees standing. People hunting in the woods came upon this path, some of them were frightened away, thinking some monster was roaming the land, eating people. Others were braver and walked along the wide path finding peeled-off meat. Soon, word was spread and people moved their camps to this strip of broken land. Every day, people gathered the meat, storing it away.

Many, many moons later, the days grew long and warmer. By and by, spring came again, and with spring, the birds came back from the South. The snow on the mountains and trees melted, turned into water, and made little rivers come back to life with the merry sound of running water.

Gradually, the wide path filled up with water from the melting snow, and started to flow in the direction taken by the big ball of meat. At first it was awfully muddy, but as time went on, it got clearer and as it did, all the broken trees were carried away to the mouth.

Many winters and summers later, white men started to come into the country, discovering lakes and rivers, and naming them. One such white man came and got the river named after him,[1] but we were here before he came and we know this much more about the Mackenzie than he does!

[1] The reference is to Alexander Mackenzie, the first white man to descend the river to its mouth in 1798. The Aboriginal name for the river, Deh Cho, is still preferred by many Native residents instead of the Mackenzie River. – ed.

6 Wind Chill in Northern Canada

M.K. THOMAS AND D.W. BOYD

Man has talked about the weather since the beginning of time. Frequently these discussions and arguments have concerned relative coldness, that is "whether last Tuesday" or, "that cold day last winter," was colder than this morning. During these thousands of years, man has invented various methods of protecting himself from the weather. He has learned to wear clothing, to find or build himself a dwelling, and more recently, to heat his dwelling in cold weather, to ventilate or cool it in hot weather, and even to control the humidity.

Man has also devised methods of measuring the weather elements. Over three hundred years ago he invented the thermometer and used it to measure the temperature of air. He found that it gave higher readings in the sun than in the shade. Now he has discovered why the readings are different and has learned to measure solar radiation, and other elements that affect his comfort, such as humidity and wind speed. However, the arguments still continue: "Is it colder in Halifax when the northwest wind pushes the temperature down to zero, or in Saskatoon on a calm, sunny day with the temperature thirty below?"

It must be admitted that coldness, for most of us, is not a simple matter of low temperature. Currie attempted to measure this coldness by using the only available instrument that would give the correct answer: the human body. Only the human body can measure how cold the weather feels. The students who helped Professor Currie at the University of Saskatchewan agreed quite well about how cold it was

SOURCE: *The Canadian Geographer* 10 (1957): 29–39. Reprinted by permission of the author and publisher.

each winter morning on their way to lectures. Their coldness sensations were plotted on a diagram using temperature and wind speed as coordinates, and isopleths of sensation were drawn and labelled with such terms as cool, cold, and bitterly cold. Currie's diagram shows how coldness depends on wind speed, but it does not take into account other factors such as humidity and solar radiation.

The coldness of the body also depends on the amount of activity of the person and the amount of his clothing. This may influence his estimate of the coldness of the weather. Currie kept these factors fairly constant by using only the reported sensations of students who were walking or waiting for transportation and who were similarly clothed. However, the sensations of a group of students at Dalhousie University in Nova Scotia would probably have produced quite a different diagram due to differences in clothing habits and to the weather that they are accustomed to.

It is obviously necessary to have some device which is much simpler than using students as an instrument to measure coldness. The rates of loss of heat from other objects must be measured and these rates assumed to indicate the rates of loss of heat from exposed flesh. This will in turn indicate whether clothing is required to keep a human being comfortable.

The same thing could be expressed somewhat differently. What is required is a measurement or estimate of some elusive characteristic of the weather which is frequently referred to as "coldness" and which is known to depend on temperature, wind speed, and other weather elements. This coldness is probably closely related to the rate of loss of heat from exposed flesh and is frequently assumed to be proportional to the measured rate of cooling of a suitable object.

Cooling rates have been measured for many objects, including thermometer bulbs of various shapes and sizes, platinum wires, and copper cylinders and spheres. From these experiments and also from theoretical considerations it is found that the rate of loss of heat (h), whether by radiation, convection, or conduction, is very nearly proportional to the difference in temperature (Δt) between the object and the surroundings. $h \propto \Delta t$. This assumes that the surroundings are all near air temperature and hence does not take account of solar radiation.

The rate of heat loss by convection depends on the wind speed (v). With streamline flow it is a linear function of the square root of the wind speed according to most authorities. $h \propto \sqrt{v} + const$.

For turbulent flow, Reynolds in 1874 found the rate of heat loss to be a linear function of the wind speed. $h \propto v + const$. The general form including both cases would be: $h = \Delta t (A + B\sqrt{v} + Cv)$.

By choosing a suitable value for the constant "A", this form could include radiation and conduction. The values of the constants "A",

"B", and "C" differ considerably in the formulae proposed by different investigators, and quoted by Court. This is to be expected because "h" also depends on certain properties of the cooling body. Some investigators have used powers of "v" somewhere between one half and one. It is obvious that the different methods of measuring the cooling power of the air lead to quite different results and this increases the difficulty of selecting a suitable formula. Much of the work on cooling power has been done at quite low wind speeds, and most of it at temperatures above freezing. Since the latter part of this paper is concerned with conditions in the Arctic, it seems reasonable that we should use a formula which is based on observations obtained at low temperatures and with moderate winds.

Siple[1] has had considerable experience in Antarctica, where, during the winter of 1941, he measured the time required for the freezing of 250 grams of water in a plastic cylinder about $2^{1}/_{4}$ inches [56 mm] in diameter and 6 inches [152 mm] long, under a variety of conditions of low temperature. He assumed that the rate of heat removal was proportional to the differences in temperature between the cylinder and the air, and that the cylinder remained at the freezing point throughout the period of freezing. The results were expressed in kilogram calories per square metre per hour, per degree Centigrade, and plotted against wind speed in metres per second.

Siple disregarded a few of his experimental results for one reason or another, and on the basis of the remaining readings he computed the formula: $h = \Delta t \ (10.45 + 10 \sqrt{v} - v)$.

Court recalculated Siple's formula using, in addition, the data which Siple had rejected. This gave the formula: $h = \Delta t \ (9.0 + 10.9 \sqrt{v} - v)$. This latter form has been adopted by the Quartermaster Corps of the United States Army and has been used by Bristow in preparing wind chill maps of the United States.

Siple's original formula has been chosen for the present paper, and hence the maps are not strictly comparable to those published in the United States. The cooling times in his experiments varied from less than an hour to about twenty-four hours, and hence either hourly or daily mean values of wind speeds and temperatures should be used. If mean values of wind and temperatures for longer periods are used, the resulting wind chill will be considerably greater than the mean of the individual wind chill values. The error will depend largely on the variation of the wind speed, since the wind chill is not a linear function of the speed. Siple himself, however, used monthly mean values to compute wind chills. Since the error always has the same sign, the values

[1] P.A. Siple, "Measurements of dry atmospheric cooling in sub-freezing temperatures," *Proc. Amer. Phil. Soc.* 89 (1945): 177–99.

will at least be comparable, and that is all that is needed. One must, of course, be careful not to compare daily values with values computed from monthly means.

Halifax with conditions such as mentioned above, a temperature of 0°F [−18°C] and a wind of 20 miles [32 km] per hour would give a wind chill of 1,600 units. Saskatoon at −30°F [−34°C] would need a 6 mile [10 km] per hour wind to give the same wind chill value. With a lighter wind or in the sun it would not be so cold.

Monthly values of the wind chill factor have been obtained directly from monthly mean values of temperature and wind speed. The length of record varies from thirty years at a few Mackenzie valley stations to five years at Alert on Ellesmere Island, the most northerly station in Canada. Further south, in settled Canada, the data are based on the standard thirty-year period from 1921 to 1950. Available data indicate that there is an area in Keewatin [Kivalliq] District that is colder in mid-winter than the Arctic coast to the north. It is in fact more than ten degrees colder than the coastal area north of Aklavik. However, lower mean temperatures, more than −35°F [−37°C] are experienced at Isachsen and Eureka in the northernmost Arctic.

Mean wind speed maps are not entirely satisfactory, since the immediate location of a station is of prime importance, and because periods of calm do much to lower the mean wind speed. The data used are based on irregular periods ending in 1954. The relatively high wind speeds over Hudson Bay and Strait are significant, and contrast sharply with the relatively low speeds in the Yukon and mid-continental areas as well as in the northernmost Arctic. The core of the high mean speed area extends along the Arctic coast of the mainland. Recent reports from new stations in this area suggest that this core may be even more pronounced than shown here.

Using these temperature and wind data, monthly values of the wind chill factor were computed for all Arctic and Subarctic weather reporting stations. While the factor was the greatest in February for several stations, 75 per cent of the stations showed the greatest monthly value of wind chill in January.

The January map (Figure 6.1) shows that the factor exceeds 1,900 kg-cal/sq.m./hr. in most of the District of Keewatin [Kivalliq]. Baker Lake with 1,980, and Chesterfield with 1,950 kg-cal/sq.m./hr. have the distinction of being the worst meteorological stations in Canada on the basis of wind chill. Isachsen with 1,840, and Mould Bay, Resolute, and Cambridge Bay with 1,800 are almost as severe as are Ennadai Lake with 1,820, and Churchill with 1,740 despite their more southerly latitudes.

It is interesting to note that some populous southern Canadian cities do have high wind chill values in January. Winnipeg with 1,490 is

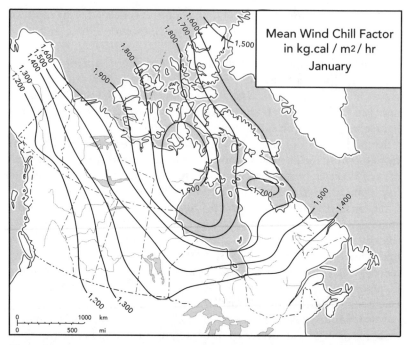

Figure 6.1

higher than any station in the relatively calm Yukon Territory, and even Montreal with 1,220 is almost as cold as Whitehorse, with 1,250. At Toronto in January the value is 1,110, at Halifax 1,000, and Victoria 820 kg-cal./sq.m./hr.

Figure 6.2 illustrates the mean wind chill factor in the mid-spring month of April. Because of the seasonal temperature lag, the wind chill values are still high in April and the pattern is similar to that of January. The "core" of the extreme values has moved north to the islands, but April is still a winter month, more severe in this area than Winnipeg in January.

During the mid-summer season, wind chill values in the far north are comparable to January values at Victoria, British Columbia. In July (Figure 6.3), the most severe area is along the northwest coast of the Arctic islands where Isachsen is 840 and Mould Bay 830 kg-cal./sq.m./hr. Values in excess of 800 are also observed in the mouth of Hudson Strait at Resolution Island, where the July mean temperature is below 40 degrees [4°C]. In comparison, the value at Toronto in July is 330 kg-cal/sq.m./hr. Since solar radiation is not taken into account, this map applies only to conditions in the shade and its value is doubtful.

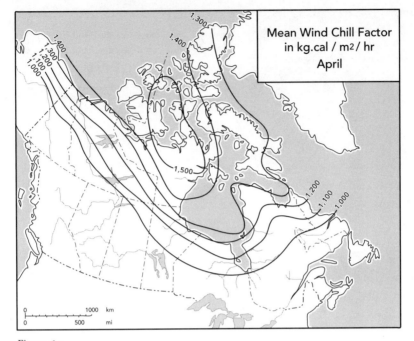

Figure 6.2

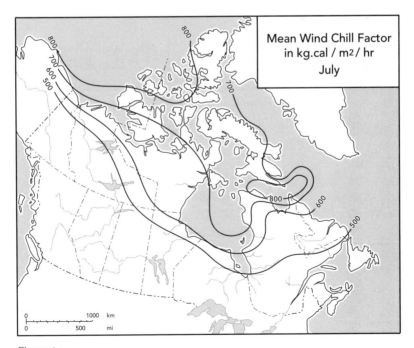

Figure 6.3

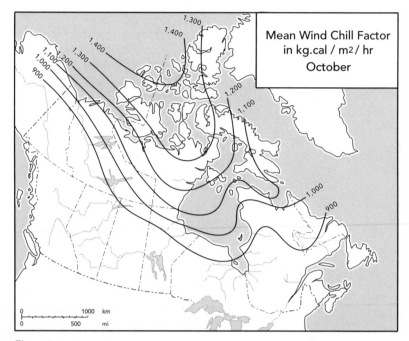

Figure 6.4

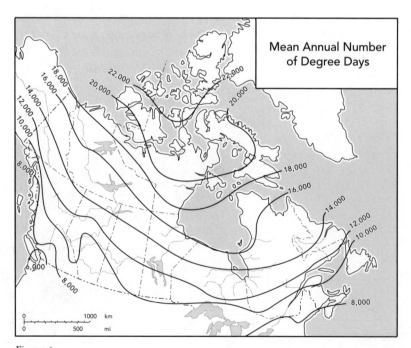

Figure 6.5

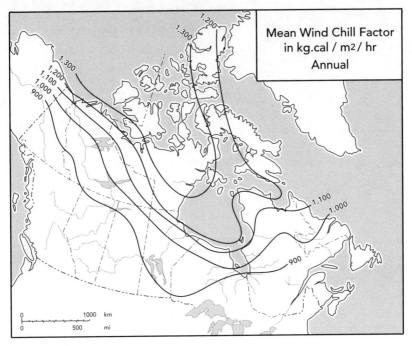

Figure 6.6

The October mean wind chill factor map is shown as Figure 6.4. This map has a pattern similar to the July map, although values in the northwest Arctic are equal to January values in southern Manitoba.

It might be interesting to compare these wind chill maps with other measures of coldness or severity of winter conditions in Northern Canada. Figure 6.5 illustrates the distribution in Canada of mean annual total degree days below 65°F [18°C]. These degree day values are used in estimating fuel consumption. Since this map illustrates an accumulation over the season the increase from south to north is quite marked and perhaps illustrates a more common idea of the difference between the Arctic and the south.

Many people are interested in extremes and often base their idea of coldness on the lowest temperature ever recorded in any location. In Canada the lowest temperature ever officially reported is −81°F [−63°C] at Snag in February 1947. Temperatures colder than −70°F [−57°C] have occurred over most of the Yukon but there is no official record of such low temperatures occurring in the Arctic. In fact record low temperatures reported from some Ontario and Quebec stations are lower than those at several Arctic stations. By comparison Toronto has had a temperature as low as −26°F [−32°C], and Ottawa −38°F [−39°C].

TABLE I

MEAN MONTHLY AND ANNUAL WIND CHILL FACTORS FOR SELECTED
STATIONS (IN KG-CA./M2/HR.)

Month	Winnipeg Man.	Churchill Man.	Chesterfield Nunavut	Baker Lake Nunavut	Resolute Nunavut	Alert Nunavut	Snag Y.T.	Toronto Ont.	Bay of Whales Antarctica	Cape Denison Antarctica
Jan	1,490	1,740	1,950	1,980	1,800	1,370	1,130	1,110	1,072	1,131
Feb.	1,400	1,720	1,920	1,870	1,840	1,490	1,100	1,110	1,335	1,273
Mar.	1,200	1,530	1,720	1,660	1,700	1,450	970	970	1,615	1,550
Apr.	860	1,300	1,480	1,450	1,480	1,250	830	770	1,722	1,750
May	630	990	1,160	1,080	1,170	1,040	600	580	1,706	1,814
June	480	780	860	840	920	820	480	390	1,660	1,880
July	390	580	670	640	810	740	440	330	1,948	1,864
Aug.	430	610	730	650	850	790	500	340	1,907	1,802
Sept.	600	800	910	890	1,060	1,040	610	480	1,934	1,800
Oct.	810	1,040	1,190	1,210	1,360	1,280	780	620	1,670	1,695
Nov.	1,110	1,400	1,520	1,520	1,520	1,370	960	820	1,374	1,426
Dec.	1,370	1,680	1,780	1,790	1,650	1,390	1,060	1,030	1,078	1,237
Annual	898	1,181	1,324	1,298	1,347	1,169	788	713	1,585	1,602

With these maps as background, Figure 6.6 shows the mean annual wind chill factor for Northern Canada. This map was constructed by averaging the monthly values at all stations and should be considered as a mean of the rates of cooling. Although the windiness and relative summer coolness of Hudson Bay is evident, the far northern stations of Isachsen, Mould Bay, and Resolute have more severe conditions over the year as a whole.

In summary, it might be as well to point out again that wind chill is a computed index based only on air temperature and wind speed. It is probably the best indication we have at present of that weather characteristic which is referred to vaguely as "coldness." It is therefore at least a rough indication of the cooling rate of exposed flesh and of whether or not clothing is necessary. Its usefulness at high temperatures is doubtful.

Monthly values of the wind chill factor are shown in Table 1 for a number of representative stations. Chesterfield and Baker Lake are the two weather observing stations with the highest wind chill values in mid-winter, followed closely by the Bay of Whales and Cape Denison, two of the coldest stations in Antarctica. Resolute is typical of several stations scattered over a large part of the Arctic. Alert, the most northerly station, is quite similar, at least in winter, to Winnipeg. Snag, the station with the record low temperature of –81°F [–63°C] is in the average winter, very little colder than Toronto. Churchill seems to have been well chosen as the location of many field testing units for the effect of winter weather on personnel and equipment.

Finally, in any study of wind chill, attention must be drawn again to conditions during the most severe month of the year and to parts of the country where these conditions are most marked. Study of the January map (Figure 6.1) focuses attention on the Churchill-Chesterfield-Resolute area. There are, of course, differences between the wind chill factor and other cold weather criteria. Yukon has the distinction of recording lower temperatures than any other area in Canada, while the northernmost islands require the most fuel for a season's heating. But the strong winds over the flat district of Keewatin [Kivalliq] and the very low temperatures in the continental interior combine to make it one of the most bitterly cold areas on earth.

7 The Ecology of Snow

W.O. PRUITT, JR

If snowflakes were rare objects, generations of graduate students would doubtless have received degrees for research into the properties and potential uses of snow. As it is, some tens of billions of these beautiful crystals of frozen water-vapour pile up in each square metre of snow. The very abundance of snow seems to have suppressed almost all but the negative aspects of getting rid of it as quickly as possible. In the literature of the sciences that ought to be most concerned there is little to suggest that snow is a major element in the environment of life. In greater or lesser amounts, however, snow covers more than half of the land area of the northern hemisphere at some time during the year.

Taiga Snow

As a result of experience with snow at varied places in the northern parts of North America, I have come to look on the snow of the Subarctic taiga[1] in interior Alaska as "typical" snow. That is, it is snow that is the least modified by external factors. There, in the great topographic bowl north of the Alaska Range and south of the Brooks Range, the snow arrives early in the fall and remains until late spring. For a goodly part of the winter it is virtually unaffected by incoming solar radiation, since this region is only two degrees south of the Arctic

[1] The dominantly coniferous Subarctic.

SOURCE: Canadian Society of Wildlife and Fishery Biologists, *Occasional Papers, No. 1*, M.T. Myers (ed), (Dept of Biology, Univ. of Calgary, Oct. 1965), 1–8.

Circle. The peculiar meteorological conditions prevailing there cause a standing inversion to be common, thus the snow is little affected by wind. The result is a snow cover that piles up loose and fluffy, modified only by the heat and moisture which rise through it from the soil below. The thermal inversion is also responsible for another phenomenon which has ecological importance – the deep cold of the calm, dense, air that accumulates in the great, flat valleys of the Tanana and Yukon Rivers, and is also responsible for the "ice-fog" that rises from any source of free moisture. The snow piles up on the tree branches and the surface is rough with crystals, lying just as they fell and undisturbed by wind.

Taiga snow is the interface between the warm, moist soil and the dry, very cold, Subarctic air; it is in fact an ecotone between drastically different environments. Although the instruments in a standard Stevenson Screen may measure an air temperature of, say, –40° or –50°F [–40° or –46°C], the air immediately above the snow may be as low as –70°F [–57°C]. But once measured underneath the surface of the snow, the temperature rises dramatically, until at the base of a fully-developed taiga snow cover the temperature hovers not far below freezing and with the addition of a moss layer under the snow, the temperature may be only slightly below freezing. Not only is the subnivean environment relatively warm and moist, it is also markedly stable.

The presence of such a warm, moist, and stable environment is of extreme importance to small mammals, plants, and invertebrates. It has been shown by many experiments that small mammals (mice, voles, and shrews) cannot withstand the supranivean environment of the Subarctic winter. They are physiologically incapable of producing enough heat to offset the loss to the cold, dry air and their mass-to-surface relationship does not allow them sufficient insulation. Thus, if it were not for the snow cover, large areas of the northern taiga would be lacking in small mammals. This would have far-reaching effects, since the small mammals are the main herbivorous base of the ecosystem, and many carnivorous mammals and birds depend on them for food. Nevertheless, because the thickness of the snow cover fluctuates from year to year, there are yearly differences in the severity of the bioclimate in which the small mammals live, so there are yearly differences in the numbers of small mammals.

The autumnal decline in air temperature is a fairly regular event, governed by the regular decrease in incoming solar energy as the days shorten. The onset of a snow cover is more fortuitous, being governed by the precise succession of meteorological events that bring the correct mass of moist air into contact with cold air. Thus, a snow cover may arrive quite early in October one year, but may not arrive until late

November the next year. When there are only a few centimetres of snow on the forest floor of the Subarctic taiga, one may see many signs of small mammal activity, but when the snow cover reaches a thickness of fifteen to twenty centimetres, there is a dramatic decrease in activity on the surface. Such a thickness of snow is sufficient to insulate the soil against fluctuations in air temperature. When this thickness arrives, I call it the *hiernal threshold*, for it is the true beginning of winter for the small creatures of the forest floor.

Not only is the snow cover variable in time, but it also varies in space. As the snow flakes fall, many of them are caught on the needled branches of the coniferous trees of the taiga. Thus there is a "snow shadow" formed around the base of each spruce tree. As the winter progresses the snow shadow continues and increases in sharpness. The Forest Eskimo of the Kobuk Valley, in northwestern Alaska, call these bowl-shaped depressions in the snow at the base of trees *qaminaq*. We have seen how small mammals prosper with deep snow and decline in numbers with little snow, so it is no surprise to learn that individual red-backed voles (*Clerthrionomys*) avoid the *qaminaq* and more often frequent the parts of their home ranges that have thicker snow cover. Their home ranges have vacuoles, as it were and these vacuoles are the *qaminaq*.

A thick protective blanket of snow is not without dangers, however. The relatively warm subnivean environment allows a certain amount of bacterial action to continue, even in mid-winter. This results in the production of carbon dioxide under the snow. If the snow cover is thick enough, and particularly if there are dense, relatively impervious layers within it, there may accumulate in certain low spots enough carbon dioxide to be harmful to the small mammals. The voles have developed a behavioural adaptation to counteract this situation. They construct ventilator-shafts up through the snow cover to the surface.

In regions of especially thick snow cover, the small mammals are virtually immune to predation by carnivorous birds during the time when the snow cover is present. The only time that they are exposed to predation is when they come up their ventilator-shafts to the upper air. The distinguished Russian naturalist, A.N. Formozov, has shown that it is such activity on the part of the small mammals that allows certain species of small owls to overwinter in parts of the Eurasian taiga. Without the voles caught at their ventilator-shafts they would have insufficient food for survival in the region.

There are two main classes of mammals in the taiga – those that are large and live above the snow cover, and those that are small and are forced to live beneath the protecting blanket. There is one mammal, the red squirrel, that lies just on the borderline between the two size-groups of mammals. Red squirrels are active above the snow most of the time, but when the air-temperature falls below −25°F [−32°C] or −30°F [−34°C], they vanish from the scene. This is a critical temperature for them and when it arrives they leave the supranivean environment of the moose and hare and join the voles in their warm subnivean environment. Thus they have the best of both worlds.

There are three main ways in which mammals adapt to snow. One is by using the snow as a blanket to avoid the deep cold. A second adaptation is that of the moose – the possession of long legs or stilts; even this adaptation is not perfect, since the snow may sometimes become too thick for stilts. The places where moose overwinter are also places where the snow cover is thinnest. The third way of adapting to snow is to float over it. The lynx and the snowshoe hare are the best examples of this method. A lynx appears to be a large animal, but when skinned out it is seen to be not much larger than a small dog. Most of the animal is fluffy insulating fur and the large snowshoe feet. These large feet enable it to float on the surface of the snow, even when galloping. The snowshoe hare is the classic example of a floater. In closely related species in Eurasia, *Lepus timidus* and *Lepus europaeus*, the snowshoe hare possesses more bearing surface on the hind feet, even though it is smaller than its southern relative. However, sometimes even flotation fails. The snow may be so light and fluffy that even the snowshoe hares sink into it. Then they change their behaviour and become "trailers," following each other's trails. Each time an animal passes, the snow gets packed a little bit more. Eventually, a hard trail is formed along which the animals can move freely.

Caribou are animals that are closely associated with snow. In 1957–58 I studied their ecology for the Canadian Wildlife Service. That winter we flew many hundreds of miles in light aircraft at low elevations, carefully plotting on maps the locations of overwintering caribou, and also where the animals were *not* present. We found that most of the area had no caribou. Then there were areas of light concentrations, within which were areas of heavy concentrations.

At this time I set out a series of snow stations. I measured the thickness, hardness, and density of each layer of the snow, and also the grain size and type, and the temperature of each layer. When I plotted the results on maps and compared them to the distribution of caribou, I found that the areas of heavy caribou concentration had snow that was light, soft, and thin. Hardness ranged from 6.5 to 60 gm./sq.cm. for forest stations and 50 to 700 gm./sq.cm. for lake stations, while density varied from 0.13 to 0.20 for forest stations and 0.13 to 0.32 for lake stations. Thickness varied from 19 to 59 cm. For the areas without caribou, the snow could sometimes be soft, but it could also be very hard, dense, and thick. Hardness varied from 35 to 7,000 gm./sq.cm. for forest stations, and from 150 to 9,000 gm./sq.cm. for lake stations, while density varied from 0.16 to 0.92 for forest stations, and 0.17 to 0.92 for lake stations. Thickness varied from 19 to 82 cm. Remembering that the density of freshwater ice is 0.92, it is clear that the non-caribou layers had ice layers in the snow cover, whereas the caribou areas did not.

Thus we see that caribou have thresholds of sensitivity to the hardness, thickness, and density of the snow cover. The threshold of hardness sensitivity is approximately 50 gm./sq.cm. for forest snow, and 500 gm./sq.cm. for lake snow. The threshold of sensitivity for density is approximately 0.19 or 0.20 for forest snow, and 0.25 or 0.30 for lake snow. The threshold of sensitivity for thickness is approximately 60 cm. When these thresholds are exceeded, the caribou react by exhibiting a migratory type of appetitive behaviour until they encounter snow of lesser hardness, thickness, or density.

Let us examine some of the ecological aspects of the snow which collects on trees. In my work in the North I have found that the "official" meteorological words for snow are woefully inadequate to describe its phases. Consequently I have turned to the languages of the native peoples of the North, the Eskimos [Inuit] and the Indians, for words which represent those phases of snow which are important to animals and plants. Of all the native languages I have examined, that of the Kobuk Valley Eskimo [Inuit] (the "Forest Eskimo") of northwestern Alaska appears richest in snow terms. Their word for the snow that collects on trees is *qali*, and for the snow that collects on the ground is *api*.

In the windless taiga of central Alaska, *qali* assumes great ecological importance. It is one of the agents initiating forest succession. If spruce departs from the vertical it is doomed to breakage, someday, because it will accumulate *qali*. When a tree breaks, adjacent trees become susceptible to *qali*-breakage and the "glade" grows until it is sufficiently large that wind circulation prevents massive accumulation of *qali*. In the glade, the broken spruces die and the rain of dead needles chokes out the feather-mosses on the forest floor. Thus seeds have a good site for germination. Deciduous trees invade – alders, birches, aspens, and willows. These trees mature and die and, in their leaf litter, young spruces can germinate. They mature and the spruce forest is eventually restored to the site, to await further *qali* breakage.

Qali is also of direct economic importance to man's activities. Power lines are frequently broken, either by *qali*-broken trees or by a heavy accumulation of *qali*. Near Fairbanks, the local Rural Electrification Cooperative has met the challenge by flying a helicopter along the most vulnerable lines. The rotor blast cleans the *qali* from the power lines.

In summer the taiga may be a mass of greenery – primarily alders and young birches. In winter these trees and shrubs are bent over by *qali* accumulation. This is their way of adapting to the presence of *qali*.

The spruce stands straight and tall and resists *qali*; alders and birches are limber and bend with the *qali* and recover in the spring. When the trees are bent over by the *qali*, their tender growing-tips are brought within range of the snowshoe hares, which feed extensively on them. This is a very important source of food for the hares. When the trees are bent over, snow-caves form under them. In very cold weather, even the hares avoid the infinite heat-sink of the night sky, by finding refuge in these caves. In the spring the alders and birches spring back vertical again, with the hare-barked twigs high in the air, and at this time of the year one may see many signs of the winter-feeding by hares. The relationship of hares and shrubs is reciprocal: the plants furnish the hares with food but the hares return the food to the soil by their fecal pellets, which accumulate in quantities around the shrubs utilized most heavily.

In a winter when the snow comes early and accumulates gradually all winter long, the hares are constantly elevated on top of the snow surface to reach fresh supplies of food. But in a winter when the snow remains at a constant thickness, or when it even settles and decreases in thickness, the hares are unable to reach for fresh sources of food higher up and they turn to unpalatable foods, such as spruce. Such snow conditions may result in a decline in the hare population by spring.

In January, the birches shed their seeds onto the snow surface and many resident birds, such as redpolls, utilize this food source. Eventually, however, another snowfall covers this seed layer. Mice and shrews then tunnel up through the snow and mine out the seed layer.

Tundra Snow

So far we have concerned ourselves with taiga snow. Let us now examine the ecology of the hard, wind-moved tundra snow. Tundra snow is characterized by this factor of having been moved by the wind.

There are two phases, in the physical sense, to the snow in an Arctic tundra region. The snow cover consists of those particles which are not picked up by the wind (*api*), and of hard wind-worked particles which have become consolidated into a mass (*upsik*).

Above the snow cover, in the air, on the surface of the snow cover, and sometimes incorporating even the top layers of the snow cover itself, is another phase. This is the moving snow or *siquoq* which, depending on the wind direction and force, is either consolidated into a succession of drift forms, or moves along and above the snow surface. Because of variations in the force of the wind transporting them, and because of nivographic details, the particles comprising the *siquoq* become stabilized for varying periods of time and form drifts.

The sequence of drift types appears to be as follows:

Snow particles are released from suspension in the air whenever the speed of air movement is not sufficient to support them. Thus, snow accumulates in microtopographic depressions, streambeds, and behind obstructions (which may themselves be nivographic details). Later winds of greater force or different direction may scour these spots and re-deposit the particles elsewhere. On a flat, relatively unobstructed surface many of these particles advance in groups. A group assumes a characteristic arrowhead shape with the point upwind, a gradually sloping upwind face, and a lee slope which is abrupt and concave laterally. At the tang of the arrowhead the thickness of the drift is greatest. These drifts are known popularly as barkhans but more accurately, in Eskimo [Inuktitut], as *kalutoqaniq* (a barkhans is technically a sand drift of this shape while *kalutoqaniq* refers to this precise snow drift type). *Kalutoqaniq* migrate downwind as the particles are exposed on the windward face, are moved over the surface of the drift, and then are temporarily immobilized on the steep lee slope. Whenever the wind slackens, the *kalutoqaniq* become consolidated through the processes of sublimation and re-crystallization.

Later winds, if of sufficient force, will erode away the *kalutoqaniq*, producing sculptured forms which have great beauty but which are exceedingly difficult to traverse. The sculpturings are widely known by the terms *zastrugi* (Russian) or *skavler* (Norwegian), but are more accurately known as *kaioqlaq* (Eskimo [Inuit]). *Zastrugi* or *skavler* refer to surface sculpturings in general. *Kaioqlaq* refers to large, hard sculpturings while the word *tumarinyiq* (Eskimo [Inuit]) refers to small *zastrugi* or "ripple marks."

Kaioqlaq eventually may be eroded away completely and the particles regrouped downwind again into *kalutoqaniq*. A late stage of *kaioqlaq* is the formation of overhanging drifts or *mapsuk*. The windward point of a ridge of *kaioqlaq* is eroded faster at base level than above it, thus forming the characteristic anvil tip which points upwind.

Thus this succession of drift forms may be diagrammed as follows:

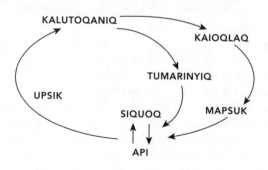

Valleys of small streams become completely filled with a thick mass of wind-blown snow, and these drifts may not melt until late in the following summer. Such drifts are known by the Russian word *zaboi* and may be of considerable ecological effect. They retard plant growth and, in those spots where they do not melt until late in the summer, their presence may prevent certain species from living. In extreme cases they may prevent all plants from growing on the site where they form. These bare spots are then subject to intense cryopediological processes. In some spots on the tundra where snow has completely filled a small stream valley, eddy currents may scour out the snow and produce a cavity that may assume tremendous size. This is one of the "traffic hazards" of tundra travel. These scoured spots are known to the Eskimo [Inuit] as *anmana*. Such scoured spots can be seen even in the taiga when there is a light wind.

Now, how do animals react to the varied types of tundra snow? Some animals, such as the caribou, react by moving. Most caribou leave the tundra during the snow season and migrate to the taiga where the snow is softer and less dense. Some caribou remain on the tundra and these groups may be found in either of two situations: first, there may be caribou where, because of the topography, even on the open tundra there are spots where the snow is relatively soft; second, in those regions of the tundra where the winds are so strong that virtually all the snow is blown away from the vegetation; these spots are known as good caribou hunting grounds, for they regularly support a population of overwintering animals.

On the tundra, some resident birds, such as the ptarmigan, are able to find small pockets of soft snow and dive into them and use them as a blanket for protection from the cold winds. Some mammals, such as the Arctic ground squirrel, which hibernate, choose their hibernation sites where thick *zabois* form. There they make "escape holes," and the tunnel down to the snow may be nearly twenty feet long. The animal thus spends the winter where it is relatively warm. The tundra snow cover also protects small mammals, such as voles and lemmings, from predation.

To sum up this discussion of the ecology of snow we can do no better than to quote Professor Formozov:

Analysis of the factors described leads to the conclusion that, in order to study the winter ecology of mammals and birds in regions with snowy winters, zoologists and biogeographers must have available, without fail, in addition to the data furnished by meteorological stations, numerous specialized measurements and descriptions of the snow cover and its structure in various habitats and environmental types. The study of snow cover is necessarily conducted simultaneously with systematic calculations of distribution, numbers, and characteristics of the vital activities of animals, since elucidation of their varied reactions allows the possibility of judging the positive or negative influence of the snow cover of a certain strength, and structure on the winter conditions of existence of the fauna.

The study of the ecological role of the snow cover and its structure requires long-term observations which will give the answers to a series of questions that are important for protection of valuable animals and rational planning for their utilization, to work out predictions of the numbers of harmful rodents, predictions of the probability of damage to winter crops, fruit trees, and cultivated shelter belts, etc. These observations are best carried out at specialized stations, which conduct, in addition, simple experiments on animals, carried out in open pens arranged under the open sky.

8 Permafrost Map of Canada[1]

R.J.E. BROWN

Permafrost, the ground that remains frozen throughout the year, is widespread in northern Canada and greatly influences developments in this vast region. Any soil or rock, whose temperature remains below 32°F [0°C] is considered to be in a perennially frozen or permafrost condition. The existence of this phenomenon is related to the cold climate, which over the ages has caused permafrost to accumulate to depths exceeding one thousand feet in the Arctic. This perennially frozen ground occurs further south in the Subarctic but it is not found everywhere beneath the ground surface as in the Far North. It is vitally important to newly-discovered mining areas and other activities rapidly developing in northern Canada. Understanding the permafrost, its distribution, characteristics, and effects on construction, is absolutely essential to successful northern development.

Research in Canada on the distribution and nature of permafrost, and engineering problems associated with it, began nearly twenty years ago. Since its inception in 1947, the Division of Building Research of

---

1 Copies of the original map from which the map included as a supplement to this issue was made are available from the Geological Survey of Canada, Dept. of Energy, Mines, and Resources, Map 1246A, First Edition, and from the Division of Building Research, National Research Council, Map 9769.

It has not been possible to include either of these maps in this volume for technical reasons. The map provided by Dr Brown presents some critical permafrost boundaries. For the latest permafrost map of Canada, see the 5th edition of the *National Atlas of Canada*, plate 2.1. – ed.

SOURCE: *The Canadian Geographical Journal* 76, no. 2 (Feb. 1968): 56–63. Reprinted by permission of the author and publisher.

the National Research Council of Canada has considered the problem of building in the North to be one of its major responsibilities. Early studies showed that one of the special problems of engineering construction in northern areas arises because of the presence of permafrost. As a result, a Permafrost Section was formed within the Division in 1950. It soon became evident that many research needs could be fulfilled only by actual field investigations – which prompted the Division to establish, in 1952, a Northern Research Station at Norman Wells on the Mackenzie River, ninety miles [145 km] south of the Arctic Circle. In 1962 this operation was moved to the Inuvik Research Laboratory of the Department of Indian Affairs and Northern Development as permafrost research projects developed at the new town of Inuvik in the continuous permafrost zone. Another field station was established in 1965 in the new nickel mining town of Thompson, located in the southern part of the permafrost region.

Throughout this period and at the present time, the Permafrost Section, now known as the Northern Research Group, has carried out investigations in many parts of northern Canada, from Ungava and Baffin Island in the east to the Mackenzie River and Yukon Territory in the west. Knowledge of permafrost distribution in this vast region is a vital prerequisite to successful construction. The Division is collecting information continually by various means, including field mapping surveys and studying the technical literature.

Distribution and Occurrence of Permafrost

Permafrost is not confined to northern Canada but underlies 20 per cent of the world's land area, being widespread in North America, Eurasia, and Antarctica. It is a highly important factor in the development of Siberia, and like Canada, about one-half of the Soviet Union is underlain by permafrost. Indeed, these two northern neighbours share most of the permafrost territory in the northern hemisphere.

The permafrost region is divided into two zones – continuous and discontinuous zone. In the continuous zone, the climate is so cold, with mean annual air temperatures below about 17°F [−8°C] – compared with 42°F [6°C] at Ottawa – that permafrost occurs everywhere beneath the ground surface. It varies in thickness from more than one thousand feet [305 m] in the northern part of the zone to about two hundred feet [60 m] at the southern limit of the zone. The surface layer of ground, termed the "active layer," which freezes in winter and thaws in summer generally varies in thickness from about 1½ inches to 3 feet [4–90 cm].

In the discontinuous zone, which is Subarctic and not so cold, a host of terrain factors contribute to the existence of areas and layers of

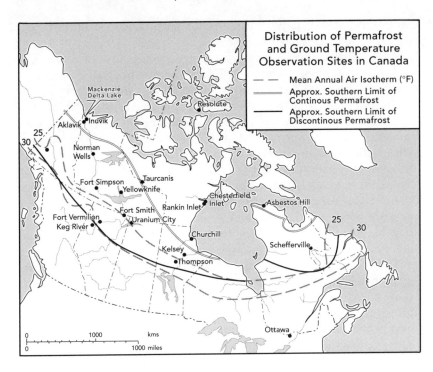

Distribution of Permafrost and Ground Temperature Observation Sites in Canada

unfrozen ground. Between the northern limit of this zone and the belt where the mean annual air temperature is 25°F [-4°C], permafrost is widespread and varies in thickness from two hundred feet [60 m] to about fifty feet [15 m]. Southward, permafrost becomes patchy and is only a few feet or even inches thick at the southern limit of its distribution where the mean annual air temperature is about 30°F [-1°C]. The depth to the top surface of the permafrost, termed the "permafrost table" is extremely variable ranging from about two feet to more than ten feet [0.6-3.0 m]. The active layer extends to the permafrost table where the latter lies within about five feet [1.5 m] of the ground surface. In areas where the permafrost is the product of past colder climates and lies at greater depths, the zone of annual freezing and thawing will usually not reach it.

In the mountainous regions of Canada, permafrost occurs at high elevations far south of the limit shown on the map. In northern British Columbia it is estimated that permafrost may be encountered above an altitude of four thousand feet [1,200 m]. Southward, the lower limit of permafrost rises progressively to an estimated elevation of between six thousand and seven thousand feet [1,800-2,100 m] at the 49th parallel. The distribution of permafrost in the mountains is further complicated by the fact that it is more widespread and thicker on north-facing slopes, which face away from the sun.

Temperatures in the permafrost range from a low of about 5°F [-15°C] in the High Arctic to 23°F [-5°C] at the boundary separating the continuous and discontinuous zones. Southward, the temperature increases progressively to a few tenths of a degree below 32°F [0°C] at the southern limit of the permafrost region. A broad relationship exists between the air temperature and the temperature of the permafrost. Observation in many parts of Canada indicate that the temperature of the permafrost is roughly 6°F [14°C] warmer than the mean annual air temperature at a given location, mainly because of the winter snow cover. Local variations in climate and terrain cause variations but a value of 6°F [14°C] can be used as an average figure. Therefore, it is possible to predict roughly the temperature of the permafrost where the mean annual air temperature is known.

The broad pattern of permafrost distribution is determined by climate, but local terrain conditions such as relief, vegetation, drainage, snow cover, type of soil and rock, and geological history are responsible for the extent and thickness of permafrost, and variations in thickness of the active layer. Relief influences the amount of heat from the sun reaching the ground surface. A north-facing slope receives less sun than a south-facing slope, causing variations in permafrost distribution as in mountainous regions. Vegetation, particularly the moss cover, so characteristic of northern regions, shields the permafrost from the thawing effects of summer air temperatures. The presence of water bodies and poor drainage causes thawing of the permafrost. Heavy snow in the late fall and early winter inhibit winter frost penetration but snow on the ground late in the spring delays thawing of the underlying ground. The type of soil and rock influences the active layer, causing it to be thicker in gravel, sand, and rock than in silt and clay. During the Ice Age virtually all of Canada was covered with glaciers, greatly affecting the past and present distribution of permafrost.

Engineering Aspects

Permafrost is an important consideration in engineering work in northern Canada. One feature of perennially frozen ground particularly significant in engineering construction is ice, which occurs frequently, but not always, in permafrost as layers, coatings around particles, and large blocks or wedges. Although frozen soil provides a strong and firm base for a building or other structure, it may when thawed lose its strength to such a degree that it will not support even light loads. Considerable settlement can occur when the ice in the soil melts, usually resulting in serious damage or even complete failure of the structure. Ground movements sufficient to cause damage to structures can result also from frost heaving caused by winter ice buildup in the active layer. Another feature of permafrost is its sensitivity to heat from the sun or a structure which will cause thawing and loss of stability. Third, it is impermeable to water and impedes drainage.

Prior to construction, site investigations must be carried out to gain adequate knowledge of permafrost conditions for the design and operation of engineering structures. Air photographs, together with information on local climate and terrain conditions, can provide much useful information for a preliminary appraisal of an area. Next, field studies are conducted to determine the extent and nature of permafrost. Variations in the depth of the permafrost table and the thickness of the permafrost over a potential site are determined during this survey. These investigations are particularly important in the southern fringe of the permafrost region where frozen ground occurs in scattered islands and is near thawing.

Field studies must include a program for examining and sampling subsurface materials at the site. This work can be carried out by probing, hand augering, or drilling methods, or by the excavation of test pits. Determination of ice content is particularly important. For example, at Aklavik in the Mackenzie Delta, the top ten feet [3 m] of perennially frozen silt were found to contain 60 per cent of ice by volume. The serious engineering implications are evident, for if this frozen soil thawed there would be a settlement of six feet [2 m] in the top ten feet [3 m] of the ground.

Ground temperature measurements are also an important part of these studies. This is particularly critical in the southern fringe area, where the permafrost temperatures are within one degree of 32°F [0°C]. Further north, the ground temperatures are several degrees lower and consequently several degrees of warming can be tolerated without adversely affecting the strength of the foundation soil.

When general site conditions have been evaluated, further detailed investigations are normally required at the location of individual buildings or structures. The results of these will indicate the approach to be taken in foundation design and the construction techniques to be used. There are four possibilities: disregard permafrost conditions, preserve frozen conditions for the life of the structure, remove frozen conditions or material before construction, thaw frozen ground with the expectation of subsequent ground settlement taken into account by foundation design.

Permafrost can be neglected when structures are located on sands and gravels or bedrock, containing no ice, and conventional design and construction are possible. In the continuous permafrost zone, particularly where fine-grained soils with high ice content occur, every effort must be made to preserve the frozen condition. In the discontinuous zone, it may be convenient to remove the frozen material by thawing or excavation and to replace it with well-drained material not susceptible to frost action; standard foundation designs can then be used. For some structures, in either the continuous or discontinuous zones, it may not be possible to prevent thawing of the ground during the life of the structure, and settlement must therefore be anticipated and taken into account in the design.

Preservation of frozen conditions can be accomplished by either ventilation or insulation; the former is commonly used with heated buildings. Foundations are well embedded in the permafrost, and the structure is raised about the ground surface to permit circulation of air beneath to minimize or prevent heat flow to the frozen ground. Pile foundations placed in steamed or drilled holes have proved well suited to this method, and have been used extensively in northern Canada and elsewhere.

Where pile placing may be difficult, as in very stony soils, alternative foundation designs may prove more economical. Insulation to prevent or reduce thawing of the underlying frozen soil may be achieved by placing a gravel pad on the ground surface where the structure is to be erected. This method is generally limited to small buildings that can tolerate some movement.

For the construction of highways, railways, and airstrips where the ventilation techniques cannot be applied, the insulation method must be relied on to preserve the underlying permafrost. Normally, fill methods are used throughout, and disturbance of the surface cover is kept to a minimum. Cuts through hills are avoided where possible. Proper drainage must be provided to prevent accumulation of water, which would thaw the underlying permafrost, and cause ice buildup, which can block a road during the winter.

Excavation of frozen ground can be difficult and costly because normal excavation techniques are much less effective in permafrost. For some structures, however, it may nevertheless prove economical to excavate the frozen soil, replacing it with coarse-grained material, not susceptible to frost action, on which the foundation can be built. Again, adequate drainage must be provided to take care of seepage water. The procurement of large quantities of fill for road or airstrip construction presents its own problems; suitable sources must be located, cleared of vegetation, and allowed to thaw well in advance of construction operations.

Permafrost complicates the provision of water and sewer services. Only limited year-round sources of water are available because many lakes and streams freeze to the bottom during the winter, and water-bearing layers are only occasionally encountered in permafrost. Normal methods of sewage disposal into the ground are generally prohibited because of the imperviousness of the permafrost. Distribution systems are generally located in insulated boxes (utilidors) on or above the ground surface because of freezing of pipes if they are placed in the active layer or in the permafrost, or problems resulting from excessive thawing of the frozen ground by the contents of the pipes.

The thawing effects on permafrost becomes particularly critical when dams and dykes are constructed on permanently frozen ground and large areas are covered by impounded water. The rate at which thawing will take place and the depth to which thaw will penetrate beneath the water and the water-retaining structures, are of prime importance in the design of their foundations. According to circumstances, the underlying frozen ground can be excavated and the structure placed on bedrock, the frozen condition can be retained by natural or artificial refrigeration, or the embankment can be built up as settlement occurs when the permafrost thaws.

This brief review of the properties and behaviour of permafrost indicates its great significance in the development of northern Canada. Not only do severe climatic conditions hamper all constructional work, but engineering structures may have to rest on soil and rock having properties quite different from those encountered elsewhere in the country. Investigations of the distribution of permafrost in Canada and its engineering properties are continuing. A revised edition of the permafrost map will be issued in a few years incorporating the results.

9 Reconnaissance Vegetation Studies on Western Victoria Island, Canadian Arctic Archipelago

S.A. EDLUND

Introduction

During the summer of 1982 Terrain Sciences Division, Geological Survey of Canada, initiated a Quaternary mapping project on western Victoria Island, west of 110°W (82 D, F-H; 88 A, B; parts of 77 B, C, F, G; 78 B). As part of this group, I studied the flora of the area and the relationships between plant communities and surficial materials. Fieldwork was done from three camps: Cape Wollaston, with J-S. Vincent; Natkusiak Peninsula, with D.A. Hodgson and J. Bednarski; and near Mount Bumpus, Wollaston Peninsula, with D.R. Sharpe and M.F. Nixon (Figure 9.1). During the latter part of July and early August we conducted helicopter traverses to the three areas while based at Holman and in August I conducted detailed studies of plant communities around Holman.

The extensive calcareous tills, ice contact, and glaciofluvial deposits, which cover most of western Victoria Island (Fyles, 1963), provide an opportunity to study calciphilous plant communities on chemically and texturally similar substrates over an elevation interval of more than 500 m and along a 550 km north-south transect, from Richardson Islands along the south coast to Peel Point in the northwest (Figure 9.1).

The glacial deposits are derived from the dominantly calcareous bedrock of Victoria Island described by Thorsteinsson and Tozer

SOURCE: *Current Research, Part B, Geological Survey of Canada,* Paper 83–1B, 75–81 (extract), 1983. Reproduced with the permission of the Minister of Public Works and Government Services Canada, 2001, and courtesy of the Geological Survey of Canada.

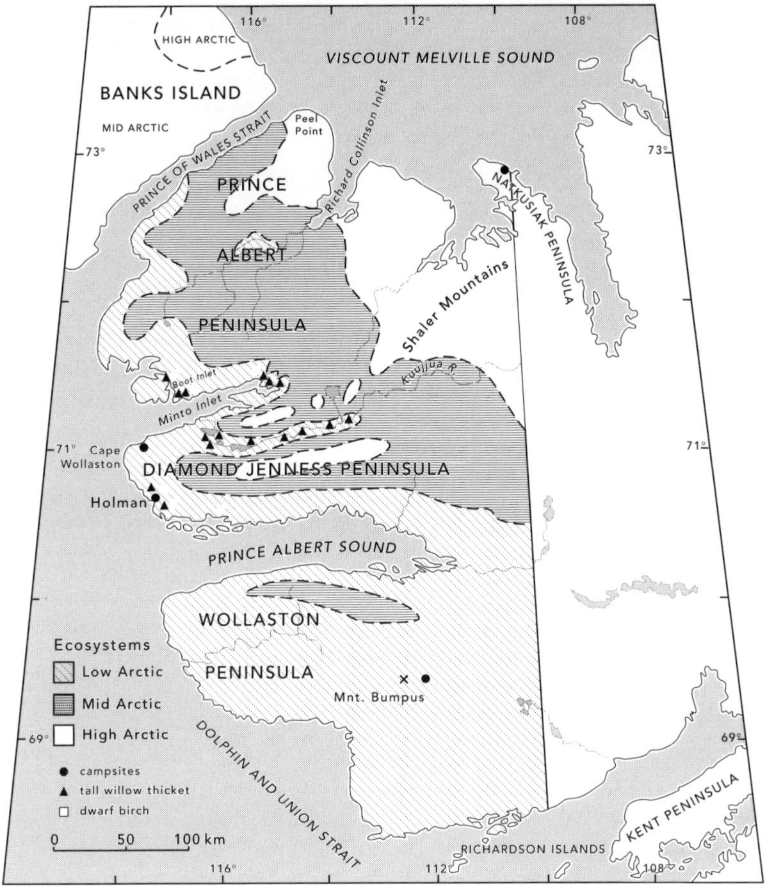

Figure 9.1
Western Victoria Island showing Arctic ecosystems, locations of campsites and of tree-size willows and dwarf birch.

(1962). Noncalcareous extrusive basalt lava flows occur in the central Shaler Mountains; where this type of rock outcrops, the weathering products are predominantly felsenmeer. Only rarely (in local pockets) does the weathered basalt form the gravel and sand required to support vascular plants; well vegetated sites in areas underlain by basalt commonly are veneered by calcareous tills.

Arctic Ecosystems

Low, Mid, and High Arctic ecosystems (Polunin 1951; Young 1971) occur on western Victoria Island. The Arctic ecosystem begins at the limit of coniferous trees and ends in the region where climate no longer permits plants to survive. The progression from Low to High Arctic ecosystems is marked by a decrease in species diversity, major changes in the life form of the plant communities, and a decrease in the per cent plant cover (Figure 9.2). In describing the ecosystems of western Victoria Island, the two extremes – High and Low Arctic ecosystems – are compared and contrasted and then the intermediate Mid Arctic ecosystem is described.

HIGH ARCTIC ECOSYSTEM

The High Arctic ecosystem is dominant in the Queen Elizabeth Islands north of the map area (Edlund 1983a). On western Victoria Island it occurs along the north coast and farther south at high elevations in Shaler Mountains, Prince Albert Peninsula, and Diamond Jenness Peninsula (Figure 9.1). The flora and plant communities in these areas are comparable to those on southern Melville Island (Edlund 1982b), southern Bathurst Island (Edlund 1983b), Cornwallis Island (Edlund 1982c), and Prince of Wales and northern Somerset Island (Woo and Zoltai 1977). Flora of this ecosystem is the least diverse found on western Victoria Island (less than one hundred vascular species), and continuous plant cover is restricted to lower slopes and wetlands.

The vascular component of the plan communities at all but the wettest sites is dominated by dwarf shrub communities which commonly have less than 25 per cent cover but locally reach 50 per cent cover. The calciphile *Dryas integrifolia* (mountain avens) is the most common dwarf shrub, although Arctic willow (*Salix arctica*) is also common in many communities. Two of the most common herbaceous associates are purple saxifrage, *Saxifraga oppositofolia*, and the sedge, *Carex rupestris*, both calciphiles.

Continuous vegetation is generally confined to moist lower slopes and lowlands found locally near the north coast. Sedge meadows with a large grass component are common on these wet areas but are relatively low in graminoid abundance and general diversity, as are other herbaceous species. Arctic willow is common on raised moss hummocks in these wet meadows.

LOW ARCTIC ECOSYSTEM

The Low Arctic ecosystem on western Victoria Island is found on Wollaston Peninsula and northwards along the coast, up Kuujjua River valley, to southwestern Prince Albert Peninsula (Figure 9.1). The rich flora (from 150 to more than 200 vascular species) is found in communities that have a nearly continuous vegetation cover on all but the coarsest and driest materials. *Dryas*, including *D. integrifolia* and *D. punctata*, is again a major component of the vascular plant stratum of

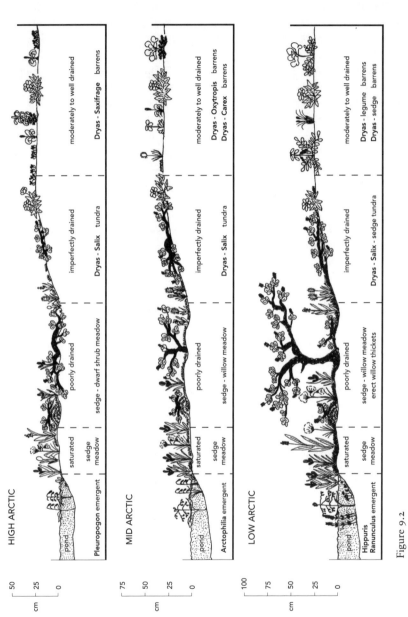

Figure 9.2
Catena showing plant communities of the Low, Mid, and High Arctic ecosystems on the different moisture regimes of calcareous deposits of Victoria Island.

the plant communities on well to moderately drained materials, with Arctic willow a common dwarf shrub associate. Herbaceous associates include a variety of legumes, such as *Oxytropis, Hedysarum*, and *Astragalus*, as well as *Artemisia, Potentilla, Kobresia, Carex*, and various grass species.

The wetlands are generally continuously vegetated with dense and diverse sedge meadows having an abundance of graminoid species and a substantial component of shrubs. The shrub component includes several dwarf shrubs as well as thickets of semi-erect and erect shrubs, commonly willows, which reach 25 to 50 cm in height and may form a nearly continuous low canopy. Heaths, such as *Cassiope tetragona* (Arctic heather) and to a lesser extent *Vaccinium uliginosum* (blueberry), *Rhododendron lapponicum* (Lapland rosebay), *Arctostaphylos rubra*, and *A. alpina* (bear berry), occur with dwarf willow on raised hummocks in wet sedge meadows and on sheltered slopes. Dwarf birch (*Betula glandulosa*) occurs locally in this ecosystem as well, generally in sheltered locations (Figure 9.1).

An unusual feature within this zone is the presence of sites with tree-sized willows in the Minto Inlet area. (Figure 9.1). Willows as high as 1.5 m have been reported near Holman (Porsild 1955; T. Washburn personal communication, 1982); tall willows have been noted in a river valley near the head of Minto Inlet (Peterson et al.1981). Nevertheless, it was surprising to see the size of the felt-leaved willow (*Salix alaxensis*) in many sheltered valleys around Minto Inlet, Boot Inlet, along terraces of Kuujjua River and dunes near its mouth, as well as in extremely sheltered niches near Holman. They form thickets, 1.5 to 5 m in height, with trunk diameters of 5 to 12 cm. Other erect willow species, such as *S. lanata* spp. *Richardsonii* (Porsild and Cody 1980), which usually grow no higher than 0.25 m elsewhere on Victoria Island, also reach heights of up to 1 m in some of these sheltered valleys. The rich Arctic herbaceous flora found nearby or as an understory to the thickets, is similar to the flora of communities near tree line, 400 km to the south.

The dense thickets have developed from shoots from a relatively few willows; preliminary tree ring analysis shows that the largest shoots are fifty years old. Numerous dead trunks, similar in size to the willow shoots sampled, were found within the thickets, but so too were new suckers. Thus, in some places, present day conditions are suitable for continued growth of the thickets.

The willow thickets represent vegetational and floristic oases. They probably form as a result of the special microclimatological conditions prevailing in some deep valleys where steep walls, composed of dark gabbro, absorb and reradiate heat into the valleys. One afternoon

(August 1, 1982), at a site at the head of Minto Inlet, a temperature difference of more than 20°C existed between the plateau (1°C) and valley floor (22°C). Protection from winds, availability of nutrients and moisture, as well as deep snow conditions during winter probably also contribute to the unique flora of these special sites.

MID ARCTIC ECOSYSTEM

The Mid Arctic Ecosystem occurs on central Prince Albert Peninsula, and at moderate elevations in Shaler Mountains and on Diamond Jenness Peninsula, and in a small area of northern Wollaston Peninsula (Figure 9.1). As in the Low Arctic ecosystem, dwarf shrubs, particularly *Dryas* species, dominate all but the wettest areas. The diversity is less than 150 species, however, and fewer types of legumes are associated with these communities than is the case for the Low Arctic ecosystem. Common associates are several *Oxytropis*, *Kobresia*, and *Carex* species, *Pedicularis lanata*, and *Parrya arctica* – all of which are also common in the Low Arctic ecosystem. But the rich diversity of Low Arctic herbs is absent.

Wetlands are characterized by sedge meadows in which the woody component consists of prostrate shrubs, primarily *Salix arctica*, but do include other dwarf shrubs such as *S. reticulata* and *S. polaris*. *Cassiope* is the only heath species, growing in sheltered spots. Species of willow that are normally semi-erect and erect are present, but in this zone they usually take on a prostrate form. The exceptions occur locally where some gnarled individual branches of *Salix lanata* and *S. alaxensis* reach about 25 cm in height.

Comparisons with other Areas in the Canadian Arctic (Figure 9.3)

Western Victoria Island is floristically similar to Banks Island to the west. Banks Island, like Victoria Island, is largely blanketed by calcareous glacial deposits (Vincent, 1982), although on Banks till masks neutral to weakly acidic bedrock. Banks Island, therefore is also vegetated by calciphilous plant communities, predominantly *Dryas integrifolia*, barrens, and wet sedge meadows (Vincent and Edlund 1978), and is characterized by all three Arctic ecosystems. On western Victoria Island, which extends farther south than Banks Island, the Low Arctic ecosystem is more extensive; on Banks Island, it occurs only in the southernmost part of the island and the only low willow thickets are found primarily in Masik River valley and De Salis Bay lowland (Figure 9.3). Nowhere on Banks Island do willows attain heights seen in the Minto Inlet area, and while dwarf birch is locally common on

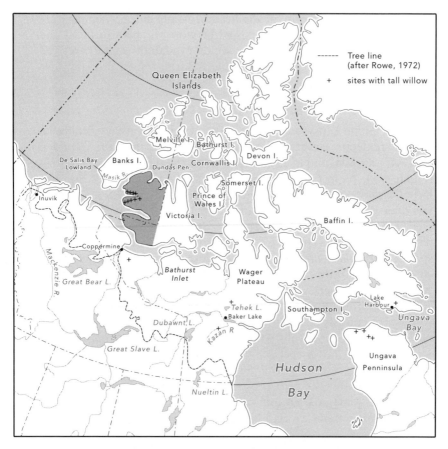

Figure 9.3
Location of tall willows in northern Canada and their proximity to tree line.

Victoria Island, only one clump on a south-facing slope in Masik River valley has been reported on Banks Island (Kuc 1970).

On Banks Island, the Mid Arctic ecosystem is dominant whereas on western Victoria Island Mid Arctic-type vegetation is much less common. The High Arctic ecosystem on Banks Island is confined to the northeastern plateau (Figure 9.1) and the foggy, low-lying northwest coast adjacent to the permanent ice pack whereas on western Victoria Island it occurs on the northern coast and in mountainous areas.

The High Arctic plant communities of northwestern Victoria Island are most similar to those of the southern part of Dundas Peninsula, Melville Island (Edlund 1982b), north of the study area, which is also covered by calcareous till probably derived from materials from Victoria Island (Hodgson et al., in press). Similar plant communities also develop on Bathurst and Cornwallis islands on materials derived from local, moderately calcareous bedrock. Woo and Zoltai (1977) found similar communities on calcareous substrates on Prince of Wales Island and northern Somerset Island.

The flora and plant communities of western Victoria Island are quite different from those found on materials derived from noncalcareous rocks of the Canadian Shield. In north-central District of Keewatin [Kivalliq], south of the study area, the Low Arctic communities are dominated by heath species such as *Cassiope tetragona* (heather) and *Ledum palustre* ssp. *decumbens*, (Labrador tea), and commonly possess a thick, nearly continuous carpet of foliose, squamulose, and fruiticose lichens, particularly *Cladonia* and *Alectoria* species, and mats of mosses such as *Rhacomitrium* and *Polytrichum* (Edlund 1982a). A similar dense and diverse cryptogamic stratum is not present on the calcareous gravels of Victoria and Banks islands; neither are heath-dominated communities common, but species characteristic of heath are found locally in hummocky wetlands and in areas having wet, snowpatch conditions where the local reducing environment of the soil results in a lowered pH (6.8 to 7.2). In spite of these distinctions, which include a general lack of overlapping dominant species, physiognomic similarities exist between the plant communities on western Victoria Island and northern District of Keewatin [Kivalliq]. The Low Arctic ecosystem in both regions has nearly continuous vegetation, and the vascular plant stratum of all but the wettest soils is dominated by dwarf shrubs. For example, on the highest parts of the Wager Plateau (Edlund 1982a) of northern Keewatin [Kivalliq] (Figure 9.3), the vascular plant stratum is dominated by dwarf shrubs, similar to that on Wollaston Peninsula, Victoria Island. At lower elevations surrounding Wager Plateau, semi-erect willows, particularly *Salix phylicifolia* and *S. alaxensis*, and dwarf birch (*Betula*) appear generally with less than 10 per cent cover; in the southern part of the plateau around Tehek Lake (north of Baker Lake), dwarf shrubs are a major ground cover, and in sheltered valleys and along some river terraces, thickets of willow up to 80 cm in height are found. This physiognomy is similar to that found in the Low Arctic ecosystem of Victoria Island.

Within the Low Arctic ecosystem, to find willow thickets of a comparable size to those in the Minto Inlet area, one must look farther afield (Figure 9.3). The Bathurst Inlet and Tree River areas south of Victoria Island, not far from tree line, harbour such thickets in some sheltered valleys (W. Blake, Jr and B.G. Craig, personal communication, 1983). Tree-sized willows are also found south of Baker Lake in the Kazan River area of Keewatin [Kivalliq], between Ford Lake and Thirty Mile Lake (A.N. LeCheminant, personal communication, 1983). Maycock and Matthews (1966) described *Salix alaxensis* thickets reaching 4.6 m in height in northern Ungava Peninsula, Quebec and they refer to reports by J.D. Soper of 3.7 m-high willow (probably *S. plainfolia*) near Lake Harbour, southern Baffin Island. All such communities are extremely local and rare. It appears that the occurrence of the willow "forests" at Minto Inlet are the northernmost ones known in North America.

Bioclimatic Zones

Victoria Island has several zones of biogeographical and bioclimatological importance. The High Arctic ecosystem, which extends across the Queen Elizabeth Islands to the north, has its southern limit on Victoria Island. The High Arctic ecosystem on Victoria Island lies entirely within the richest subzones of the High Arctic, where dwarf shrubs are the major vascular plants (Edlund 1983a). Farther north, in the Queen Elizabeth Islands, dwarf shrubs are no longer dominant. The boundary between the zone where dwarf shrubs are dominant and that where dwarf shrubs occur only locally may be thought of as a 'mini-forest line'. The northernmost limit of dwarf woody species is the 'mini-tree line'.

All of Victoria Island lies well within the 'mini-forest zone'. Dwarf shrubs are dominant on moderately to well drained substrates. They are absent from plant communities only in areas where local snowbeds persist well into mid and late July; in such places, the communities are floristically similar to those seen farther north above the 'mini-tree line'.

This concept of broad bioclimatic zonation of the ecosystems can be applied to other parts of Victoria Island as well. The southern boundary of the High Arctic ecosystem is, in effect, the limit of erect and semi-erect shrubs – an 'erect shrub limit'. Throughout the Mid Arctic ecosystem, erect and semi-erect shrub species are locally present in poorly drained areas but are less than 25 cm high and have low percentages of cover. In the Low Arctic ecosystem, the semi-erect and erect shrubs of the low shrub forest zone generally reach 0.5 m high and commonly form thickets on poorly drained, sheltered sites; these represent a local erect shrub forest. The tree-sized willow thickets in valleys around Minto Inlet represent isolated tall willow-forest zones in an area far removed from tree line.

Further research may tie these biogeographical observations with climatological parameters, possibly coincident with isotherms such as the mean July isotherms, as suggested by Edlund (1983a) for the central Queen Elizabeth Islands.

Acknowledgments

Excellent field support was provided by the Polar Continental Shelf Project and R. Frost, pilot with Quasar helicopters. Special thanks are given to my colleagues, J. Bednarski, D.A. Hodgson, F.M. Nixon, D.R. Sharpe, and J-S. Vincent who allowed me to work from their field camps, provided useful discussions about the surficial materials, and brought plant observations and samples from areas I personally did not visit. I wish to thank D.R. Sharpe and J.V. Matthews, Jr who critically reviewed this report.

References

Edlund, S.A. 1982a. Plant communities on the surficial materials of north-central District of Keewatin. Geological Survey of Canada, Paper 80–33.

– 1982b. Vegetation of Melville Island, District of Franklin: eastern Melville Island and Dundas Peninsula. Geological Survey of Canada, Open File 852.

– 1982c. Vegetation of Cornwallis, Little Cornwallis, and associated islands, District of Franklin. Geological Survey of Canada, Open File 857.

– 1983a. Bioclimatic zonation in a High Arctic region: central Queen Elizabeth Islands. In Current Research, Part A, Geological Survey of Canada, Paper 83–1A, 381–90.

– 1983b. Vegetation of the Bathurst Island area, Northwest Territories. Geological Survey of Canada, Open File 888.

Fyles, J.G. 1963. Surficial geology of Victoria and Stefansson Islands, District of Franklin. Geological Survey of Canada, Bulletin 101.

Hodgson, D.A., Vincent, J-S., and Fyles, J.G. Quaternary geology of central Melville Island, Northwest Territories. Geological Survey of Canada, Paper 83–16. (in press)

Kuc, M. 1970. Vascular plans from some localities in the western and northern parts of the Canadian Arctic Archipelago. Canadian Journal of Botany 48, no. 11, 1931–8.

Maycock, P.F., and Matthews, B. 1966. An Arctic forest in the tundra of northern Ungava, Quebec. Arctic 19, 114–44.

Peterson, E.G., Kabzems, R.D., and Levson, V.M. 1981. Terrain and vegetation along the Victoria Island portion of a Polar Gas combined pipeline system. Report of Polar Gas Environmental Program (Toronto). Prepared by Western Ecological Services (BC) Ltd.

Polunin, N. 1951. The real Arctic: suggestions for its delimitation, subdivision and characterization. Journal of Ecology 39, 308–15.

Porsild, A.E. 1955. The Vascular Plants of the Western Canadian Arctic Archipelago. National Museum of Canada, Bulletin 135.

Porsild, A.E., and Cody, W.J. 1980. Vascular plants of Continental Northwest Territories, Canada. National Museum of Canada, Ottawa.

Rowe, J.S. 1972. Forest Regions of Canada. Department of the Environment, Canadian Forestry Service Publication no. 1300.

Thorsteinsson, R., and Tozer, E.T. 1962. Banks, Victoria and Stefansson Islands, Arctic Archipelago. Geological Survey of Canada, Memoir 330.

Vincent, J-S. 1982. The Quaternary History of Banks Island, N.W.T., Canada. Géographie physique et Quaternaire. 36, no. 1–2, 209–32.

Vincent, J-S., and Edlund, S.A. 1978. Extended legend to accompany preliminary surficial geology maps of Banks Island, Northwest Territories. Geological Survey of Canada, Open File 577.

Woo, V., and Zoltai, S.C. 1977. Reconnaissance of the soils and vegetation of Somerset and Prince of Wales Islands. NWT Northern Forest Research Centre, Information Report NOR–X–186.

Young, S.B. 1971. The vascular flora of St. Lawrence Island, with special reference to floristic zonation in the arctic regions. Contributions from the Gray Herbarium of Harvard University, no. 201, 11–115.

10 Climate and Zonal Divisions of the Boreal Forest Formation in Eastern Canada[1]

F. KENNETH HARE

The boreal forest formation is the great belt of coniferous forest stretching across the Subarctic latitudes of Eurasia and North America. Perhaps because of its inaccessibility and unsuitability for permanent rural settlement, the boreal forest has been little studied by geographers in North America, though Russian and Scandinavian workers have devoted much time to it. The phytogeographer and the ecologist have been concerned primarily with the southern part of the formation, which is the main home of the lumbering industries of Canada, Scandinavia, and the USSR. The greater part of the formation, beyond the limit of merchantable trees, has received scant attention.

The idea that the boreal forest consists of an endless repetition of muskeg and forest, with little difference from one place to another, is a popular delusion. In both Eurasia and eastern North America the formation resolves itself naturally into three or four zonal divisions, each with its characteristic type of cover and each standing in a definite and predictable relationship to climate. This paper is devoted to a study of these zonal divisions in Canada east of the Ontario-Manitoba boundary, and particularly in the great Labrador-Ungava Peninsula, where the author himself has worked.

[1] The author wishes to express his gratitude to Dr Ilmari Hustich, pioneer of forest studies in Labrador, to Dr Pierre Dansereau of the Department of Botany, University of Michigan, and to Mr Harry Lash of the Department of Geography, McGill University, for advice and assistance in the preparation of this paper.

SOURCE: *The Geographical Review* 40, no. 4 (Oct. 1950): 615–35. An Appendix Table of Climatological Stations has been deleted from the original version. Reprinted by permission of the author and publisher.

Labrador-Ungava offers many advantages to the student of zonal structure of the boreal forest. The peninsula is essentially a tilted peneplain, relatively low in latitude 58 degrees (near the Arctic tree line) but rising southward to a general level approaching three thousand feet [900 m]; some fifty to one hundred miles [80–160 km] north of the St Lawrence. Toward the river it falls in a spectacular and highly complex escarpment, the Laurentide scarp, which has yet to appear on most contour maps of North America. The southward rise of altitude has the effect of offsetting to some extent the normal southward rise of temperature. The thermally controlled zonal divisions of the boreal forest are hence widened and can be studied over much greater distances than in Russia or Finland. The same is true on a smaller scale between Hudson Bay and Lake Superior.

A further advantage springs from the fact that the entire eastern half of the Canadian boreal forest has an abundant precipitation. Control of growth by temperature, well known to be the usual climatic control of natural vegetation in high latitudes, is thus manifested in full measure. Drought effects such as those reported by Marie Sanderson in western Canada are unknown.

The boreal forest formation in east Canada, as elsewhere, is bounded on the north by the tundra. On the south, it passes into a mixed forest formation, the Great Lakes–St Lawrence Forest of Halliday or the Lake Forest of Weaver and Clements. A similar mixed forest borders the Russian boreal forest west of the Urals. As with the zonal divisions of the boreal forest itself, both the southern and northern limits are readily determinable in climatological terms.

Composition

The chief associations of the formation are dominated by white spruce (*Picea glauca*), black spruce (*P. mariana*), larch or tamarack (*Larix laricina*), and balsam fir (*Abies balsamea*). The jack pine (*Pinus banksiana*) is also an important element in the western half of the region. Tree lines of these species are given in Figure 10.1, but a discussion of their climatic relations is not attempted in this paper.

The black spruce, white spruce, and tamarack have almost identical northern tree lines, and all three range throughout the formation in North America, though the tamarack is rare in Alaska. Their relative abundance, however, varies greatly. In Labrador-Ungava, black spruce is overwhelmingly the most common. In northern and central districts it occurs either alone or in association with tamarack: in the south and southeast black spruce-balsam fir is the principal association.

Balsam fir and jack pine have northern tree lines for which no exact climatic equivalent can be found. The jack pine, for example, has made

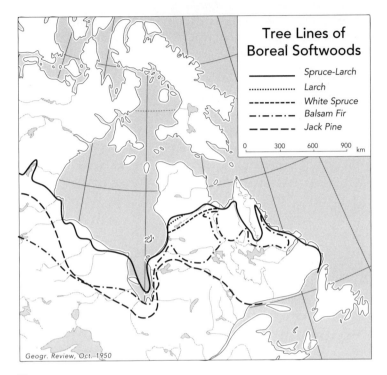

Figure 10.1
Tree lines of boreal softwoods. Note absence of the jack pine from much of Labrador-Ungava. The two spruces and the larch have coincident tree lines in most places

little progress into Labrador-Ungava, in spite of an apparently favourable climate.

Associated with the dominant coniferous species is a small group of hardwoods having a widespread distribution in the boreal forest in both North America and Eurasia, though the actual species differ between the two land masses. The North American representatives are the white birch (*Betula papyrifera*), balsam poplar (*Populus balsamifera*), aspen (*Populus tremuloides*), and certain alders (*Alnus* spp.). None of these trees form part of the climax, but all are important elements in the successions.

Birch and aspen attain their greatest significance in areas recently burned. Huge areas of birch-aspen associes[2] extend today over the fire devastated parts of Labrador-Ungava and northern Ontario. Beneath

2 An "associes" is the successional equivalent of "association." The birch-aspen communities of the boreal forest are short-lived, yielding place to the coniferous forest that is climax in the formation; hence the use of "associes" rather than associations.

the pale-green foliage of the hardwoods, spruce, larch, and fir seedlings grow rapidly, and in a few decades the climax coniferous association is re-established. White birch and aspen also occur as individual relict trees in the coniferous associations, though aspen does not appear in northern Labrador-Ungava.

Balsam poplar occurs widely throughout the region, though it is always local in distribution. The alders are found almost exclusively along watercourses and lake shores, forming impenetrable thickets through which landings can hardly be forced.

Along the southern margins of the formation the boreal forest is invaded by isolated individuals or groves of species proper to the Great Lakes–St Lawrence mixed forest formation. Among the softwoods, these include the white cedar or arbor vitae (*Thuja occidentalis*), white pine (*Pinus strobus*, a dominant species of the mixed forest), and red pine (*P. resinosa*). The hardwoods include black ash (*Fraxinus nigra*), yellow birch (*Betula lutea*), and bigtooth aspen (*Populus grandidentata*). The cedar and black ash invade the boreal forest deeply, but the others penetrate only a short distance. At the southern limit of the boreal forest, white and red pines and sugar maple become dominants.

Forest Types

The boreal forest formation is not readily divisible into associations, as is the deciduous forest. Partly because of the small number of species, partly because the region has been heavily glaciated and hence suffers from deranged drainage and highly variable soils, the forest exhibits structural types rather than fixed associations. These types differ as to the spacing of trees, the layering of the vegetation, and the nature of the ground cover. Many of them are definitely not "climax" and probably represent successional stages that must soon give way to a higher form of forest. The principles of succession in the boreal forest are, however, too vaguely understood to allow a genetic classification. Accordingly the forest types are ranked as equals in the following discussion, regardless of their status.

The classification of forest types used here is based on a study by the Finnish ecologist Hustich.[3] He has studied both the Labrador and the Scandinavian-Finnish boreal forest, and his classification for the Labrador-Ungava region can be cross-referred to both the Russian Taiga and the forests of western Canada as studied by Raup. Hustich's detailed subdivisions are not considered here, and new English terms are suggested for the main types. These are three in number:

3   Ilmari Hustich, "On the Forest Geography of the Labrador Peninsula: A Preliminary Synthesis," *Acta Geographica* 10, no. 2 (1949): reference on 36–42.

1   The *close-forest type* ("Moist Series" of Hustich) is a continuous stand of closely spaced trees, in Labrador-Ungava usually a black spruce-balsam fir association. Such stands occur on well-drained land with a water-retentive soil and hence with abundant but not excessive moisture. The ground vegetation is rich in mosses, especially feather mosses, and there are some characteristic small herbs such as bunch berry (*Cornus canadensis*) and wood sorrel (*Oxalis montana*).

2   The *lichen-woodland* type ("Dry Series" of Hustich) consists of open stands of trees with a thick and beautiful floor of lichens. Black and white spruce, tamarack, and jack pine all occur in such woodland, though the spruces are overwhelmingly the dominants. The lichen floor consists of a layer several inches [cm] thick of the pale-gray, purple, orange, or green fruiting bodies of *Cladonia*, the genus to which the so-called "reindeer mosses" belong. The trees are from two to twenty-five yards [2–23 m] apart. Black spruce tends to assume a beautiful "candelabrum" form in the lichen-woodlands. This type occurs only on drier sites in the south but is widespread in central and northern districts. Various subtypes of lichen-woodland probably cover more than 60 per cent of the Labrador-Ungava plateau and are widespread throughout the formation in North America and Eurasia. The term "woodland" is used to suggest the wide spacing of the trees.

3   The *muskeg* type ("Wet Series" of Hustich) occurs on badly drained ground and is variable in appearance. Black spruce and to a smaller extent, tamarack are the typical trees. Both are slow-growing and slow-reproducing, appearing like gaunt sticks often largely devoid of branches or green leaves. The wet ground is covered by sphagnum mosses and certain shrubs, of which Labrador tea (*Ledum groenlandicum*), leatherleaf (*Chamaedaphne calyculata*), and a heath (*Kalmia angustifolia*) are representative.

These clear-cut forest types occur throughout the formation in eastern Canada. The zonal divisions about to be discussed are definable in terms of the relative frequency of the forest types. This important principle has only recently emerged as the basis of division in the boreal forest.

The Zonal Divisions

We come now to the first of the two main purposes of this article – the definition of zonal divisions within the formation. The argument closely parallels that of Hustich, who laid down major forest regions for Labrador-Ungava in 1949. The basis of definition and the precise lim-

TABLE 1
PROPOSED ZONAL DIVISIONS OF THE BOREAL FOREST IN LABRADOR
UNGAVA

Zonal division	Hustich's term	Dominant forest type
Forest-Tundra Ecotone	Forest-Tundra	Thin lichen-woodland in valleys; pure tundra on interfluves
Open Boreal Woodland	Taiga	Lichen-woodland
Main Boreal Forest	Southern Spruce Forests	Close-forest
Boreal-Mixed Forest Ecotone		Close-forest containing Great Lakes-St Lawrence indicators

its of the divisions proposed here differ significantly from his. In the next section the climatic relations of each of the divisions are reviewed.

South of the tundra four zonal divisions are proposed. These are arranged in north-south sequence in Table 1.

The proposal zonal boundaries in Labrador-Ungava are shown in Figure 10.2. Where these differ from those of Hustich, the differences are based on an inspection of aerial photographs, on flights across the boundary zones, and in a few cases on ground traverses. The map is in any case provisional and will demand continual revision.

The dominant forest types refer to areas of well-drained soil. Throughout the formation areas of poor drainage are covered by treeless swamps and muskegs that differ little from zone to zone. The detailed character of the cover in each zone is beyond the scope of this report, but some further description is necessary.

The *forest-tundra ecotone* extends across northern Labrador-Ungava from the Hudson Bay coast to the Atlantic. Here it is truncated by the pure coastal tundra, which runs along the Atlantic shore to the Strait of Belle Isle. Along the line of the Torngat uplift, running north-south just east of the George River, the ecotone is narrow and is displaced southward by the greater altitude. The forest-tundra ecotone[4] is the zone in which associations of the tundra and boreal forest formations intermingle. The boreal forest is represented by long strings of lichen-woodland along the chief rivers, but the interfluves are covered by pure tundra entirely free of trees. Permafrost is widespread throughout the zone. The northern limit is the Arctic tree line. The southern,

4 The term "forest tundra" is a literal translation of the Russian term *lyesotundra* for the same belt in European Russia and Siberia. "Forest-tundra ecotone," a preferable form in English, was introduced by J.W. Marr, "Ecology of the Forest Tundra Ecotone on the East Coast of Hudson Bay," *Ecological Monographs* 17 (1948): 117–44.

the line along which the lichen-woodland covers interfluves as well as valley floors, is known with some confidence from traverses by Rousseau on the George River, Low on the Kaniapiskau River, and Polunin at Lac Bienville. The present author was also able to traverse the boundary by air near the headwaters of the Whale River.

The *open boreal woodland* as a term seems preferable to Hustich's "taiga," since the latter is applied by Russian ecologists to the entire formation and is so understood by geographers everywhere. As the present term implies, this zone is dominated by enormous stretches of the lichen-woodland forest type; tall and well-developed spruce (more rarely other conifers) stand several yards apart in a sea of *Cladonia*. On the wetter ground muskeg supervenes, with stunted trees, Labrador tea, and sphagnum. Close-forest types are absent over most of the zone but become abundant on steeply sloping ground near the southern boundary. This boundary – one of the most significant economic limits on the continent, as it is the virtual northern limit of lumbering – is defined as that along which close-forest exceeds lichen-woodland in area.

The open boreal woodland presents one of the most picturesque, colourful, and extensive landscapes of the continent and is equally important in the Eurasian boreal forest. The beauty of its *Cladonia* floor, which retains the impress of footprints for years and whose pastel shades defy the colour film, is still largely unknown to North Americans, since convenient routes nowhere penetrate its solitudes.

The *main boreal forest* is far better known, for it yields more than 90 per cent of the pulpwood cut of eastern Canada. Along its entire length it is penetrated by railways, logging roads, and power-generating rivers. In eastern Canada, it extends from north of Lake Superior across southern Labrador-Ungava to Anticosti Island and Newfoundland. Large outliers cover inland Gaspé, the highlands of New Brunswick, and parts of Maine.

This zone is largely covered by close-forest associations of black spruce and balsam fir east of Lake St John, white spruce and balsam fir to the west, thus approaching the traditionally accepted boreal climax. Lichen-woodlands occur only on dry soils and appear to be a late stage of the xerosere.[5] Muskeg, with black spruce and tamarack as dominants, is again common, especially on the dreary plains south and west of James Bay.

The northern limit of this zone was first defined by Halliday, who consolidated traverse records from many transverse valleys crossing the boundary. Hustich accepted Halliday's line with few exceptions. An accurate determination of its position on the Romaine River by

---

5 The succession of covers achieved as the forest extends over dry surfaces like rocky outcrops or sand plains, which abound in this region.

H.N. Lash and N. Drummond showed, however, that Hustich's line was too far north. The position given on Figure 10.2 incorporates their results.

An important outlier of the main boreal forest covers the lowlands around the head of Lake Melville and in the Hamilton Valley. This favoured region of Labrador has much close-forest, though lichen-woodland is extensive on sand plains and gravels. It was formerly believed that the richness of this vegetation sprang from deeper and more fertile soils developed on the Proterozoic sediments contained in the basin. It is now obvious, however, that this outlier is a climatic effect, a subject treated below. A revised version of Halliday's map includes another outlier, the middle Kaniapiskau Valley, on the strongly folded Proterozoic sediments of the Labrador Trough. Here the vegetation consists of lichen-woodland and is included in the open boreal woodland of Figure 10.2.

The southern limit of the main boreal forest is the line along which the white pine-maple associations of the Great Lakes-St Lawrence mixed forest formation replace the spruce-fir of the boreal forest. The position of this boundary as shown on Figure 10.2 is taken without change from *Native Trees of Canada.*[6] The Lake St John basin forms a conspicuous enclave of Great Lakes–St Lawrence associations within the main boreal forest.

The ecotone between these two formations is less easy to define than the forest-tundra. On Figure 10.2 the northern edge of the ecotone is taken as the tree line of white and red pine. However, certain elements of the Great Lakes–St Lawrence forest, notably the white cedar and black ash, extend well beyond this line.

Climatic Relations

Climatic correlation has been impossible in the past because of the lack of inland climatological stations. Not until 1937 was a station established in the interior of Labrador-Ungava, a region equal in area to the United States east of the Mississippi and south of a latitude of 40° North. Since then, however, many stations have been opened by the Canadian Department of Transport and the United States Air Weather Service, and a rudimentary climatological network is now in operation. In 1947 the author began preparation of a report on the climatology of the region, and extensive use has been made below of materials gathered during this investigation.

[6] *Native Trees of Canada,* Forest Service Bulletin 61, 4th ed. (Ottawa, 1949), see map inside covers.

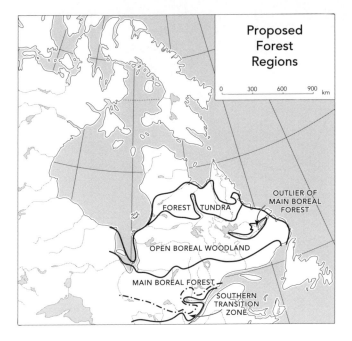

Figure 10.2
Proposed zonal divisions of the boreal forest in Labrador-Ungava. The term "region" is applicable because the map was originally drawn to establish natural regions in the peninsula; "zonal division" is a better term as applied to the boreal forest as a whole. The classification closely follows Hustich, but the terminology and the position of boundaries are revised. "Southern transition zone" refers to the ecotone between the boreal and Great Lakes–St Lawrence forests.

Raw climatic data have little application in ecoclimatology. Effort must be made to find means of combining and integrating the elements into indices having a more direct applicability to ecological problems. The best available system is C.W. Thornthwaite's classification of 1948,[7] and it is used here. An account of the climates of Canada as a whole according to this new classification system has already been published by Sanderson. A considerably denser network of stations has been used in the preparation of Figures 10.3 and 10.4, however, and these maps differ in detail from those of Sanderson.

Annual potential evapotranspiration is the function used by Thornthwaite to establish the degree of thermal efficiency possessed by

7 C.W. Thornthwaite, "An Approach Toward a Rational Classification of Climate." *Geogr. Rev.* 38 (1948): 55–94.

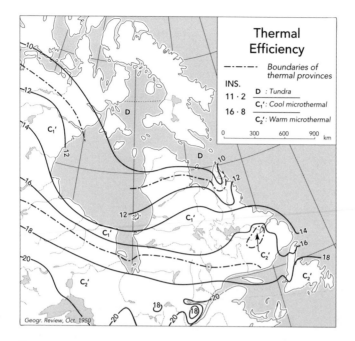

Figure 10.3
Thermal efficiency, expressed in terms of potential evapotranspiration (in inches),
according to Thornthwaite's classification of 1948.

a climate. It is an accumulating logarithmic function of monthly mean
temperatures, regarded as expressing thermal efficiency on the basis of
a presumed analogy with the control of growth rates by temperature.

Figure 10.3 shows annual potential evapotranspiration over eastern
Canada and the boundaries of the thermal provinces suggested by
Thornthwaite. The D'C_1' boundary runs north of Baker Lake, across
the Ungava Peninsula from Portland Promontory to Payne Bay, and
across the northernmost part of the Torngat massif. The C_1'/C_2' boundary (separating cooler and warmer microthermal provinces) runs from
the Hayes River near Gods Lake across James Bay to the northern tip
of the Long Peninsula of Newfoundland. Thus the greater part of
Labrador-Ungava falls into the cool microthermal province (C_1'). All
the southern districts lie in the warm microthermal province (C_2');
mesothermal climates do not occur within the area of Figure 10.3
though they are found in the St Lawrence lowlands near Montreal.

The records from Goose Bay airport show that an important outlier
of warm microthermal climate occurs around the head of Lake
Melville and the lower Hamilton Valley. Fragmentary records from
North West River and the Hamilton Valley confirm that this is an

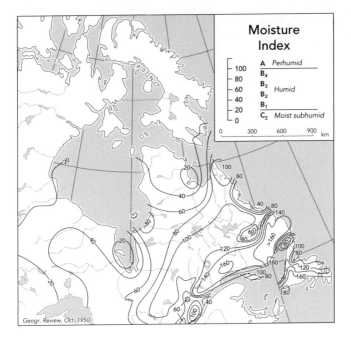

Figure 10.4
Moisture index, according to Thornthwaite's classification of 1948. Note high values typical of southern Labrador-Ungava, low values along western margin.

extensive area, but its form can at present only be sketched in relation to the terrain.

The moisture index, the other main element in Thornthwaite's classification, is shown in Figure 10.4. Labrador-Ungava, Newfoundland, and northern Ontario have an abundant well-distributed precipitation and rank almost exclusively as humid or perhumid. The southern half of Labrador-Ungava has indices of more than 100 (A; perhumid), as does most of Newfoundland. Highest values occur along the Laurentide scarp belt, just north of the Gulf of St Lawrence, and in southern Newfoundland. Conspicuously drier areas include the structural depressions of Lake St John and Lake Melville, and also the Atlantic coastal strip from Nain to Cape Harrison. Farther north and west indices are lower but everywhere exceed 20, except in the James Bay and Ungava Bay depressions. This is in striking contrast to the condition reported for the western boreal forest by Sanderson, who found indices ranging from below −20 to about +20, i.e. between arid and moist subhumid.

There is some doubt about the extent of drought in the James Bay region. The isopleth of 20 in the centre is based on values computed for Fort Albany (17), Moose Factory (19), and Fort George (33). It is possible, however, that these values are too low, perhaps because of faulty exposure of the snow-measuring site. More recent observations at Moosonee, a first-order station staffed by trained professional observers, indicate a heavier winter snowfall than at nearby Moose Factory, and the index stands at 74. The point is academic, since almost the entire district is covered by muskeg in which bad drainage upsets the normal moisture cycle.

Correlations between Climate and the Zonal Forest Divisions

It now remains only to establish the relation between the climate distributions and the zonal divisions of the forest. Such attempts were previously made by Halliday and Villeneuve for various parts of the region, but in both cases before data were available from the interior of Labrador-Ungava and before Thornthwaite's new system was published. Villeneuve's maps do not extend beyond the 51st parallel.

Elsewhere, a good deal of work has been done on the growth conditions of the coniferous trees composing the boreal forest climax. With few exceptions, all these investigations have suggested that midsummer temperatures control growth rates and that precipitation is largely ineffective as a control. Thus Hustich reported that the width of the annual rings in Scotch pine (*Pinus sylvestris*) at Utsjoki, Lapland, was closely correlated with July mean daily temperature; vertical growth of the trees was likewise related to the July temperature of the previous year. Neither vertical nor radial growth was related to variations in summer precipitation. Similar results were obtained by Erlandsson and other Scandinavian workers for Scotch pine and other boreal forest conifers. Giddings stressed the dependence of the spruce on summer temperature in both Alaska and the Mackenzie Valley. The widely accepted view that the northern forests are governed in their growth by temperature, and that precipitation is everywhere adequate to supply the needs of the growth possible under such cool conditions, was also accepted by Thornthwaite in his earlier climatic classification.

The present investigation amply confirms this view. A comparison of Figures 10.3 and 10.4 with Figure 10.2 shows at once that there is an obvious correlation between the zonal forest divisions and thermal efficiency (that is, potential evapotranspiration); the interdivisional boundaries tend to follow the isopleths of potential evapotranspiration. On the other hand, there is no obvious correlation between the moisture provinces and the forest divisions. No evidence whatever has been found to suggest any control of the forest structure by the moisture factor, other than the effects of poor drainage in the muskeg. Table

TABLE 2
FOREST DIVISIONS AND POTENTIAL EVAPOTRANSPIRATION IN
LABRADOR-UNGAVA

Division	Typical Value of P-E Along Boundaries (inches)	Dominant Cover Type
Tundra		Tundra
	12.0–12.5	
Forest-Tundra Ecotone		Tundra and lichen-woodland intermingled
	14.0–14.5	
Open Boreal Woodland		Lichen-woodland
	16.5–17.0	
Main Boreal Forest		Close-forest with spruce-fir associations
	18.5–19.0	
Boreal–Mixed Forest Ecotone		Close-forest with white and red pine, yellow birch, and other non-boreal invaders
	20.0	
Great Lakes–St Lawrence Mixed Forest		Mixed forest

2 shows how close is the correlation between thermal efficiency and the forest divisions. With trifling exceptions, the interdivisional boundaries follow closely the isopleths of potential evapotranspiration suggested in Table 2.

The Arctic tree line, the southern limit of the tundra, nowhere reaches Thornthwaite's theoretical potential evapotranspiration value of 11.2 inches [285 mm]. Near the mouth of the George River, which affords a good migration route northward, a thin stand of black spruce and tamarack reaches the value of 11.5 inches [292 mm], but elsewhere the tree line lies between 12 and 12.5 inches [305-318 mm]. Marr has shown, however, that white spruce is actively invading the tundra near the Great Whale River, and it may well be that there has not been time since the Wisconsin glaciation for the forest to attain its climax tree line.

Near Richmond Gulf on the Hudson Bay coast and south of Hebron on the Atlantic coast, well-developed black and white spruce groves stand well north of the isotherm of 10°C (50°F) for the warmest month.

The narrow strip of coastal tundra fringing the Atlantic coast of Labrador as far south as Belle Isle was formerly thought to be the reflection of the chilling effect of the Labrador Current and its pack ice. Climatological stations directly on this coastal tree line, however, show that the thermal efficiency is adequate to support lichen-woodland

(Hopedale, 15.3 inches [389 mm], Cartwright, 15.3 inches [389 mm], Belle Isle, 14.5 inches [368 mm]). Evidently the temperature is not alone responsible for the lack of trees.

The open boreal woodland extends between the potential evapotranspiration values of 14.0–14.5 inches [355–368 mm] and 16.5–17.0 inches [419–432 mm]. Its southern limit coincides with Thornthwaite's suggested divide between warmer and cooler microthermal climates (C$_1$'/C$_2$'). In other words, the forest-tundra ecotone and open boreal woodland correspond with the cool microthermal province.

Wide variations in moisture index occur within this division, without any apparent effect on the vegetation. Since moisture is abundant everywhere, it is of interest to speculate as to the origins of the curious structure of the dominant lichen-woodland, with its widely spaced but fully developed trees and its lichen floor that requires little moisture. Farther south lichen-woodland is definitely a drought type, confined to sandy or gravelly soils.

It may well be that the widespread character of this dry type in the open boreal woodland results from physiological drought, as Schimper has called it. Where frost in the soil is still unthawed in July, the season of peak growth, the trees can derive moisture only from the topmost layers of the soil and hence are driven to assume a horizontally developed root system. Competition between neighbouring trees must then mean that the space between individuals has to be greater.

The main boreal forest occupies the range of potential evapotranspiration between 16.5–17.0 inches [419–432 mm] and 20.0 inches [508 mm]. It is invaded by indicators of the Great Lakes–St Lawrence formation as far north as the 18.5-inch [470-mm] and 19.0-inch [483-mm] isopleths. The southern limit, 20.0 inches [508 mm], is faithfully followed from Lake Superior to Gaspé; the 20.0-inch [508-mm] isopleth even curves around the Lake St John lowland, with the little enclave of mixed forest mentioned above. It is to be noted that Thornthwaite's microthermal-mesothermal boundary (C$_2$'/B$_1$') is 22.4 inches [469 mm]: the main boreal forest and the Great Lakes-St Lawrence formation thus meet in the middle of the warm microthermal province.

The small outlier region of warm microthermal climate in the Lake Melville-Hamilton River region coincides reasonably well with the detached area of main boreal forest in the same districts. Though poor in species, the well-developed close-forests of this region offer one of the largest untapped reserves of pulpwood in eastern North America. The region has a thermal efficiency similar to that of the forests now being cut near Clarke City and on Anticosti Island.

The Remaining Areas

Although the delimitation of zonal divisions has not yet been undertaken beyond Labrador-Ungava, a few comments may be made concerning other parts of eastern Canada.

Native Trees of Canada includes a revised version of the Halliday map of Canadian vegetation. It distinguishes between a "northern transition" zone and the main boreal forest; and in Labrador-Ungava the line very nearly coincides with the boundary between the open boreal woodland and the main boreal forest. From the vicinity of James Bay west to Manitoba the boundary continues to lie between the 16.5-inch [419-mm] and 17.0-inch [432-mm] isopleths of potential evapotranspiration. The regions farther north in the Hudson Bay lowland are too little known to permit further correlation. The islands of Newfoundland and Anticosti lie wholly within the range of potential evapotranspiration found in the main boreal forest, with the solitary exception of the Long Peninsula of Newfoundland. Both islands are largely covered by close-forest, with the spruces and balsam fir as the dominants. There seems no doubt that they lie within the main boreal forest division. The warmest area of Newfoundland (potential evapotranspiration 19–20 inches [483–508 mm]) lies along the railway line from St George Bay to the Avalon Peninsula. This thermal efficiency corresponds with the boreal-mixed forest ecotone on the mainland, and it is interesting to note that many non-boreal trees occur in isolated localities (for example, white pine and red maple, *Acer rubrum*). The south coast of Newfoundland is chilled by the onshore prevailing winds crossing the offshoot of the Labrador Current that moves westward along the coast. Much of the high ground of the south is covered by treeless moss barrens. In many cases the lack of trees is an effect of altitude, but moss barrens occur also at low support high forest. Their origin has not been explained.

A preliminary glance at the Russian taiga has shown that zonal divisions comparable with those defined above for Labrador-Ungava have almost identical relationships with climate. This encourages the hope that it will ultimately be possible to extend the present review to the entire extent of the boreal forest formation. It will be of particular interest to see whether anything of the same zonal structure is revealed in the boreal forest of western Canada, where Sanderson has reported the retarding effects of drought. It may well be that the dependence of growth on thermal efficiency so strikingly confirmed in the humid east breaks down in the drier west.

11 Organic Terrain and Geomorphology[1]

NORMAN W. RADFORTH

Those who specialize in geomorphology do not accept landform as a static phenomenon, and recognize that land conformations prevailing at any one time are transient symbols of reference for purposes of classification. The dynamic concept is favoured and it is evolution of form that inspires inquiry. No doubt this is why the Arctic and Subarctic make an attractive panorama. The ice cap retreated and left new configurations that rested on the old. Nothing is more fascinating than a journey from the fresh landforms of the North to the old ones of the south to observe how time has consorted with forces to wear the land into senescence.

But wear and tear are not inevitable, for in vast areas of the land, new depth is added. I refer not to that mineral increment that climate translocated from one place to another. I mean organic deposition that has the inherent power to grow – a new landscape crowned with living plants growing on their ancestors beneath. This combination creates new aspects of form.

[1] The writer is indebted to the Defence Research Board for financial support for his studies on aerial interpretation of organic terrain and to the National Research Council for sponsorship of the work on ground studies. The special study on confined muskeg is supported by the Waterways Experiment Station of the US Army Corps of Engineers through the writer's interest in trafficability studies.

SOURCE: *The Canadian Geographer* 6, nos. 3–4 (Winter 1962): 166–71. Reprinted by permission of the author and publisher.

Organic Land

SURFICIAL FEATURES

Those whose business it is to read the character of the land want to know how to distinguish organic terrain, or muskeg as it is frequently called in Canada, from mineral terrain. This has never been my concern because mineral terrain has been outside my ivory tower. Recently I have given the matter some thought, for I have been impressed by those who without formal training can pick out organic from mineral terrain, especially when they are airborne.

Their first symbol of recognition is hidden in the vegetation. To find it they must fly at somewhat less than five thousand feet [1525 m], if they are novices. They discover, if mineral terrain is in line of sight and is tree-covered, that the tree crowns will vary usually markedly in level. Also, if the trees are deciduous there will be characteristic admixtures of form. When organic terrain comes into view there are no convoluted admixtures, and tree crowns form no longer in multiple level but in a single plane. This plane may be either sloping, curved, or horizontal.

Where there is treeless mineral terrain, the cover is changeable as to proportions of tall shrubs, low shrubs, and herbaceous plants except where the landform is constant over a large area, for example, prairie. On treeless organic terrain the cover is also changeable but less heterogeneous. Configurations of over-topping tall or dwarf shrubs are delineated against a background of herbaceous cover, which is mainly grass-like.

To attempt further separation of mineral and organic terrain by reference to cover requires partial reference to land conformation. Thus, characteristically, grass-like cover for the prairie is on a gently rolling surface, whereas for organic it is on a flat expanse.

The principle of cover-form heterogeneity for mineral and homogeneity for organic terrain is also better understood when basic land conformation is included in the consideration. When muskeg is confined, there is zonation of homogeneous cover-form but often there is local overgrowing at the zone interfaces and the total effects stimulates heterogeneity on mineral terrain. Usually, however, there is no difficulty in detecting the organic terrain because the delineated margin of the muskeg is observed to coincide with the edge of a land depression.

It has been reported elsewhere that those natural attributes of muskeg cover which lend themselves to ready identification are stature, habit, and presence or absence of woodiness. Use of them results in the designation of nine classes of cover, which usually occur in combinations. On application of the rule that if a given class of cover does not exceed 25 per cent it is not significant in symbolizing prevailing character, no more than three classes will appear in combination and often fewer.

Combinations of classes are known as cover formulae and are portrayed by grouping appropriate letters chosen from A to I, the nine symbols of cover class. Typical examples of cover formulae are AEI (trees over fifteen feet [5 m] tall overtopping low shrubs less than three feet [1 m] occurring with mosses – not lichens), FI (non-woody sedge to grass-like growth with mosses), and HE (lichen-like habit with low shrubs less than three feet [1 m] tall, and so on. The predominating cover class (tallest overtopping layer having greatest coverage) is always on the left of the formula with those on its right in descending order of predominance.

The number of formulae that occurs is happily few by comparison with the number of mathematical possibilities. This, coupled with the fact that there is much muskeg, means that in nature there must be frequent recurrence of formulae. This is indeed the case and, furthermore, some show a much higher frequency of occurrence than others; for example, EI and FI occur much more frequently than do ADF or BDE.

Since cover formulae integrate to provide three dimensional air-form patterns, the latter once identified can be extrapolated back to cover; for example, Marbloid, a 30,000-foot-altitude [9,000-m] air-form pattern is constituted largely of HE, EH, and EI and not FI, DFI, which are basic to Dermatoid, another 30,000-foot [9,000-m] air-form pattern, and not AEI or ADE, which are basic to Stipploid.

TOPOGRAPHIC FEATURES

Surficial cover for organic terrain has greater significance than it does for mineral landscape. The reason is that muskeg cover forms the culminating element to a whole series of buried elements that collectively convey ordered post-Pleistocene vegetal history. The history explains the structural composition in the vertical dimension of the peat, which facilitates comprehension of difference in organic terrain. There is no equivalent for this to assist in the interpretation of muskeg-free terrain.

Botanical organization of organic land masses plays such a dominant role in imparting character that examination of it in relation to geomorphic manifestations seems reasonable. Like botanical phenomena, the topographic ones are repetitive and not at all fortuitous in their occurrence. Also, evidence of correlation between cover and conformation is inescapable. For instance, in one family of cover formulae characterized by EI and EH, closely applied mounds occur. Their bases are usually much wetter than their peaks, which rise to one or two feet [30–60 cm] in height. With another family of cover classes, that represented by FI, the mounds become very widely scattered or if they coalesce they do so to form ridges in tortuous or crescentic parallel arrangements. Whenever D cover class enters the cover formula, intermittent traps or depressions with vertical sides intervene with high frequency. Usually, this designates a drainage course and almost invariable the lower end of the gradient terminates in terrain with FI cover, in which not mounds but hummocks characteristically appear.

Superimposed upon this micro-topography is a macro-pattern. Thus, where mounds occur, the terrain is at higher elevation than is ridged or hummocked terrain. Sometimes the elevation reaches a height of twenty to thirty feet [6–9 m], a common feature in semi-marine climates where the condition is known as "raised bog land." This macro-feature is common in Labrador and Newfoundland. On the other hand the topographic feature sometimes defines the air-form pattern. The anastomosing ridges of Reticuloid air-form pattern, the lace bogs referred to by Allington and Sjörs, are usually of FI or FBI cover unless the water table is low or raised bog effect becomes superimposed and then BDE may form the framework of the reticulum.

Because of the relationships already suggested between topography and vegetal cover, presence of topographic features can be interpreted once air-form pattern is known.

ICE EFFECT

So far, topographic differences have been expressed as suggesting relationship to cover, which in turn reflects structural relationships established in the vertical dimension of peat. These relationships can be either accentuated or sometimes altered by an ice factor.

There are three conditions where freezing temperatures are influential. First, there is permafrost, which the writer rightly or wrongly regards as terrain that is frozen indefinitely. In this circumstance, the elevations are accentuated and peat plateaus occur. Probably because of the ice that becomes incorporated into peat year in and year out, the plateaus rise to a height of ten or twelve feet [3–4 m] above the surrounding terrain.

Another condition of frozen ground also occurs perennially but from two to several years, not indefinitely. Temperatures arranged in this order of time produce the condition designated by the author as "climafrost." Obviously climafrost and permafrost can occur together, but it is significant that the former may occur in permafrost-free country, a situation which has been found and studied in the organic terrain south of Waboden, Manitoba. It is climafrost that is largely responsible for ice polygons, common features in Marbloid country. It is also important where perforations in subsurface ice typify FI-covered muskeg.

Finally, there is active frost, perhaps better defined as "seasonal frost," for it gives rise to winter ice. Perhaps the best example of it is ice-knolling, which is associated with the organic factor to produce mounds. The process that forms the subsurface condition known as ice-knolling in mounds and the vertical growth of iceblocks in peat plateaus likens to the ice dynamics in pingos.

The ice factor in muskeg contributes to its own special kind of topography in organic terrain, but it is a hidden topography in that it is beneath the surface. The link between the ice and the organic factors is always real. One very significant example of relationship involves not the peat but the living cover – the H, or lichenaceous cover. Aerial inspection on flights from south to north reveal EI changing to EH and finally HE. The latitude at which H makes its appearance is now known to bear some relationship to the southern limit of permafrost.

THE WATER FACTOR

In recent work, the study of hidden topography has taken on another aspect which does not concern ice. Structurally weak peats have been discovered beneath FIE cover in small masses averaging about eight feet [2 m] in diameter, and surrounded by stronger peat with FEI cover. Analysis suggests that the weak areas were once open water – small ponds in an open expanse of floating peat. Thus paleo-topographic features may in some cases have occurred only to become obliterated in the course of time.

In the case described, the organic factor has taken over from the water factor but often the reverse may be the cause of topographical differential. Thus, in northern shallow peat plateaus erosion caused by local drainage produces irregularities. Fringe irregularities caused by seasonal melt water at the edges of peat plateaus are not as common or as accentuated in the south as they are in the north. Melt water may also encourage shift in position of ponds and of water courses that arise from time to time, particularly where FI and DFI abound.

The mechanical factor of water erosion and the biotic factor combined do more than climate in the control of vegetal succession. Where drainage change does not exist, usually there is no change in the botanical history of the peat, which means that the culminating flora is the same in composition as the ancestral ones. There is therefore no climax in the classical sense that the expression "climax" conveys.

Mineral Land

There is of course topographic control resulting from conformation of the mineral terrain lying beneath the organic.

In these circumstances, the water factor, if it arises as a result of organic differentials, is secondary and the primary effect of water is a function of contour of the mineral sub-layer. In heavily forested muskeg where the peat is usually relatively shallow, drainage is always towards DFI cover. Where the formula is ADE (in contrast to AEI) there will be intermittent shallow depressions and the mineral sub-layer will be gently rolling. Where ADF occurs the depressions become contiguous, and extensive flooding can be expected for a period lasting well into the summer.

At the macro-topographic level, one might expect controls imposed by mineral terrain on muskeg formation to be impressive. There is evidence to support such expectation, but discriminatory relationships are hard to appreciate because every landform encountered in North America may support muskeg so far as the author can confirm. Whatever their genesis, hillsides or mountainsides may be muskeg-covered. Whatever age or form of a major drainage system, it too can support muskeg. Sedimentary or glaciated flats serve equally well as foundation for muskeg. Glacial features, beach-lines, Precambrian exposures, or land surface depressions may all become muskeg-covered.

It is usually easy to designate the mineral landform occurring beneath organic overburden. It is the determination of mineral soil type that is difficult. There are as yet no objectively established rules which enable an observer to predict mineral sub-layer. There is a useful hypothesis in which it is proposed that contemporary cover of the muskeg bears relationship to type of mineral aggregate beneath the peat. Although there is now much evidence to support this claim, it has been derived by empirical procedure and is not yet conclusive. At this time the statement will be appreciated that it is almost invariably the case that the A and the B families occur over coarse aggregate mixed with silt and sand. If the formula is AH, it is safe to claim that the mineral matter beneath will be sand. Commonly the H and the E families prescribe for sandy silt except for HE in which case the constitution of mineral matter is usually gravelly sand. The D family relates to silty terrain and the F family to clay. An explanation for these phenomena has yet to be supplied.

Confined Muskeg

Kettle holes, and depressions that simulate kettle holes, examples of which are found on the Canadian Shield, require special study for organic terrain interpretation.

Analyses and distribution of cover formulae support the classical claim that zonation, oriented with respect to the margin of the water body, characterizes the depression. On the other hand, the writer finds that the successional concepts conveyed in most textbooks on geomorphology are very much in need of qualification. For instance, when peat commences to form at the edge of a glacial lake, it is not always the case that the initial cover is sedge-like. The so-called xeromorphic condition which is constituted of ericoids in company with mosses may usurp the privilege of the sedge-like plants as the initial colonizers. There are other cases where exceptions become the rule, and certainly evidence suggests that confined muskeg cover does not conform to the climax formation theory for it appears at the time of writing that depth of organic terrain and possibly depth of depression will turn out to be very influential in controlling cover type. Certainly this, combined with other factors not yet adequately designated or understood, will prove to outweigh climate in cover control and peat development.

Application of Organic Terrain Studies

Forestry and agricultural industries, together with engineering development of one aspect or another, stand to benefit from the results of organic terrain studies. Devices for terrain interpretation can be fashioned to afford prediction pertinent to application of almost any kind. Prediction procedures based on cover and air-form pattern have been tested on a circumpolar basis and the results are most encouraging. For Canada, off-the-road access is the most significant requirement for northern development, and it is most encouraging to realize that for any given vehicle, not only can routes across organic terrain be selected for which it can be calculated that the operation contemplated can succeed, but also, the operational costs can be estimated. This is achieved in one preliminary aerial survey that is attractively inexpensive.

But it is not the application of the studies that encouraged the writer to present this paper at this time; it is the realization that a sister science of paleoecology can throw some light on geomorphic principles and provide an extension for geomorphic study.

ABORIGINAL PEOPLES

12 The Fragments of Eskimo[1] Prehistory

WILLIAM E. TAYLOR, JR

A reader of Arctic history and ethnology might conclude that the friendly Arctic was sometimes deadly. Yet, in turning to archaeology, the reader learns that man has survived in Arctic America for some five thousand years and that the Eskimo's ancestors prevailed over a vast expanse from easternmost Siberia to the Strait of Belle Isle and to Denmark Strait between Greenland and Iceland.

Although many pieces are yet to be found for the complex jigsaw puzzle of Eskimo prehistory, archaeologists can outline the general nature of the picture by fitting in place its available fragments. This reveals the fifty centuries of Canadian Eskimo prehistory readily dividing into four major periods or stages. First there was the Pre-Dorset stage of nomadic hunters who drifted across the deglaciated Canadian Arctic from Alaska. Archaeologists, somewhat given to polysyllabic locutions, refer to the Alaskan parent of Canadian Pre-Dorset as the Cape Denbigh Flint Complex of the Arctic Small Tool tradition.

The Denbigh Flint Complex, best known from the Alaskan side of Bering Strait and from the Brooks range of northern Alaska, contains a long list of chipped chert and obsidian tools, such as microblades, end scrapers, side scrapers, knife blades, and the most delicately fash-

[1] Though "Inuit" is now the preferred word for the present Aboriginal inhabitants of the Arctic, "Eskimo" is still widely accepted in an ethnological sense, referring to the ancient ancestors of the present-day Inuit. – ed.

SOURCE: The Beaver, Outfit 295 (spring 1965): 4–17. The material presented here forms the first part of the longer original version. Reprinted by permission of the author and publisher.

ioned inset side blades and points for hafting in lances, spears, arrows, and harpoon heads. Denbigh Complex sites also include a high percentage of burins, a distinctive and specialized chipped stone tool used for slicing and perforating such hard materials as bone, caribou antler, and walrus ivory. The Denbigh Complex people were seasonal nomads, many of whom summered on the coast hunting seal, probably with the aid of boats; others lived in the interior where they stressed caribou hunting. Various kinds of evidence, including radiocarbon dating, paleoclimatology, and geology, suggest that the Denbigh Complex or something very closely akin to it, existed in northern and western Alaska at about 3000 BC, and probably it persisted there from about 3500 BC to 2500 BC. Although some Denbigh traits recall still earlier Indian cultures far to the south in the interior of North America, many more reflect a relationship, perhaps old and indirect, with recently discovered Paleolithic and Mesolithic cultures of the Far East and with the early "Neolithic" of Siberia. Because no human skeletons, and of course no traces of the language of the Denbigh hunters, have been found, one cannot readily conclude that they were Eskimos. They did, however, have an Eskimo way of life, that is, a distinctive culture and economy adapted to treeless country, and further, some of their objects persist in a slightly altered form in much later, clearly Eskimo, sites. As J.L. Giddings, the discoverer of Denbigh, recently wrote, "Regardless of how we designate them, these Denbigh people appear to be in a direct line of cultural continuity with Eskimos."[2] Also, interesting studies in the relatively new and exciting field of lexico-statistical dating suggest the Eskimo and Aleutian languages may, as a unified language family, be at least five thousand years old.

Faced with the many, sometimes startling, developments in Eskimo archaeology over the past ten years, Arctic prehistorians are sometimes struck mute by caution or made unintelligible to the reader because of confusion and indecision. Nevertheless, it seems likely that the earliest "proto-Eskimo" of northern America derived their culture from Siberia and originally migrated from there. Perhaps the ancestors of Denbigh people drifted eastward along the southern edge of the former Bering Strait land bridge from what is now the southeastern coast of Siberia. If that were the case, the camps of those wanderers, now over eight thousand years old, are long since submerged under the cold waters of the Bering Sea.

Whatever the origin of the Denbigh people, they and their descendants were well equipped to survive in the tundra world. The success of their Arctic adaptation appears clearly in the archaeological record

[2] *The Archeology of Cape Denbigh* by J.L. Giddings, Brown University Press, Providence (1964), 243.

of their migrations, for the descendants, harvesting the game on which their lives depended, spread eastward across northern Alaska, the central Canadian Arctic, and the eastern islands to Greenland. Eventually they reached at least as far as northeastern and southwestern Greenland, Ungava Peninsula in northern Quebec, and down through the Barren Lands and the west coast of Hudson Bay to Churchill, Manitoba. Carbon 14 dating suggests they reached northeastern Greenland by 2000 BC. In the Canadian Arctic, this eastern development from the Denbigh threshold is called Pre-Dorset culture and it persisted over a large area until about 800 BC. In southwestern Greenland, Danish archaeologists have found a late variant of Pre-Dorset, called Sarqaq, which lasted there until about 500 BC.

When lumping Denbigh, Pre-Dorset, Sarqaq and a few other regional variants all together, archaeologists call the lump the Arctic Small Tool tradition for it is spread all across the tundra top of the continent, is characterized throughout by very small, carefully and delicately chipped stone tools, and lasted perhaps for three thousand years. With comforting consistency these sites produce the burins, microblades, scrapers, knife and weapon points, and side blades by which the prehistorian recognizes the tradition. Some of the Canadian sites include the small subrectangular depression left by the semi-subterranean winter houses of Pre-Dorset people, or the ring of boulders that secured the bases of their skin tents in summer camps. Occasionally charred and split stone cobbles mark the old hearths where they burnt greasy bones and scrub vegetation. The nature and the location of Pre-Dorset sites indicate that these people lived in small, widely scattered, nomadic bands, moving seasonally to exploit various game resources. They used toggling harpoons, spears, lances, and the bow and arrow in hunting caribou and seal. Very likely fish and summer birds appeared on their menu and probably wide-eyed children heard yarns of encounters with bears, wolves, muskoxen, and walrus.

The second period in Canadian Eskimo prehistory is that of the Dorset culture. It derives its name from Cape Dorset on Baffin Island for it was from the Hudson's Bay Company post there that the first collection of Dorset period material was sent to Ottawa. Collected by Eskimos, it was received some forty years ago by Diamond Jenness in the National Museum of Canada. Although Eskimo archaeology had barely begun and, despite the fact that the collection was completely mixed up, Jenness, in a brilliant feat of archaeological detective work, managed to isolate the diagnostic specimens and to prepare the original definition of the culture (Jenness 1925). The abundant work of the past forty years has confirmed and amplified Jenness' then revolutionary interpretation.

Like many others so long silent, the Dorset people had scant effect on the clattering, embattled course of man's history. Nevertheless they occupied a large part of the earth's surface and did so for an impressive number of centuries. The Dorset culture existed approximately from 800 BC to AD 1300 and spread from Bernard Harbour and Melville Island in the west to eastern Greenland and the northwest part of Newfoundland Island. In fact, the Newfoundland sites are some 2,400 miles [3,900 km] and 2,300 miles [3,700 km] respectively from those in northeast Greenland and those at Bernard Harbour in the Canadian Western Arctic – roughly the same as the mileage from Montreal to Los Angeles or Winnipeg to Tegucigalpa in Honduras. Within the Dorset area, sites seem most abundant in the Hudson Strait–Foxe Basin region. Although Dorset material occurs down the east side of Hudson Bay to the Belcher Islands, it has not been found on the bay's west coast; nor does it seem to occur in the Barren Lands interior west of the bay. So far the only inland find of Dorset sites have been at Payne Lake near the centre of Ungava Peninsula.

The origin of Dorset culture has long been a question of hot scholarly debate despite the cold silence of the subject matter. Before the Pre-Dorset period was discovered scarcely a dozen years ago, some archaeologists claimed Dorset was derived by migration from Alaska while others argued it was basically an Indian way of life that was carried from the Great Lakes–St Lawrence Valley area. A volley of recent reports on very early Dorset sites and on Pre-Dorset sites along with new results of radiocarbon dating has led to general agreement that Dorset culture developed first within the Canadian Eastern Arctic from the Pre-Dorset culture, for many culture traits are shared by the two, and other Dorset tool types are clearly evolved from Pre-Dorset prototypes. Further the Pre-Dorset and Dorset peoples lived very similar kinds of lives with the same adaptation, economy, and settlement patterns. Nevertheless a few Dorset traits, lacking Pre-Dorset antecedents, may have been acquired by cultural diffusion from the western Subarctic, and from early Indian groups in southeastern Canada. The National Museum of Canada's 1963 Arctic field survey re-opens the possibility that the change from Pre-Dorset to Dorset involved the spread of some ideas eastward from Alaska because that survey extended the known range of Dorset occupations some 450 miles [725 km] westward to Bernard Harbour, where a rather early Dorset culture site was examined.

Like their predecessors, Dorset people lived in small seasonally-nomadic bands with little camps of skin tents in summer, sheltering in winter in small clusters of partly-underground pit houses. Some of these winter houses seem to have had skin roofs. Dorset man may have used, indeed might have invented, the snow house. They hunted seal, bearded seal, walrus, and caribou; they fished extensively using stone traps and barbed spears; spears were also used in bird-hunting. Heavy spears, lances, and toggling harpoons were used against the larger animals. Since there is very little evidence of domesticated dogs in this culture, Dorset people may have man-hauled their small ivory-shod sleds. Although they seem to have had skin boats, nothing is known precisely of the boat type. The Saga of Erik the Red mentions Skraeling skin boats or canoes propelled by staves or paddles. In the northern part of Newfoundland island, where the observation was likely made, such might well be a reference to Dorset culture kayaks. Needle cases and an abundance of delicate bird-bone needles suggest that Dorset people wore tailored fur clothing. They had, albeit a smaller model, that traditional Eskimo hallmark, the blubber-burning lamp, carved from soapstone, which provided heat for the dwelling, light, a means of drying clothes, and an answer to their humble cooking needs. Implements made of antler, ivory, bone, or driftwood were tipped or edged with chipped and sometimes polished stone blades of chert, quartz, or quartzite. Such implements generally reflect a Pre-Dorset heritage, but another category, blades of ground and polished slate, seem to have no adequate Pre-Dorset precursors, and thus may reflect Alaskan influences, or may have been learnt from Indians to the south, some of whom used that technique.

The most excitement in excavating a Dorset site comes when someone unearths one of the small delicate carvings in ivory, antler, or bone, that characterizes Dorset art. These rare pieces, shaped with consummate skill with stone tools, range from about four inches [10 cm] to as little as three-eighths of an inch [10 mm], and sometimes weigh only a small fraction of an ounce [a few grams]. Such figurines, often precisely realistic, sometimes of sophisticated abstraction, usually depict animals, birds, fish, humans, or mythical beasts. Sometimes a complete specimen represents only a part, such as a walrus head, a caribou hoof, a human face, or a gull's head. A second category of art is ornamentation, commonly of short lines, confidently set on various objects, often as a skeletal motif on the figurines.

The recent and very distinctive art of the Angmagssalik Eskimo of the east coast of Greenland shows a number of traits similar to those found on Dorset culture carvings and this leads to the speculation that Angmagssalik art may have perpetuated some Dorset period art styles. If that were so one might wonder whether or not Angmagssalik culture was in some part derived from Dorset culture, or whether perhaps it was a blend of Dorset and later cultures.

Until a few years ago nothing was known of Dorset people's burial practices or skeletons. Recently archaeologists have discovered stone vault graves, stone-lined pit graves, and small gravel mound graves containing grave goods and red ochre. The skeletal remains have been very poorly preserved, but what little there is suggests that the Dorset people were physically typical Eskimos. The much-debated work of linguistics leads me to think that the Dorset population spoke some old variant of the Eskimo language. Thus we may conclude for the Dorset period that the general picture is of Eskimo culture, and although not all the usual Eskimo traits are present, the picture does not fall from its frame.

Around AD 900 Dorset culture began to be crowded off the Arctic stage as the third major period of Canadian Eskimo prehistory pushed in from Alaska. Between about 900 and 1300 a vast, thin drift of population spread over Arctic Canada and coastal Greenland almost completely burying the Dorset. Although there is some evidence of contact and mutual influence between the older Dorset and the emigrating Alaskans, the overall view is of nearly complete replacement. This third period, the Thule culture, persisted until about AD 1750. Thule evolved directly out of Birnirk culture, of the north Alaskan coast, and Birnirk, in turn was a product of a long evolutionary trend of Eskimo culture stages in the Bering Strait region. Although some part of that evolutionary lineage extends back to the old Denbigh Flint Complex of 3000 BC, diffusion from Siberia and from northwestern North America along with local modifications and inventions must have played a significant part in that still little-known progression.

This seems the place to inject a necessary rebuttal to Tryggvi Oleson's surprising recent revival and extension of Duason's odd ideas on Eskimo prehistory and protohistory in Canada and Greenland. Oleson argued that the old Dorset Eskimo and Greenlandic Norse groups blended, both racially and culturally, to produce the Thule culture and its people, that subsequent to this proposed origin Thule culture and people spread throughout Arctic Canada and west into Alaska. That speculative reconstruction embodies so many errors that a critic might despair of listing them in detail. Perhaps it will suffice to say that there is no evidence of a Norse-Dorset blending such as Oleson requires, that there is no evidence that the Canadian Thule culture people were racially blended with Caucasoids, and that there is no reason to believe that Thule culture began earlier in Greenland than in Alaska. The archaeological record shows but very scant evidence of Norse-Dorset contact,[3] let alone a cultural blending, that Thule culture people were pure Eskimo in racial type, and that the earliest Thule sites occur in the

3 In 1977 a carved Thule figurine of a man wearing thirteenth-century Norse clothing was discovered on Baffin Island. The following year a piece of medieval chain mail and a Viking ship rivet were found on a small island off the central east coast of Ellesmere Island. These and other discoveries provide tantalizing evidence of an early Norse presence in Canada's Arctic, probably on exploration or hunting voyages. – ed.

west, not in Greenland. The evidence to support the Oleson-Duason views would of necessity be archaeological and yet I am sure no Arctic archaeologist would support their speculations. Certainly none has presented an appraisal of Canadian Eskimo prehistory compatible with that attempted by Oleson.

Beginning not later than AD 900, Thule migrants gradually wandered eastward from northern Alaska along the Arctic coast and northeastward through the High Arctic islands reaching northwest Greenland perhaps about AD 1100. Subsequently, in Greenland Thule peoples came under the influence of and into close contact with the Viking settlers on that island's southwest coast. As one Thule wave washed onto the Greenland shores, another carried southeasterly crossing Hudson Strait to flow south down the east coast of Hudson Bay to the Belchers and down the Labrador coast to the Strait of Belle Isle.

Although Thule hunters harvested caribou, seal, walrus, birds, and fish like their Dorset predecessors, and had a basically similar tundra-adapted way of life, there were marked significant differences between the two. To begin, the Thule had a more effective cultural adaptation to the Arctic: there is only scant evidence of domesticated dogs in Dorset culture but Thule people had dogs, a valuable aid in hunting and, harnessed for sled hauling, a means to increase the range and rate of travel; a second vital advantage for Thule was its possession of the full range of gear for hunting the great baleen whales, a major food supply not available to the Dorset people. Indeed, whaling more than anything else distinguishes Thule culture from earlier and later Canadian Arctic culture periods.

The Thule people were typical, indeed classic, Eskimo in their culture, their language, and their physical type. Skeletons from the several grave finds place them clearly in the distinctive racial sub-group of modern Eskimos. All Eskimologists agree that they spoke the Eskimo language. Their way of life is fully within the Eskimo pattern. Parts of tailored fur clothing, including the parka and skin boots, preserved in the permanently frozen soil of Arctic sites, are quite like recent Eskimo dress. They used kayaks, umiaks, sleds and sled dogs, whips, harpoons, spears, lances, fishing gear, and bows and arrows of typical Eskimo type. The same may be said of tools as of weapons. The women's kit, needles, needle cases, the ulu, the soapstone lamps and pots, and wick trimmers that have been excavated have close counterparts in recent Canadian Eskimo culture; and so do Thule culture adzes, drum parts, snow knives, dippers, seal scratchers, snow goggles, sealing stools, snares, drying rack fragments, snow beaters, bow drill parts, and snow probes. Even the amulets and the toys such as the *ajaqaq* (a cup and ball variation using pierced bone and pin), wooden dolls, model boats, weapons, and utensils echo the commonalty of Thule culture and recent Canadian Eskimo life as the later was seen by the early European explorers and whalers. Indeed, the arrow that struck Martin Frobisher in the buttocks as he fled to the beach of Frobisher Bay was delivered from a Thule bow by a Thule culture Eskimo whose unsuspecting ancestors had come all the way from Alaska for the event. One may safely conclude that racially and culturally the modern Canadian Eskimo descended from the old Thule culture population.

Its winter villages reflect the more effective Arctic adaptation of Thule culture as compared to Dorset. Thule winter villages commonly contain six to thirty rather large solid houses made of stone slabs and sods set over a whale-bone framework; these have a cold-trap entrance passage, raised flagstone sleeping platforms, a flagged floor, and various little storage cubicles, food bins, and pantries. Usually they are partly underground, often set into a gently sloping gravel hillside facing the sea. Dorset houses, on the other hand, are usually less elaborate, smaller, and in winter clusters of only about three to fifteen in number. Like the Dorset people, Thule Eskimos used skin tents in summer and in winter they built snow houses, perhaps only for temporary camps.

As Thule replaced Dorset over a vast area, that replacement must have taken some time and undoubtedly ideas were exchanged between the contending cultures. Eskimo folk tales include numerous accounts of the Tunit or old people who were, in fact, the Dorset population. It seems likely that Thule people learnt of the snow house from Dorset for it is not an Alaskan trait. The same might apply to soapstone lamps and pots for prehistorians have not found these in Birnirk, the Alaskan culture ancestral to Thule. Also some Thule types of harpoon heads suggest the copying of Dorset harpoon head styles. Nor was the borrowing in one direction only for it seems that the latest stage of Dorset house types incorporates the cold-trap entrance passage copied from the new Thule subdivisions. It must have been a valuable innovation since both Thule and Dorset people faced a worsening phase of climate after AD 1100 in the eastern Arctic.

Coming down the west coast of Greenland, Thule wanderers soon contacted the Vikings who had begun settling in southwest Greenland a few score years before. The Vikings had considerable influence on the Thule Eskimo in Greenland and, over the generations, much contact with them. But there too, deteriorating climate (and a lack of concern in Europe) squeezed the Norse settlements out of existence, so that in the fifteenth and sixteenth centuries the diminished remnant of Norse Greenlandic culture blended with its Viking-influenced Eskimo matrix.

The fourth and final stage of Canadian Eskimo archaeology is that of the recent Central Eskimo which can be dated from the eighteenth century. The recent people derive directly from the Thule culture population; but there are differences and they are largely a result of a gradual collapse in, and virtual end to, the Canadian Eskimo hunting of baleen whales which had been an economic mainstay of the culture. When whale-hunting declined, the large permanent villages of sturdy winter houses were abandoned, for a more nomadic life was required now that the people became increasingly dependent on the more scattered herds of seal and walrus. Thus there was a gradual shift to the snow-house on the sea ice as the customary winter residence. Further, the "Little Ice Age," a time of harsher climate from 1650 to 1850, seems to have forced a withdrawal of population from the northernmost Canadian islands – Ellesmere, Devon, Somerset, Cornwallis, and Bathurst – that the Thule people had settled, east to Greenland and south to the south coast of Victoria Island, Boothia Peninsula, and Baffin Island. That "Little Ice Age," bringing more extensive ice cover and shorter seasons of open water, may partly explain the decline in Thule culture whaling. The whales' summer range might also have been shrunk by a decreasing depth of sea passages caused by the continuing post-glacial rise of the land. Third, the diligence of European whalers in northern waters may have reduced the supply of whales available to Thule harpooners. In fact very little is known of this transition except, of course, the matter of whaling and the very minor changes in styles of harpoon heads and other fragments whose study writes the prehistory just summarized.

The other change from Thule to recent Central Eskimo culture rests in the introduction of European goods and ideas. That package of steel needles, so generously given, exploded into trading stations, missions, governments, a DEW line – and even sore-backed archaeologists wandering about in a vast past picking up pieces lost, left, and forgotten like the tribes that used them.

13 The Inuit Sea Goddess

DARLENE COWARD WIGHT

The most powerful of the spirits in the Inuit shamanic belief system is known variously in different areas of the Canadian Arctic as Sedna, Taleelayuk, and Nuliajuk. She was the guardian of the animals and mistress of both land and sea. Ruling through *tornaq*, or evil spirits, she made sure that all souls, both human and animal, were shown the respect that the ancient rules of life demanded. She could make the animals either visible and easy to hunt, or invisible so that humans went hungry. The powerful sea goddess could also punish people for breaking taboos by causing them to become sick or injured. It was the duty of the Inuit *angakok* (shaman) to visit Sedna through soul travel to determine what wrong had been done. The *angakok* would have to placate Sedna by combing and braiding her dishevelled hair, or by overpowering her to force her to release animals for the hunt or to call away her *tornaq*. In a carving by Peter Nauja Angiju (Puvirnituq), the sea goddess grapples with a man, perhaps enacting the struggle between a shaman and the guardian of the animals. This relationship is also the subject of the whale bone carving *Shaman Summoning Taleelayuk to Release the Animals*, by Manasie Akpaliapik of Arctic Bay. However, in this work, the Inuit *angakok* sings magic songs to persuade the sea spirit to allow her creatures to be captured by hunters.

The origin myth of Sedna is told in different forms in different areas. Essentially, it is the story of a young girl who defied convention by refusing to take a husband. In anger, her father forced her to marry a

SOURCE: *Inuit Art Quarterly* 14, no. 2 (summer 1999): 34–35. Reprinted by permission of the author and publisher.

dog-man and live with him on an isolated island, where she gave birth to a litter of dog-children. Feeling sorry for his daughter, the father killed the dog-husband, leaving the girl with no husband to hunt for her family. To save them, she put some of her children in a sealskin boot which drifted out to sea. They eventually landed to become the ancestors of the white people. The remaining children were put in another boot which drifted to land where they founded the North American Indian race.

In Baffin Island versions of the story, Sedna then married a northern fulmar, or *kakoodlak* (bird of the storm), disguised as a handsome man. In shame for her mistake, she followed her new husband to an island to live in seclusion. Again, her father took her away in his boat, but the fulmar caught up with them and caused a huge storm by flapping its wings. To save himself, the father threw the girl overboard, but she clung to the side of the boat. So he chopped off her first finger joints, and they turned into seals as they fell into the water. Again she grasped the side of the boat, but the father hacked off the second joints of her fingers which became walrus. Still the girl hung on, and the last joints became whales. She then sank into the sea to become a spirit, the mother of the sea animals. The father was swept down to join his daughter where he became a dog in a house at the bottom of the sea.

14 The Northern Athapaskans: A Regional Overview

C. RODERICK WILSON

Although the region strikes most southerners as definitely inhospitable, people have lived in the Western Subarctic longer than in any other part of Canada. Bone tools discovered in the Old Crow region of the Yukon have been widely accepted as indicating human presence some 25,000 years ago. Some archaeologists think that other artifacts found in the region are much older. Whatever dates are ultimately demonstrated, Amerindian people have clearly lived in the region from "time immemorial."

The physical environment of the region can be characterized as the zone of discontinuous permafrost in western Canada. The southern half is boreal forest (spruce, fir, and pine, with some poplar and white birch) and the rest is primarily a transitional zone of boreal vegetation intermixed with patches of lichen-dominated tundra. (As a cultural region the Western Subarctic in places extends north of the tree line into the tundra because some Subarctic societies made extensive use of tundra resources.) In absolute terms the region receives little annual average precipitation (about 40 cm), yet it has two of the largest river systems on the continent and innumerable lakes. The ground is snow-covered six months of the year, and few places have more than fifty frost-free days. Temperature extremes typically range from lows of −50°C to summer highs of over 20°C.

SOURCE: R. Bruce Morrison and C. Roderick Wilson, from *Native Peoples: The Canadian Experience* (2nd ed., Toronto: McLellan & Stewart, 1995), 225–31. Copyright © 1995 by R. Bruce Morrison and C. Roderick Wilson. Reprinted by permission of the authors and Oxford University Press Canada.

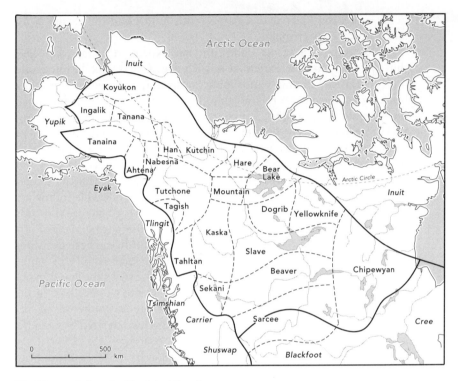

The Western Subarctic: Aboriginal Peoples

Large game animals, particularly caribou, constituted the primary resource here for Aboriginal peoples. Moose, goats, sheep, and even bison were locally important. Small animals (especially the snowshoe hare), fish, migratory waterfowl, and grouse were of secondary importance.

The entire Western Subarctic culture area is inhabited by Indians speaking a series of closely related Athapaskan languages. Linguists believe that these languages were undifferentiated as recently as 500 BC and that from their "ancestral" homeland they expanded further west in Alaska and eastward into the Northwest Territories and then southward. Some, notably the Apacheans, became physically separated from the others and ended up thousands of miles to the south, but for Northern Athapaskans, because of limited linguistic variation, communication is still possible across considerable distances. This implies frequent communication between people from different localities throughout the prehistoric past. Correlatively, the ethnographic evidence is that few Northern Athapaskan groups had significant relationships with non-Athapaskans. The exceptions are the Chipewyan (with Cree), the Hare and Kutchin (with Inuit), and the Kaska (with Northwest Coast groups).

In this context it seems significant that Northern Athapaskan kinship systems could be extended in such a way that usually one could find a "relative" even in bands quite far distant, and hence a legitimate basis for establishing a relationship. In general these people developed social strategies characterized by great flexibility and informal institutional arrangements. For example, leadership among the eastern Northern Athapaskans was largely situational: people were listened to or followed not because they had the power to make people obey but because they had demonstrated an ability to lead in that particular activity. An outstanding hunter might attract a considerable following; nevertheless, he had no permanent power. The overall picture thus was of individuals, family groups, and even larger groupings making short-term decisions about where and how they would live. These decisions were based on a large number of factors: the local supply of game, reports from elsewhere, degree of satisfaction with fellow band members, which relatives lived where, and so on. In the western part of the region, the existence of clans correlated with a more complex and formal social life; nevertheless, that life also was characterized by remarkable social flexibility.

Another key to Aboriginal Athapaskan society is its egalitarianism (autonomy and self-reliance are closely associated). This was true even among the most westerly Athapaskans who, like their maritime neighbours, had "chiefs." As McClellan and Denniston (1981: 384, 385) point out, these positions were conditional, and stratified rank was not possible for them until the fur trade period. The staples necessary for life were in general equally accessible, and the skills necessary to transform them into finished products were shared widely in the community. Although some people were more competent than others, and more respected, there was no basis from which an exclusive control over goods, and hence people, could develop. Even in the realm of spiritual power, the ultimate basis for any success and an aspect of life in which people varied conspicuously, the possibility of becoming a shaman was in principle open to all.

The initial contact of Canadian Athapaskans with Europeans was consistent with the general trend of the frontier moving from east to west. Direct trading contacts were made by the easternmost group, the Chipewyan, in 1714, while none of the westernmost tier of Canadian Athapaskans had direct contacts prior to 1850. There was not, however, a single frontier: Athapaskans in southern Alaska had trading contacts from 1741.

Helm (1975) makes the following points concerning the early contact process: European goods and diseases usually preceded direct contact with traders; the first-contacted tribes obtained guns and expanded their fur-collecting operations at the expense of their western neighbours; trading posts were welcomed by those living in the "new" territories; traders tended to act as peacemakers in this newly competitive context. The latter three points particularly contrast with Russian-Native relationships in Alaska.

The fur trade had numerous consequences, and it constituted a social revolution. The introduction of new goods, especially guns and traps, is salient. New social roles, such as trading chiefs, and changing social relationships are also striking. Perhaps most startling are features that are now identified as traditional, such as dog sleds, which were introduced during this period. Somewhat simplified, the guns, traps, and sleds enabled people to engage in old and new bush activities more efficiently. However, the new elements made their own demands: a dog team in a year would eat thousands of pounds of fish, thereby increasing the "need" to obtain new goods such as fish nets and to harvest bush resources at a higher level. Further, an innovation frequently had a cumulative impact. The effect of the gun, an early trade item, in contributing to higher harvest levels becomes most evident after 1900. Nevertheless, these early changes were limited in scope, and the continuity with the past was evident. The ensuing way of life was remarkably stable for many groups for almost a century and a half. Even the general acceptance of Christianity resulting from the activities of Oblate and Anglican missionaries was notable for people's continued adherence to old beliefs as to new.

The impact of introduced diseases is not clear. Aboriginal population levels are not well known, nor are the consequences of the various outbreaks. Are they isolated events, as a narrow reading of the sources would indicate, or do they represent merely the documented cases of widespread epidemics? What is evident is that a number of new diseases (smallpox, scarlet fever, influenza, measles, venereal diseases, and tuberculosis – in rough chronological order) became common and that there were at times very high mortality rates, both because the new diseases were in themselves devastating and because in their wake small groups might well starve to death. Nevertheless, the recuperative powers of the populations were such that the region as a whole appears not to have experienced significant demographic change until recently. With the advent of modern medicine some forty years ago, the Yukon and Northwest Territories [and Nunavut] have experienced the highest rates of population increase in Canada.

For most of the area, the fact that Native societies had become incorporated into Canada did not become a social reality until the early 1950s. There were two major exceptions. Much of the Beaver Indians' traditional territory, the Peace River country of northern British

Columbia and Alberta, was arable. Farmers began displacing Natives circa 1890. Even more dramatic was the Klondike gold rush. Dawson alone had grown to 25,000 by 1898, while the Han "tribe," occupying adjacent portions of Alaska and the Yukon, numbered only about 1,000. The rush, and the government and commercial presence it created, resulted in an early marginalization of Native people in the Yukon to a degree that the people of the Northwest Territories generally have still not experienced.

A few generalizations about contemporary life are in order. First, for much of the region, most adults over about age forty have personal memory of what life was like before significant government-industrial presence. For some the "traditional life" is a still present reality. Second, as both Cruikshank and Ridington demonstrate, even for severely impacted groups, truly significant features of traditional life remain important.

Third, the treaties continue to be living documents. As Asch notes, there is active debate over how Treaties No. 8 and No. 11 should currently be interpreted, but both sides agree that they help define the constitutional debate on what it means to be Native in those regions. A unique element of this debate is Judge Morrow's 1973 investigation of the circumstances under which Treaty No. 11 had been signed in 1921, unique because at the time there were signatories to the treaty still alive and able to testify in court that the treaty had been everywhere presented as a gesture of friendship and goodwill and not, as a literal reading of the written text would indicate, a means of extinguishing Aboriginal title.

A fourth point is that Athapaskans have attempted both to maintain themselves as a distinctive people and to accommodate the powerful forces for change thrust upon them. The primary mechanism in the Northwest Territories for doing this, the Dene National Assembly, is profoundly rooted in traditional values. Membership, for instance, is open to all Natives, including Metis and Cree, as equally part of the community of people. Although they have had victories, they have not had the same success in establishing political legitimacy as have the Inuit in the eastern part of the Territories [and Nunavut].

Our last point follows from this comment. The Inuit case is unique for Canadian Natives; they are numerically dominant in their region. The Dene, like most Natives in Canada, must find mechanisms other than a territorially based legislature for political accommodation. Presumably that means some form of limited, shared, or joint arrangement with Euro-Canadians. Their recent history demonstrates a willingness to seek workable solutions; will Euro-Canadians be as open to accommodation?

References

Helm, June, et al.. 1975. "The contact history of the subarctic Athapaskans: an overview." In Clark, A. McFayden, ed., *Proceedings: Northern Athapaskan Conference*, vol. 1. Canadian Ethnology Service Paper no. 27. Ottawa.

McClellan, C., and Denniston, G. 1981. "Environment and Culture in the Cordillera." In Helm, June, ed., *Handbook of North American Indians, vol. 6, Subarctic*. Washington, DC: Smithsonian Institution.

15 The Northern Algonquians: A Regional Overview

JENNIFER S.H. BROWN AND C. RODERICK WILSON

The Eastern Subarctic is sometimes referred to as the Northern Algonquian culture area because the entire region is occupied by a branch of the widespread Algonquian-speaking peoples. The surviving languages form two series of closely related dialects that can be grouped into two languages, Cree and Ojibway. Our usual terminology does not reflect this understanding very well: terms like Naskapi and Montagnais imply the status of being separate languages, rather than being two of the nine dialects of Cree. In any case, Cree and Ojibway seem to have developed independently from Proto-Algonquian, possibly separating about 3,000 years ago. The relationship between the dialects of these languages is quite complex, in part because in historic times (and earlier?) whole groups of people have shifted from one dialect to another.

The inhabitants of any region must come to terms with their environment. The Eastern Subarctic is characterized by long winters, short summers, and a continental climate. The generally cold climate is also related to the jet streams which, passing from west to east, tend to draw Arctic high-pressure air masses to the southeast. In spring and summer, intensified sunlight decreases the dominance of Arctic air, so seasonal contrasts are strong. Minimum/maximum daily mean temperatures in the Severn River drainage in northern Ontario, for example, range from between −29°C and −19°C in January to between 11°C

SOURCE: R. Bruce Morrison and C. Roderick Wilson, from *Native Peoples: The Canadian Experience* (2nd ed., Toronto: McLellan & Stewart, 1995), 143–9. Copyright © 1995 by R. Bruce Morrison and C. Roderick Wilson. Reprinted by permission of the authors and Oxford University Press Canada.

and 21°C in July. But even hot summer days may soon be followed by frost, and variations from the average can be considerable in either direction.

Precipitation in much of the area is relatively light. Total annual precipitation in northern Ontario averages only about 60 cm, most of it coming in summer thunderstorms. However, the climate east of James Bay is much affected by Hudson Bay. Air currents in fall and early winter pick up moisture from Hudson Bay to dump it along the eastern shores and inland. From midwinter to early summer, the Bay remains ice-covered, depressing temperatures and delaying the coming of spring in lands to the east. As a consequence, this area experiences very heavy snowfall and cold temperatures.

The presence of such extreme climatic conditions in these latitudes was difficult for Europeans to accept. When in 1749 the Hudson's Bay Company faced a parliamentary inquiry into its conduct, critics complained that it had not established agriculture and colonies around Hudson Bay and asserted that company representatives must by lying about the climate; after all, York Fort was on the same latitude as Stockholm, Sweden, and Bergen, Norway; and the Severn River was on a level with Edinburgh, Copenhagen, and Moscow. The critics' ignorance was pardonable, however. Fuller understandings of the effects of large-scale and even global weather patterns on the region have only recently been developed. Current research, in fact, is drawing on Hudson's Bay Company journals from the 1700s and 1800s to trace the regional weather patterns.

Northern Algonquian have therefore long been adapting, with a success that startled their early, ill-equipped European visitors, not only to cold, but to unpredictable and extreme climatic conditions. A late spring, for instance, would mean late break-up of lakes and rivers for travel, and late arrival of migrating geese and other birds important as food. Less moisture than usual meant, among other things, less snow cover, meaning in turn less shelter and lowered survival rates for some basic food sources such as ptarmigan, hare, and other ground-dwelling animals. Drying of streams is a serious impediment when the movements of people to different seasonally used food resources (between winter hunting camps and summer fishing spots, for example) depend on canoe transport. Excess precipitation would bring floods, mud-filled portages, and swollen rapids dangerous to small vessels.

The land forms, rocks, and soils of the Algonquian Subarctic support many forms of life, but they, too, pose challenges and constraints. The Canadian Shield is the single topographic feature that has most influenced the shape of North Algonquian life. Even a casual traveller sees how this rough rock base, polished clean in places by glaciers and overlain in other places by glacial clays, sand, and gravel, provides the

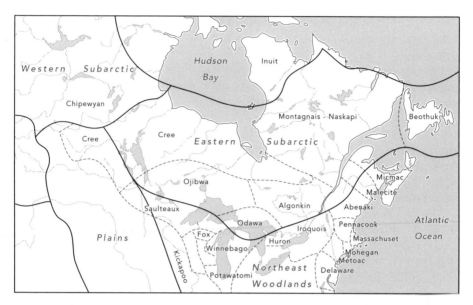

The Eastern Subarctic: Aboriginal Peoples

contours for countless lakes, streams, and swamps – ideal habitat for beaver, muskrat, and other animal species long important for food and furs. The French who reached Ontario in the 1600s found that the Hurons valued the trade furs and leather they received from the Algonkin, Nipissing, Ottawa, Ojibway, and others who made their home in the Shield region. And these Algonquian groups in turn valued their trade with the Hurons, prizing in particular Huron cornmeal.

The Shield country in central Ontario and Quebec is transitional between temperate and Subarctic. The observant traveller notices, going north, that the mixed deciduous trees of the south yield increasingly to evergreens – white and red pines mingled with spruce – then to a predominance of black spruce. Continuing northward the landscape changes again. The rocks of the Canadian Shield mostly disappear. The Hudson Bay lowland – a spruce-dominated forest on poorly drained, clayey soil – covers a vast area west and south of the bay. The growing season is short and intense throughout the Algonquian Subarctic. People intensify their activities, taking advantage of the open waters, the fisheries and waterfowl, and such plants as blueberries, which can only be harvested for a few short weeks.

The landscape presents limits of various kinds to its occupants, and anthropologists and other Western scientists are only now beginning to appreciate the extent to which Aboriginal peoples interacted with the environment. Europeans, for example, have always described the forests they found in America as "virgin," "primeval," "wilderness," and so on. In contrast, the forests were not only occupied, but their productivity was actively managed and maintained. We have long been aware that one could manage game directly by varying the intensity of hunting; it is now clear that Native peoples also managed game levels indirectly by manipulating the environment, primarily through the selective use of fire.

Small, carefully located and timed fires were extensively used to hasten new growth in the spring, which would attract desired animals and birds, foster desired plants such as blueberries and raspberries, create a more varied habitat that would support larger numbers of animals, and open up areas for travel and hunting. Some species that benefited from controlled burning were moose, deer, beaver, muskrat, bear, and ducks. Other species, notably caribou, require the mosses and lichens of mature, "climax" forests. Where caribou was the preferred basic resource, as it was in the northerly parts of the region, the use of fire was lessened. The choice was not simply a matter of food preferences but ultimately one of social organization, since the strategies for hunting solitary and herd animals vary substantially. In either case, the forest was not simply something provided by nature.

Changing patterns of human activity have also been major determinants of Northern Algonquian life. Because the following chapters are strongly historic in orientation, the historic context for the region will be limited to two generalizations. First, despite the fact that large numbers of its contemporary inhabitants pursue lifeways that are seen as strongly "traditional," this region had an extremely long period of contact with Europeans. Almost certainly its southeastern reaches along the St Lawrence River were visited by Bretons before 1500, and by 1670 the Hudson's Bay Company had initiated trade in the more northerly Hudson drainage. First European contact for the Cree in the northwestern corner of the Eastern Subarctic was only a few years prior to the first direct contact for their closest neighbours in the Western Subarctic, the Chipewyan. As a whole, contact in the east of the region was substantially earlier – in some cases more than 350 years earlier.

Partly because of the length of the cultural contact, people's lives over the years have changed substantially. Again, this is most true of the south and of the coast, but it also places the Eastern Subarctic as a region in contrast to the Western Subarctic. The region saw very early missionizing of its people, early exposure to new diseases, and generally greater involvement in the fur trade than similar zones further west. As an extreme example, by the end of the era of competitive fur exploitation between the Hudson's Bay Company and its Montreal rivals (1763–1821), in some areas the dominant resource bases, caribou and moose, had been virtually exterminated. A consequence of this longer and more intense period of cultural contact in the Eastern Subarctic than in the West is that we are less sure of what the eastern Aboriginal life patterns and beliefs were. On the other hand, in the East there is substantial historic (European) documentation of events dating from the early seventeenth century, for which there is no parallel in the Western Subarctic.

There follow three final points of general relevance. First, there is always the problem of sources – which voices have spoken to us and why. Native voices are rarely heard from the documentary record, and, when they are, they are often reported at second or third hand.

A second matter is that of homogeneity versus local variability. A superficial observer sees a great sameness: Algonquian hunters inhabiting a vast, cold, mainly spruce-covered region. In fact, local variations – the shape of a lake, the slope of the land, pockets of soil – produce considerable range of micro-environments. Local food resources, particularly game, are not evenly distributed. People have reacted in even more complex fashion. Furthermore, their lives have never been solely subsistence-oriented. They evolved distinct social traditions, world views, and cultural and religious patterns that had their own dynamics of variability and conformity.

Superficial generalization is accordingly to be avoided. The challenge is to get beyond the simple traditional stereotype of hunters in wigwams or the more modern one of isolated northern villages, to begin to know the Northern Algonquians as complex, diverse human beings whose lives have their own historic richness and vitality. We close with a series of contemporary facts that have implications worth pondering. Cree children who first learn to read in Cree, reading stories produced in their home community, later learn to read English better than their older siblings did. Some families ensure that some children learn the old bush skills and send others to university. The Grand Council of the Crees is rooted in tradition; it is also a contemporary political innovation of the first order.

HISTORICAL PERSPECTIVES

16 The Identification of Vinland

ALAN COOKE

During the past two hundred years, few informed persons have doubted that about AD 1000 the Norse attempted to colonize a part of the New World that they called Vinland. But the location of Vinland and the meaning of the word itself have long been subjects of dispute among scholars, and the lack of incontrovertible evidence of Norse occupation has permitted a proliferation of hypotheses. Among the earliest in this matter were Torfæus (1705) and Forster (1784), but the initiation of spirited controversy may be credited to Rafn (1837), who first forcibly drew attention to the fact that it was the Norse who, some five hundred years before Columbus, discovered the New World.

Excavation has now revealed house sites in northernmost Newfoundland that, to competent judges, appear neither Indian nor Eskimo, but typically Norse. This immensely satisfying discovery is the achievement of the Norwegian explorer, scholar, and writer, Helge Ingstad, and his archaeologist wife, Anne Stine Ingstad. Ingstad began his search for Vinland by a careful study of the voluminous literature. The evidence is derived chiefly from Adam of Bremen, an ecclesiastical historian, who wrote about 1075; from Icelandic sagas written during the thirteenth and fourteenth centuries; from certain early maps, especially the late sixteenth century delineation of the northern regions by Bishop Sigurdur Stefansson; and from modern knowledge of weather and ice conditions, ocean currents, and the techniques of navigation of a thousand years ago.

SOURCE: The Polar Record 12, no. 80, 583–7. Reprinted by permission of the author and publisher.

The sagas that mention Vinland are not consistent in all details, but that is not surprising in accounts written two centuries or more after the event. The *Grœnlendinga saga* gives the fullest account of the voyages to Vinland which began about 986, when Bjarni Herjolfsson was driven off course during a voyage from Iceland to Greenland and came in sight of new land. An accident prevented Erik the Red, first colonizer of Greenland, from leading an expedition of land-hungry Greenlanders to explore the new land, and the honour fell to Erik's son, Leif the Lucky. Following Bjarni's course in reverse, Leif sighted first a land of flat stones, which he named Helluland; further south, he found a well-timbered coast, which he named Markland; and to the southeast, he came upon a pleasant country, where his expedition passed the winter. This he named Vinland.

The next summer, Leif returned to Greenland. A year later, his brother Thorvald led a second expedition to colonize Vinland. Using the houses built by Leif as base, he made in the first summer a long exploration westward; in the second summer, he explored east and north, probably along the coast of Markland. Here the Norse met natives for the first time, whom they called Skrælings, and by a Skræling arrow Thorvald died. After passing a second winter in Vinland, his party returned, with their sad news, to Greenland. Another of Erik the Red's sons, Thorstein, set out to retrieve Thorvald's body from its Markland burial, but he was obliged to give up the plan after a storm-tossed summer at sea, and he died the next winter in Greenland. His widow, Gudrid, married Thorfinn Karlsefni, who followed the course of his wife's brothers-in-law, taking with him to Vinland three ships, 160 men, some with families, and livestock. But, after further explorations and a battle with the Skrælings, they, too, withdrew. A son, Snorri, was born to Gudrid in Vinland, the first American of European parents, so far as record tells. Leif's half-sister, Freydis, made another attempt to settle Vinland, but her venture ended in civil discord and murder.

The sagas do not dwell on later voyages to Vinland and Markland, perhaps because they had become commonplace and lacked the epic quality of those made by Erik's vigorous and bold children. In other Icelandic records, there is mention of the departure of a missionary bishop to Vinland in 1121 and of a wood-gathering trip from Greenland to Markland in 1347. Nicolo Zeno quotes a fisherman's story that seems to show the presence of Norsemen in Vinland or Markland as late as 1354.

But where was Vinland? There is good reason to suppose that Helluland was some part of present-day Baffin Island and that Markland was some part of the coast of Labrador, perhaps the handsomely forested region of Lake Melville and Hamilton Inlet. Ingstad has assumed that the saga directions and the descriptions of time-distance relations mean what they say – an assumption that few of his predecessors in the Vinland search have cared to make – and that, therefore, Vinland must lie fairly far north in Newfoundland; within two days of Markland and a short summer's sail from Greenland. He has accepted, as a few recent scholars have, the suggestion made by the Swedish philologist Soderburg in 1888 that *vin* may be taken to mean pasture, and that Vinland was remarkable in Leif's eyes, not for its wines or vines, as Adam of Bremen and the saga-writers after him supposed, but for its grazing potential. This theory Ingstad has elaborated in a massive study, *Landet under leidarstjernen* (1959).

In 1960 Ingstad began a survey, by boat and airplane, of the coast from Rhode Island north to Newfoundland. In northern Newfoundland, a resident of L'Anse-aux-Meadows gave him news of nearby ruins, which, during the summer of 1961, Fru Ingstad began to excavate. From the beginning, their close resemblance to Norse remains in Greenland was promising. The Ingstads continued excavation during the summers of 1962, 1963, and 1964, assisted by scientists from Norway, Sweden, Iceland, Canada, and the United States. Their labours revealed several house sites. One of them measures about 20 by 15 metres, and is composed of a great hall and four connecting rooms. Nearby, they found a smithy with a stone anvil and many hundred pieces of slag, bits of iron, and bog iron; they found, also, a pit for the manufacture of charcoal and, in the neighbourhood, a rich deposit of bog iron. The techniques evidently used for making charcoal and for hot-forging iron were unknown to primitive Indians and Eskimos, and to post-Columbian occupants of the region they would probably have been obsolete. A dozen Carbon-14 dates from the site cluster around 1000. The first and, so far, the only recognizably Norse artifact, a soapstone spindle whorl, was uncovered just at the end of the 1964 season. The scarcity of Norse artifacts is disappointing but not surprising, for most will have decayed long since in the wet and acid soil. Over the diggings, the Government of Newfoundland has erected shelters, and it has declared the site an historical monument.[1]

It cannot be asserted positively that L'Anse-aux-Meadows was the New World home of Leif Erikson himself, but the geography of the region so nearly matches the saga descriptions of Vinland that identity seems very likely. How much further south the Vikings sailed, or how far westward into the continent they penetrated cannot yet be stated with any confidence. There is no reason to suppose they did not

---

1 In 1977 Canada designated L'Anse-aux-Meadows a National Historic Site, and in 1978 the United Nations named it a UNESCO World Heritage Site. – ed.

explore far, both south and west, and there is reason to hope that the interest generated by the Ingstads' discovery may lead to new revelations of Norse occupation of the New World.

In the United States, heightened public interest in Norse knowledge of the New World has led to the declaration of Leif Erikson Day. In August 1964, Congress passed a joint resolution (H.J. Res. 393, introduced into the Senate by then Senator Humphrey) that authorized the president to proclaim the day each year on October 9. It was thought appropriate, in naming the commemorative date, to accord Leif a three-day precedence over Columbus, whose discovery of the New World is celebrated on October 12.

The posthumous publication of Vilhjalmur Stefansson's autobiography (1964) is a reminder that the discovery of the New World was completed by the Norse race that, by following a westward course from Norway to Ireland, Iceland, Greenland, and Vinland, had begun it. In 1915 and 1916, Stefansson, a Canadian-born Icelander, added the last major land masses to the map of the Canadian Arctic Archipelago. The kinship-conscious Norsemen of the eleventh century would, no doubt, have taken satisfaction in the knowledge that Stefansson, in common with many other Icelanders, could trace his family back through the wonderfully complete records of Icelandic history to include Snorri, the first son of Vinland.

17 Early Geographical Concepts of the Northwest Passage

THEODORE E. LAYNG

The venerable Adam of Bremen in his great ecclesiastical history of northern Europe, written about 1075, recorded for the first time the existence of Iceland, Greenland, and "islands" bearing farther away in the same direction. He was in fact giving to an uninterested Europe the highlights of the Norse odyssey to America. Do not number him however amongst the early proponents of a northwest passage. He advises that beyond Greenland "there the ocean, shrouded in mist, forms the boundary."

There is a paradox here. The "islands" west of Greenland – the Helluland, Markland, and Vinland of the Norsemen – could scarcely be out-of-bounds. Unhappy Adam. He accepted in principle the disconcerting concept of a globular world, but his thoughts still generated within that very practical image that the mapmakers of his day were wont to present. The awful ocean flowed all round, corsetting the tripartite world of Europe, Asia, and Africa, and it was scarcely within the scheme of things to bulge the tight little cartographical circle to accommodate lands not mentioned in the Bible, and therefore beyond the decent regard of men.

Meanwhile Scandinavians were ranging far and wide across the top of the globe from the White Sea to the Canadian Archipelago. Amongst their kind it is certain that navigating to the west along a northerly Atlantic route was a frequent occurrence. For the mediaeval sailor it was surely a more reassuring thought to sail the top of the

SOURCE: *North* 13, no. 4 (July-Aug. 1966). Reprinted by permission of the author and publisher.

world than to risk sailing through fiery equatorial regions to the bottom, where God alone knew how a man could avoid orbiting into space.

During the fourteenth and fifteenth centuries the old Norse route became less and less frequented. The Scandinavian people seem to have lost their zest for the sea. Interest in maintaining their monopoly of the north waned in the face of new competition from Hanseatic and English traders. There is no reason to believe however that knowledge or indeed communication with Greenland ceased entirely. The fishermen of England began to frequent Icelandic waters early in the fourteenth century, and it is easy to be persuaded that by accident or design some of their numbers would occasionally pursue a course farther west. By the middle of the fifteenth century, Portuguese caravels had begun to dominate the Atlantic, and it is not improbable that before the year 1500 they had already been searching out "islands" west of Greenland.

Whatever the case, and for whatever purpose men might search along old Norse routes amongst Arctic ice flows, these matters were soon forgotten when news spread of the rich Spanish discoveries in southern climates. From the Columbian discovery in 1497 to the mapmakers' creation of the Strait of Anian (Bering Strait) in 1566, the northwest passage, in reality totally unrealized, was to remain a figment capable of assuming a multiplicity of forms.

In fairness to sixteenth-century mapmakers it must be pointed out that experimentation with the northern terminus of the newly discovered continent was only incidental to their larger task of presenting a plausible geographical relation between America and Asia. The concept of a northwest passage would not really exercise geographers until a final cartographical divorce was obtained between America and Asia.

Geography begins with Ptolemy of course, and he had long ago (AD 150) decreed that the known world between the Canaries and his farthest east should occupy 180 degrees of longitude. After the thirteenth-century travels of Marco Polo the cartographers casually added another 60 degrees of mainland to Ptolemy's farthest east and placed Zipango (Japan) another 30 degrees beyond. The total extent, east to west, of the pre-Columbian world as shown on the Behaim globe of 1492 was 270 degrees, leaving some 90 degrees of ocean to be traversed. Nevertheless, as early as 1470 the concept of sailing due west to gain the riches of the east was a fully realized concept and awaited only the confirmation of Columbus.

Columbus, before setting out, had assured himself that Ptolemy had over-calculated the length of a terrestrial degree. He believed that Cattigara, placed on Ptolemy's 180th meridian, was actually on the

225th meridian. By his calculation Japan lay about 45 degrees closer to Europe than it was placed on Behaim's globe. Columbus therefore was quite prepared to believe that Cuba was the mainland of Asia. Much farther north, John Cabot had made a landfall on what he believed was a great northeasterly projection of Asia. And there was no man capable of offering a sound case against either. What a quandary for the mapmakers. To believe either man was to reject Ptolemy. It was impossible to represent the new discoveries as part of Asia, maintain Ptolemy's value of the degree, and crowd everything into a 360-degree Earth. The magnitude of the enigma can be more perfectly realized if we stop to consider that modern science allows over 200 degrees to place America in correct relation to her older neighbours. It is surely a first-hand intimation of bewildered frustration to study the first map forms of the New World.

The first maps of Canadian regions were particularly enigmatical. However much they may have delighted scholars through the last century, they were of little use to contemporary geographers. Maps like the Contarini of 1506 and the Ruysch of 1508 misrepresent Canada as a northeasterly projection of Asia. In fact, there was so little available knowledge of the actual northern Asia that cartographers were quite capable of equating Greenland with the Siberian Peninsula. It is easy to understand why knowledge of the north lagged; still it is a curious fact that although all other quarters of the Atlantic region were mentioned in the *Treaty of Tordesillas*, 1494, or in the papal bulls leading to the treaty, no reference was made to the northwesterly quarter.

One might expect the Portuguese to have given birth to a reasonable cartographical pattern of our northeast coast, but such was not the case. For instance, the Cantino map of 1502 shows an indefinite north-south coast, southwest of Greenland, straddling the line of Demarcation, as if the mapmaker was more concerned with establishing diplomatic rights than adding to the sum of geographical knowledge. Of the east-west line of the La Cosa map (1500) and the Oliveriana (1506) little can be said except that surely enough has been written to prove that it was indeed the most questionable line ever drawn on a plane surface.

It was Martin Waldseemuller, a German geographer working in the town of St Die, who finally found a way out of the cosmographical impasse. Apparently he accepted Amerigo Vespucci's startling concept that the new discoveries were in fact parts of a hitherto unknown continent. On his great map of 1507, the Americas are shown for the first time as separate land masses. Moreover, Waldseemuller very ingeniously contrived to introduce the new continent without unduly offending any of the current schools of thought. The old world from the Canaries to Zipango was represented within 270 degrees of longitude, the far east terminating on the right hand site of his map in a manner pleasing to the loyal forces of Ptolemy. In the 90 degrees remaining to him on the left hand side of the map, he drew, in elongated form, the new and fourth part of the world with the east coast in the longitude approved by Columbus. The latter and his followers might take this to represent the east coast of Asia if they wished, but what matter, the form of the New World, upon which others must build, had been established.

The northern terminus of the new continent, however, is still a question mark. America ends in a conventional and foreshortened north-south line, bearing the standard of Spain, in far northern latitudes. Far out in the mid-Atlantic lies an indeterminate mass of land bearing the Portuguese flag. In his map proper, Waldseemuller shows a strait separating north and south America. In an inset he represents a continuing coast line. Waldseemuller was indeed a master of compromise. Some thousand copies of his map were printed and its influence for the next half century is clearly recognizable in the mapping of the new world. Here was the true genesis of the concept of a northwest passage.

Waldseemuller had created an entirely hypothetical Pacific Ocean which no European had realized let alone seen. At the same time, he offered three possible through-ocean routes to Asia. Perhaps there was a convenient way through the region of Panama. If this were not so, one had only to seek the northern or southern extremity of the new continent and set his course for Asia.

Then, in 1513 Balboa stood "silent upon a peak in Darien" and, gazing towards the south, beheld in wonderment the great Pacific and called it the South Sea. Cortes and others immediately initiated explorations on both sides of the Central American coast line, hoping to find the place where the two oceans joined. It was almost a decade before it was clearly demonstrated that there was no meeting of the waters anywhere within hundreds of leagues of Darien. Cortes suggested a search up the northeast coast towards Bacalaos (Newfoundland). No one wanted to search along the western coast beyond California. To do so would point them to the most remote point from Europe and the known inhabited world. It was bad enough searching amidst ice flows along the northeast coast, as the Portuguese, English, and Bretons knew. And for what were they searching? Even Waldseemuller had lost his convictions about the new continent. On his map of 1516, he designated Cuba as part of Asia without even attempting to show a connection between the new discoveries and Asia.

There are a few historians who would attribute to Sebastian Cabot a remarkable knowledge at this period of the new continent and a way to Asia through the north. They would have him in 1507 or 1508 discovering Hudson's Strait and divining a passage through the Arctic Archipelago. This was presumably the knowledge he held as bait to wheedle his way to fame in Spanish hydrographical circles. But indeed it is not easy to be persuaded that Sebastian was ever in Canadian waters, let alone to believe that he was the first to demonstrate the existence of the northwest passage.

It was Johann Schoner, a German geographer and globe maker, who provided the cosmographers with a new inspiration. Although he believed that north of the new discoveries some part of China would be found – perhaps even the rich Cathay – he offered a more enticing passage by the south. On his 1515 globe he located a strait, connecting the Atlantic and Pacific Ocean, which lay between America and "Brasilie Regio." Spain was persuaded. The Portuguese had successfully rounded Africa; now the Spanish would circumnavigate America.

The Magellan expedition (1519–1523) remains one of the greatest episodes in the history of discovery. It is therefore all the more ironical that, in opening a way to the Pacific by sailing around Cape Horn, the expedition led to cartographers sealing the way by the north. The story is one of the most intriguing moves in the great trans-oceanic gamble which in the early years of the sixteenth century was being played out between the kings and politicians of Spain and Portugal. After Magellan's voyage it appeared that both countries had won their stake. The Portuguese, by sailing to the East around Africa, had tapped the riches of Asia, and now the Spanish had reached the same goal by sailing westward around the new continent. Now the question arose – where did the Line of Demarcation divide Spanish and Portuguese spheres in the Pacific?

The Portuguese were already striking out on an easterly route towards the Malay Peninsula, and the Spaniards were desperate to forestall them. At the Congress of Badjoz in 1524, they claimed that the Line of Demarcation would pass through the tip of the Malay Peninsula, thus leaving the Moluccas, the Philippines, the eastern coast of Asia, and everything east of it to Spain. It was rather a large claim, and the Portuguese were having none of it. (By modern calculations the line ran 5 degrees east of the Moluccas.) Obviously the Spanish case needed strengthening and apparently the survivors of the Magellan expedition provided the right answers at the right time. The distance in longitude, they said, between the Strait of Magellan and the Philippines was 106°30'. The Ladrones were 300 leagues further to the east, or in others words, they lay somewhere between 220° and 225° longitude.

This sort of reporting, coupled with erroneous information about the longitude of the Spanish discoveries in the region of Central America, gave rise to some new and bizarre mapping. Schoner, for instance, had been informed that Mexico City was located about 225 degrees east, that is to say, on about the same meridian as the Ladrones. His mind was surely well exercised. Obviously the Asian coast was more closely related to America than he had realized. The easiest way to straighten the matter out was to rejoin America with Asia. Balboa's South Sea was in effect a great bay. And many of the armchair geographers of Europe, including wise Giocommi Gastaldi of Italy, found Schoner's solution acceptable. There wasn't much inspiration left for seeking a northwest passage.

Fortunately the pace of discovery and speculation was too rapid to allow geography to freeze upon such a concept. About the time the Magellan expedition returned to Spain, King Francis I of France decided it was time for him to stake a claim in the New World. Like the early Tudors of England, he considered the *Treaty of Tordesillas*, reserving this right to Spain and Portugal, a scrap of paper. In 1524 he sent Giovanni da Verrazzano, a Florentine, to the coast of America with the usual injunction to search for a passage to Asia. Ranging the coast of North Carolina, Verrazzano found an isthmus, "a mile in width and 200 long," which has been identified with the great sandbar extending from Pamlico to Albermarle Sound. It may have been the poor visibility, or it may have been his great longing to see the Western Sea, but from his vantage point Verrazzano mistook the lagoon to the west for "the oriental sea ... which is the one without doubt which goes about the extremity of India, China and Cathay." In 1527 Vesconte de Maiollo of Italy produced a map obligingly giving prominence to Verrazzano's Sea, showing the narrow isthmus and giving to Canadian parts the form of a well-developed turnip. More important, he extended the Verrazzano Sea northward, pointing the way to a through passage to the Atlantic. This was indeed a make-weight against the parties of union for America and Asia. Battista Agnese, another popular Italian mapmaker, took up the vogue of the Verrazzano Sea and gradually a passage to the north of the Americas began to gain ascendancy in cartographical circles, experiencing in the process some remarkable growing pains.

One form in particular that it assumed must enter into all discussions regarding the genesis of the northwest passage. It is that display on the Gemma Frisius globe c. 1537. Unfortunately, the only legible copy of it available is a sketch drawn and reproduced by Ganong[1] some years ago. There is little doubt that the prototype of this globe was in existence before the Verrazzano Sea appeared on maps, and regardless of the validity of its sources, the concept of the passage as is

---

[1] W.F. Ganong, *Crucial maps in the early cartography and place-nomenclature of Canada*,
University of Toronto Press, in cooperation with the Royal Society of Canada, 3rd series.

shown on the globe was substituted by most mapmakers for the Verrazzanian concept. It is indeed a temptation to point to the globe as demonstrating that the genesis of the northwest passage was found upon an actual voyage penetrating into the Arctic from east to west. Who were the three brothers after which the Arctic Strait is named? There were three brothers in the Cabot family and three in the Corte-Real family. Is the *"terra per britannos inventa"* land discovered by the English or Bretons? Is the strait a confused expression for the Strait of Belle Isle? In any case mapmakers developed from the model a well-defined passage north of the continent trending southwest into the Pacific.

Undoubtedly, the Dieppe maps, the Mercator map of 1538, and later Agnese[2] maps, all of which followed upon the Gemma Frisius pattern, did much to turn English thinking towards the north.

In 1548 Gastaldi introduced an ingenious combination of Schoner and Verrazzanian mapping. He placed Verrazzano's isthmus in a relatively high latitude, with the corresponding sea as the large indentation directly north of it. True, he still managed to join America with Asia, but in 1562 he became a rabid convert to separation and, in his pamphlet *"La universalle descrittione del Mondo,"* he expressly mentioned the existence of a Strait of Anian more or less in the same location as Bering Strait. No one has yet offered a reasonable explanation of Gastaldi's master stroke, but even in retrospect it appears to rival Waldseemuller's Pacific as the best educated guess of the sixteenth century. Probably Gastaldi produced a map to accompany his pamphlet, but if so it has not survived, and the earliest dated map to show the Strait of Anian is the Zalterii of 1566.

In 1570 Ortelius, the great Dutch atlas maker, adopted the Gastaldi form and its place in contemporary mapping was assured. One small point – the Ortelius maps and most of the maps of the last quarter of the century show a very well-defined indentation more or less in the location of Hudson Bay. This may appear a nice bit of speculation since Hudson Bay didn't arrive on the scene until 1612. Perhaps it is only a recession to the north of Gastaldi's Verrazzano's Sea.

It all came about by the natural aversion cartographers have to blank spaces on their maps. But what perspicacity these sixteenth-century mapmakers showed in their final selection of a suitable form for the northwest passage. Another two centuries were to elapse before the gentlemen of the Royal Navy were able to add substance to the form.

<hr/>

2 Desliens 1541, Descelliers 1546, Harleian c. 1542.

18 Voyageurs' Highway: The Geography and Logistics of the Canadian Fur Trade

ERIC W. MORSE

Two hundred years ago, as today, the ice broke up on the Ottawa River around the first of May. In the heyday of the Montreal fur trade, the half century from 1770 to 1820, the first of May saw great activity at Lachine, eight miles above Montreal. "Brigades" of big Montreal canoes, or "*canots de maître*," each craft paddled by ten or a dozen colourfully dressed *voyageurs*, and carrying up to three tons of cargo, were loaded and began to move off for the "*pays d'en haut.*"

They went straight west up the Ottawa River to Mattawa, where the Ottawa ends its big swing down from the north. Here they headed up the Mattawa River, and paddled and portaged forty miles to its source in Trout Lake, at North Bay. Three portages over a rough divide led them into Lake Nipissing, from where it was easy going down the French River to Georgian Bay. After following the North Channel above Manitoulin Island, they portaged past Sault Ste Marie and headed out around the treacherous 450-mile [725-km] passage of the North Shore of Lake Superior.

About the end of June, after eight weeks of long days, great hazards, and unremitting toil, they found themselves at their objective, Grand Portage. This was the great central entrepot of the Canadian fur trade,

SOURCE: *Canadian Geographical Journal* 62, no. 5 (May 1961): 148–58. This is the first of three articles dealing with the geography of the fur trade routes which appeared in the *Cdn. Geog. Jnl.* A fuller version by the author was published as *Fur Trade Canoe Routes of Canada / Then and Now*, by the Queen's Printer, Ottawa, 1969. Reprinted by permission of the author and publisher.

situated on a shallow bay of Lake Superior ten miles [16 km] south of the mouth of the Pigeon River, the present international border. (Grand Portage was abandoned when the Americans made good their claim to the new boundary, and from about 1803 its place was taken by Fort William.)

Grand Portage was the objective of the big *canots de maître* from Montreal, and a rendezvous. But it was not the end of the line. The other end of the "Voyageurs' Highway" was Fort Chipewyan on Lake Athabasca. The ice in that latitude did not break up quite as soon as on the Ottawa. Though the big northern lakes often were still iced over in June, the turbulent Athabasca River usually burst winter's bonds around May 15. Fort Chipewyan about then became a scene similar to Lachine a fortnight earlier. And while the Montreal canoes were paddling westward, the Athabasca Brigade was paddling eastward to meet the *canots de maître* at Grand Portage.

The waters west of Lake Superior on the whole were smaller, and two continental divides had to be crossed. The craft used here was the North canoe – high-ended, half the capacity, and paddled by six or eight men. On the trip east they carried, not trading goods, but bales of fur. Up the Athabasca River they paddled for two hundred miles [320 km], leaving it at Fort McMurray (Waterways) to ascend the swift Clearwater. This stream they left after eighty miles [130 km] to cross the gruelling thirteen-mile [20-km] Methye (La Loche) Portage, which brought them to the headwaters of the Churchill river. The Churchill was followed for some four hundred miles [645 km] to a point north of Cumberland House where Frog Portage led them over the Sturgeon-weir River, which (for its wicked rapids) the *voyageurs* called the Maligne. This carried them down to the Saskatchewan River, which they then followed to its mouth in Lake Winnipeg.

Most of this journey from Fort Chipewyan was downstream. There followed the passage of storm-tossed, shallow Lake Winnipeg, and then the laborious ascent of the Winnipeg River and its tributary the Rainy along the border-lakes chain. This led them to the divide, and over it to either Grand Portage or Fort William, the two routes separating at Lake La Croix, a little to the east of Rainy Lake.

Along the way, the Athabasca Brigade would sometimes be joined by other less distant Brigades – those from Île à la Crosse, Cumberland House, or Red River. Actually, because of the greater distance, the Athabasca Brigade usually got only as far as Fort St Pierre at the foot of Rainy Lake, where they were met by a special detachment from Grand Portage in mid-July and were allowed to get away on their two-months' home journey before the first of August. Otherwise, the reforwarding of some of their return cargo of trading goods from Fort

Chipewyan to the outlying posts in the Mackenzie District could not be accomplished before freeze-up.

All this is of necessity a sketch and a simplification, taking no account of the fur trade's vicissitudes and the later modifications of the route. Nor does it attempt to bring in the avenue used by Canada's other fur-trading enterprise, the Hudson's Bay Company, which will be fitted into the picture later. Staked out here, however, is a water route, Canada's first and main throughway, which has probably done more to shape Canada's history and development than any other of its avenues of communication. The *coureurs de bois*, the *voyageurs*, and early explorers who first used this route are symbols of Canada's heroic or epic age; and few nations have so colourful and romantic a past. Many Canadians seem to be aware of the historical associations of this highway, without realizing that the actual route still lies hardly changed today: the scenery, the conditions of wind and current, nearly all the actual portages have scarcely altered in the three centuries since the first fur-seekers headed out from Quebec and Montreal for the *pays d'en haut*.

The physical geography of this waterway and the logistics of the fur trade are as impressive as the saga itself. In an age before air travel, in a land devoid of road or rail, on a route beset with obstructions, dangers, and difficulties, how did men burdened with hundreds of tons of fur and trade goods succeed each year practically in crossing a continent and back again in the scant five months between break-up and freeze-up? How did they get across lakes like Winnipeg and Superior in craft of birch rind? How did they overcome the Rocky Mountains? Paddling and portaging often eighteen hours a day, how were the human engines refuelled as they crossed a wilderness? Most such questions can be studied more concretely by retracing the route; it is a palatable proposition that sometimes as much history can be learned from a canoe as from a history book.

This survey is designed to cover the waterways of the fur trade, the overcoming of navigational obstacles, the logistics, and the influence of the fur trade on Canadian development.

The Waterways of the Fur Trade

It is a staggering statistic that half of all the fresh water in the world is to be found in Canada. Put into other words, there are as many miles of inland waterways in Canada as in all the other nations of the world combined. And, for a craft adapted to the conditions, these waterways are navigable. It is still possible to put a canoe into the water in practically any Canadian city, and paddle from the Atlantic to the Pacific

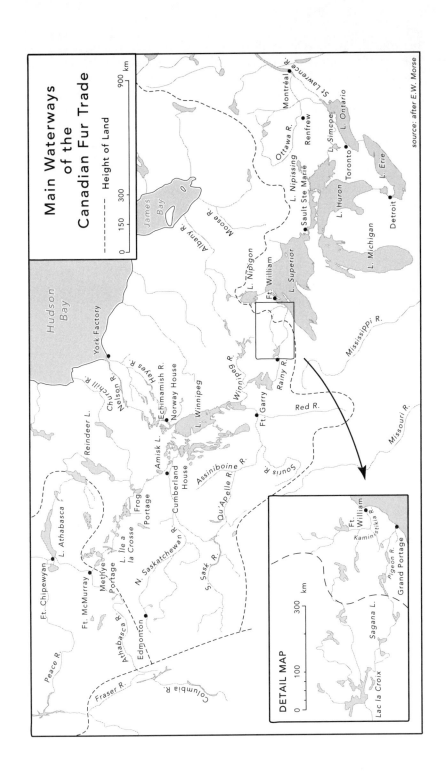

Main Waterways
of the
Canadian Fur Trade

- - - - Height of Land

0   150   300   900 km

source: after E.W. Morse

Hudson Bay

James Bay

York Factory

Montréal

Ottawa R.

St Lawrence R.

Renfrew

L. Nipissing

L. Simcoe

L. Ontario

Toronto

L. Erie

Detroit

Sault Ste Marie

L. Huron

L. Michigan

L. Superior

Ft. William

L. Nipigon

Moose R.

Albany R.

Mississippi R.

Reindeer L.

Churchill R.

Nelson R.

Hayes R.

Echimamish R.

Norway House

Amisk L.

L. Winnipeg

Winnipeg R.

Rainy R.

Red R.

Missouri R.

Ft. Garry

Cumberland House

Frog Portage

Assiniboine R.

Souris R.

Qu'Apelle R.

N. Saskatchewan R.

Sask. R.

S. Sask. R.

Fraser R.

Columbia R.

Peace R.

Ft. Chipewyan

L. Athabasca

Ft. McMurray

Methye Portage

L. Ile a la Crosse

Athabasca R.

Edmonton

DETAIL MAP

0   100   300 km

Ft. William

Kaministikia R.

Pigeon R.

Grand Portage

Sagana L.

Lac la Croix

or from the Arctic Ocean to the Gulf of Mexico. This is not, as in Europe, the result of man-made canals, but of a fantastic drainage pattern consisting of three vast, shallow basins and three great "hubs."

The water from three-quarters of continental Canada drains off through three outlets: the Gulf of St Lawrence (10 per cent), Hudson Strait (43 per cent), and the mouth of the Mackenzie (22 per cent). Furs were gathered, of course, from the other quarter of the country, but the heavy transportation of furs and goods operated almost entirely within these three drainage areas.

While it is usual to speak of drainage basins, for this story it would be more accurate to refer to these three areas as drainage *saucers*. This helps to underscore the low elevation of the rims, in relation to the vast areas encompassed. Twelve hundred miles [1,930 km] from tidewater up the Great Lakes to Fort William [Thunder Bay] is a rise of just six hundred feet [180 m]. Between the drainage areas, at no point is an interconnecting portage more than about fifteen hundred feet [460 m] above sea level; and the same holds for the several connecting gateways to the Mississippi Basin.

The combination of these three facts of Canadian geography explains much regarding the fur trade routes. Three quarters of Canada east of the Rockies presented no serious barrier to canoe travel.

This in turn is recognizably related to the presence of the worn down Precambrian Shield, covering half of Canada. The Shield with its countless systems of lakes and rivers – quite apart from its birch trees – is canoe country. Actually, for about two thousand [3,200] of its total three thousand miles [4,800 km], the Voyageurs' Highway passed along, or close to, the Shield's southern border. The Ottawa River west to Renfrew, the north shores of Lake Huron and Superior, the border lakes including Lake of the Woods, the Winnipeg River, and Lake Winnipeg – all either form, or closely flank, the Shield's southern rim. West of Lake Winnipeg the old canoe route picks up the Shield once more in Amisk Lake on the Sturgeon-weir River, and stays just inside the Shield almost as far as Lake Île à la Crosse in Northern Saskatchewan.

The presence of so many big lakes – Great Bear, Great Slave, Athabasca, Winnipeg, Lake of the Woods, Superior, and Huron – strung along the Shield's border, is another key to east-west canoe travel. These lakes all have one side (or end) in granite, and the other in sand or limestone. Nothing bears out better the newer concept of the Shield, not as a divisive wedge, but as a cohesive core of Canada.

The Voyageurs' Highway, however, was a trunk route; and Canadian geography in its bounty has provided more – an extraordinary system of branch routes, grouped around three well-defined hubs. The hubs are *Lakes Superior, Winnipeg,* and *Athabasca,* each no higher than six or seven hundred feet [180-215 m] above sea level.

Working clockwise around *Lake Superior* from Sault Ste Marie were the following water connections:

To Detroit and Niagara via Lakes Huron and Erie.
To the Lower Mississippi via Lake Michigan and the Illinois River.
To the Upper Mississippi via Lake Michigan, Green Bay, and the Fox and Wisconsin Rivers.
To the prairies via the Winnipeg River, using three different approaches: Fond du Lac and St Louis River; Grand Portage and the Pigeon River; Fort William and the Kaministikwia River.
To James Bay[1] by two different routes: Lake Nipigon and the Albany River; Michipicoten and Moose River.
To Montreal by two different routes: French and Ottawa Rivers; Lake Simcoe, Toronto, and the St Lawrence.

Working clockwise around *Lake Winnipeg* from Winnipeg were the following water connections:

To the Missouri country via the Assiniboine and Souris Rivers.
To the western prairies via the Assiniboine and Qu'Appelle Rivers.
To the Rockies via the Saskatchewan River.
To the Athabasca country via the Saskatchewan River as far as Cumberland House and thence by Sturgeon-weir, Churchill, and Clearwater Rivers.
To Hudson Bay by two different routes: via the Nelson River; via the Echimamish and Hayes Rivers.
To Lake Superior via the Winnipeg River and one of the three approaches listed above.
To the Mississippi via the Red River.

Working clockwise around *Lake Athabasca* from Fort Chipewyan were the following water connections:

To the Pacific Ocean via the Peace and Fraser Rivers.
To the Arctic Ocean via the Slave and Mackenzie Rivers.
To Hudson Bay via the Fond du Lac River, Black Lake, and the Dubawnt River.
To the Churchill River by two different routes: the Fond du Lac River and Reindeer Lake; the Athabasca River and Methye Portage.

---

1 Two other historic canoe routes from James Bay were the route through the River and Lake Abitibi connecting with the Ottawa River at Quinze Lake, and the route, followed by Father Albanel in 1671–72, connecting Rupert River with Lake St John via Lake Mistassini.

To the Pacific Ocean via the Athabasca River, Athabasca Pass, and the Columbia River.

These branch routes were feeders. They should not overshadow the basic east-west trunk routes. And in reviewing the explosive development and exploration of the Canadian Northwest during the century following La Vérendrye's penetration into the prairies, the existence of a through way should not be taken for granted. It is of interest to compare the Canadian situation with developments south of the border.

The American settlement of the West preceded Canada's, which tends to obscure the fact that Canada felt a steady pulse of east-west commerce thirty years before the Americans had even crossed the Mississippi. What contributed largely to delay American penetration of the far West was the lack of a water throughway. The great river of the American plains, the Mississippi, runs north-south. The only American river at all comparable to the Ottawa, the Winnipeg, or the Churchill was the Missouri. The Siberian fur trade also, facing the same odds, had a slower development: the main rivers of Siberia all run north and south, at right angles to the trade-flow.

A search of the map of Canada fails to show any practicable alternative to this main route. There were (sometimes transitory) short-cuts; but, squeezed between the Shield to the north and the United States border, there was simply no other main artery. This canoe route was used for war, trade, or hunting, each section of it by its own group of Indian tribes, centuries before they guided the white man over it.

An interesting postscript is the fact that, with modern engineering advantages, in building first the Canadian Pacific Railway and then the Trans-Canada Highway, we have not departed very far from the basic route. At least from Montreal to Winnipeg, these three routes – one historic, two modern – are braided together, never more than a few miles apart.

The Voyageurs' Highway was developed by the Montreal fur traders, first the *coureurs de bois*. But Montreal's was only one of two vast fur empires that shaped Canada's destiny; and in 1821 the Hudson's Bay Company swallowed its rival, the North West Company, in a merger that retained the older name.

The reason why the Hudson's Bay Company and not the North West Company came out on top was basically geographical and economic. With sea communications from their headquarters in London stretching into the heart of the continent at York Factory, the Hudson's Bay Company could lay down trade goods in the Athabasca Country at half the price of their rivals. The Hudson's Bay Company entrepot was Norway House, at the north end of Lake Winnipeg. Between there and

York Factory lay, not 3,000 [4,800 km], but only 350 miles [560 km], a relatively safe and comfortable route which went from York Factory up the Hayes River and over to the upper Nelson by a curious link, the Echimamish (the river-that-flows-two-ways). From here the goods going up were consigned to, and furs coming down collected from, either of two main depots, Fort Garry and Cumberland House. The Hudson's Bay Company instead of canoes had York boats, and instead of Canadian *voyageurs*, Orkneymen. To best their rivals, they imitated them as far as possible, recruited whom they could from defecting Northwesters, and gave their Scots employees bonuses to acquire the *voyageurs'* skill with paddle and white water.

A discussion of the waterways of the Canadian fur trade would be incomplete without some review of the later modifications, both achieved and attempted, to the main route described at the outset of this article.

Three abortive attempts were made to cut corners or to bypass a particularly rough section of the course, respectively by Edward Umfreville, David Thompson, and Sir George Simpson. Before the North West Company was ousted from Grand Portage, Edward Umfreville was commissioned in 1784 to try to get through from Lake Superior to the Winnipeg River by way of Lake Nipigon and the English River. The route proved impracticable for heavy trade and was never used. David Thompson made an attempt in 1796, just before he left the Hudson's Bay Company, to bypass the rugged Methye Portage and the shallow Methye River by a backdoor route into Lake Athabasca from the Churchill near Frog Portage. The route went through Reindeer and Wollaston Lakes and down the Fond du Lac River. Reindeer Lake is 135 miles [220 km] long and Wollaston, seventy [110 km]. Lakes that size in that latitude don't break up as early as does a turbulent river such as the Athabasca. In the author's own experiences, for instance, Reindeer Lake broke up in 1957 only on June 20, more than a month after the brigades would have left Fort Chipewyan. Great Bear broke up in 1959 on July 23. Though Thompson does not say explicitly in his journal, this was probably the reason why his attempted route was never used.

A third variation tried (on the Hudson's Bay Company route) was Sir George Simpson's effort in the 1820s to get his Athabasca brigades to go from York Factory in a direct line west to the Churchill River, by way of the Nelson and Burntwood Rivers. The *voyageurs*, however, balked at forgoing the fleshpots of Norway House and Cumberland House en route. The Little Emperor's pure logic for one rare occasion had to retreat before human nature, and the two sides of a triangle continued to be used.

No alternative to the main route ever existed in the six hundred miles [965 km] between the heel of Georgian Bay and the Lakehead, nor in the one thousand miles [1,600 km] between Rainy Lake and Cumberland House on the Saskatchewan.

However, two big detours were ultimately developed with a view to easing the *voyageurs'* back-breaking labours, and to speed the trade by introducing what crude "mechanization" became available. Without noting these and putting date-tags on them, Canada's early trade routes can be confusing. The detours were:

MONTREAL TO GEORGIAN BAY

As canals were developed on the St Lawrence, Durham boats and bateaux brought the heavier goods from Montreal up the St Lawrence, through the Bay of Quinte, over the short portage at Carrying Place, to Toronto. In 1797 the North West Company contributed to the cost of building Yonge Street, north to Holland Landing. Boats carried goods across Lake Simcoe, at first to Barrie, where the Nine-Mile [15-km] Portage brought them to Willow Creek and thence to Nottawasaga Bay. Then when the Coldwater Road was built a few years later, they went from near Orillia by road to Penetanguishine. Sailing vessels then conveyed them to Sault Ste Marie. This bypass via Lake Simcoe operated only between 1797–1821. Some goods from Montreal were even portaged around Niagara Falls and carried by ship via Detroit. This roundabout route, however, was used more especially by the American traders in Albany.

CUMBERLAND HOUSE TO ATHABASCA RIVER

Following the merger in 1821, Sir George Simpson developed the use of York boats on the Saskatchewan River, taking goods the whole distance between Cumberland House and Edmonton, in order to bypass the strenuous Sturgeon-weir-Churchill route. This necessitated a 90-mile [145-km] tote-road journey from Fort Edmonton straight north to Athabasca Landing, and a barge trip down the rough Athabasca. In pattern and in purpose this detour exactly paralleled bypassing the Ottawa River and North Bay divide by the Toronto-Lake Simcoe route. Both detours added miles to the route, but save sweat.[2] There remains only to bring the story of the Voyageurs' Highway up to date, which is practically to living memory. From 1821, the long, toilsome Voyageurs' Highway between Montreal and Lake Winnipeg was abandoned as a route for heavy trade goods, which thenceforward passed over the Hayes River route between Lake Winnipeg and York Factory.

---

[2] The detour by way of Kaministikwia, Dog, and Savanne Rivers and Lac des Mille Lacs, forced by the compulsory abandonment of Grand Portage in 1803, has been noted earlier.

From 1860, by which time American railways had got through to St Paul, even the Hayes River route was abandoned for heavy traffic, in favour of a progression of varied (rail, Red River cart, and boat) transport via Chicago, St Paul, and the Red River, to Fort Garry.

People nevertheless still had to move across a roadless nation. Passengers, mail, and "express" between the Athabasca country and Fort Garry, and between Fort Garry and Montreal, continued to use the Voyageurs' Highway. In 1870 Colonel Garnet Wolseley's army had to move west to settle the Riel Rebellion without passing over American soil. For one hundred miles [160 km] immediately west of Lake Superior, they used a modification of the old canoe route, the Dawson Trail, which cut a swampy corner off the Kaministikwia-Savanne route by going up the Shebandowan River. An old man of eighty at Fort Frances described to us in 1954 his coming out by canoe as a colonist over this section of the Voyageurs' Highway when he was a boy.

Until the Canadian Pacific Railway was completed in 1885, the Voyageurs' Highway was still the fastest way to cross Canada. An old Indian whom the author talked to on the Methye Portage in 1958 clearly remembered his grandfather telling of the busy wagon traffic over the thirteen-mile [20-km] trail in the 1880s. Dr Charles Camsell, who died only recently, in his *Son of The North* tells of coming out to school over this route in 1884.

This first in a series has intended to stress the waterways of the Canadian fur trade. The main, east-west throughway bore an important relationship to the Precambrian Shield and to the continental drainage pattern – prolific in waterways, wrapped around three hubs, with no intervening barriers.

The basic organization of the Montreal fur trade in its heyday was to have two sets of canoes (each adapted to its own waters) rendezvous halfway, swap loads, and return, thus licking the problem of covering a 6,000-mile [9,700-km] return trip in the five ice-free months. Their original route came to be modified later as bateaux, York boats, and sailing vessels were used to ease labour.

The trade-pattern of the Hudson's Bay Company, using the sea as an approach to the core of the continent, was short, simple, and economically sound. This helped them to swallow their rivals in 1821. That event took the heavy traffic off the route, but the Voyageurs' Highway continued in use for other purposes till only eighty years ago.

19 Fur Trading Posts in the Mackenzie Region up to 1850[1]

JOHN K. STAGER

The story of the expansion of the fur trade in Canada may be regarded as the first attempt to solve a perennial Canadian problem – the conquest of great distances through an organized system of communication and transportation, set within the existing economic and technological framework. The great difference between the prices paid and the prices received for furs permitted the enterprising Montreal merchants to push their canoe routes across the forested belt of North America, and by about 1775 they had reached the divide beyond which waters drained to the Arctic Ocean. At first, trading in the northwest was enormously profitable, even for the individual small operator with perhaps two or three canoes, but competition forced deeper penetration westward, increased expenses, reduced profits, and gradually brought about cooperation among the Montrealers. Perhaps the most adventuresome and farsighted among his fellows was Peter Pond who, with the combined trade supplies of several colleagues, continued north in 1778 instead of returning at the end of summer to Grand Portage, the main depot on Lake Superior. He crossed the divide at Methye Portage and established himself for winter trade on the Athabasca River about forty miles [65 km] from its mouth. Pond was

---

[1] The author wishes to thank the Governor and Committee of the Hudson's Bay Company for permission to consult and quote from the company records.

SOURCE: Canadian Association of Geographers, British Columbia Division, *Occasional Papers in Geography*, no. 3 (June 1962): 37–46. Reprinted by permission of the author.

the first to trade in the new territory of the Mackenzie drainage.[2] Pond remained in this country for a couple of seasons, then left, having obtained a good deal of local geographical knowledge and a handsome return in furs for his efforts. He returned in the summer of 1785 as a partner of the newly formed North West Company, and the next year sent his man Leroux to set up a post – Fort Resolution – on Great Slave Lake at the east mouth of Athabasca River. Pond was opposed in trade at both posts – on the Athabasca, and at Great Slave Lake – and, during the winter, his opposite number, John Ross, was killed. Such rabid rivalry brought about a cool decision, and the opposing forces from Montreal joined their fortunes in a new North West Company agreement in 1787.

That fall, Pond was joined at his Old Establishment by Alexander Mackenzie, a bourgeois[3] in the new company. A winter in company with the aging trader-explorer (Pond was nearly fifty while Mackenzie was only twenty-four or twenty-five), and seeing the map prepared by Pond, convinced Mackenzie that he could follow the river out of Great Slave Lake to its mouth at Cook's Inlet. We now know that the journey performed in the summer of 1789 took Mackenzie to the Arctic Ocean, and although it was a personal disappointment to Mackenzie, the explorer, it opened a rich new area for trade.

As Mackenzie made his voyage, Laurent Leroux crossed Great Slave Lake and travelled to Lac la Martre, about fifteen days closer to the homes of the Indians who were used to trading at the lake. This move was to consolidate further the hold on the Great Slave Lake Indian, and to prevent furs from going through middlemen (Chipewyans) to the posts on Hudson Bay. Also it encourage the local Dogribs to draw furs from the country of the Hare Indians further north. Lac la Martre [Wha Ti] was a post occupied intermittently until other posts appeared along the Mackenzie River. The first establishment on the Mackenzie River proper was built in 1796 by a clerk named Duncan Livingston on the right bank a short distance downstream from the mouth of Trout Lake River. He traded successfully for three years, and in the summer of 1789 he attempted to follow Mackenzie's lead by travelling to the mouth of the river in search of silver which had been reported to him. Unfortunately, he encountered Eskimos [Inuit] below Arctic Red River [Tsiigehtchie] and the entire party was slain – one member, James Sutherland escaped the hail of arrows but was caught and

2   In 1772 Samuel Hearne, Hudson's Bay Company, was the first white man to enter the Mackenzie drainage from the Coppermine River but his was an exploratory passage rather than a trading mission.

3   "According to North West nomenclature, clerks have charge of posts, Bourgeois of districts" (Alexander Ross, 1855) – ed.

drowned by having a stone tied to his neck and being dropped in the river. The next year, John Thompson, Livingston's replacement, abandoned the Trout River Post, and divided the Mackenzie River Indians between two depots. He built Rocky Mountain Fort on the right bank at Camsell Bend, opposite the mouth of North Nahanni River, and began trade at the outlet of Great Bear Lake, the site later occupied by Fort Franklin. Sir John Franklin claimed this latter place was occupied in 1799, the year Livingston went down the river, and it may be that he initiated trade there. In any case, in 1800, there were only two posts, widely separated on the Mackenzie River.

Soon after exploration of this new land, it became evident that it was impossible to carry furs to Grand Portage and return in one season. Instead, men from the north met others coming from Lake Superior at the height of land, Methye Portage, and they exchanged loads with trade goods returning north and the furs continuing south. This system halved the distance required to make the annual returns. Even the long canoe links were perilously close in time to the length of the open season, and were possible to maintain only because no time was required to hunt. Pemmican, traded primarily for liquor from the Plains Indians, provided the food that permitted the deep penetration to the Northwest. Moreover, the operations in the Mackenzie Valley were economically precarious since it took four years at least from the time goods were sent from Montreal until the furs were returned to pay for them. This meant great burdens of credit and interest to take away from profits. The Mackenzie region was marginal both geographically and economically.

In this early period there grew up a certain disenchantment on the part of the wintering partners with their Montreal colleagues. It led to a row among the winterers, some of whom formed a New North West Company, call the X.Y. Company, and later Sir Alexander Mackenzie and Company after the explorer who lent his new-found prestige and capital support. The new company survived until 1804 when it again submerged itself in a reorganized North West Company. From 1800 to 1804 there was fierce competition between the two groups which probably occurred in the Mackenzie River as well as in all the rest of their territories. In 1803, the X.Y. Company had four posts on the Peace River to the Northwesters' five, and they were on the Mackenzie River the next year, if not before.

In 1804 there was X.Y. Company opposition at Great Bear Lake, because the master of the North West Company Fort there was killed in a dispute over Indian allegiance. Fort Alexander at Willowlake River, on Wentzel's map (1822), was possibly an X.Y. Company post of this period. There are other posts, shown on Wentzel's map and on Franklin's map (1828), which may have been built during the period of

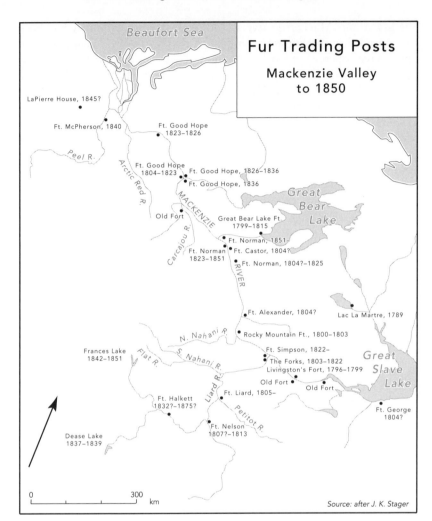

Fur Trading Posts

Mackenzie Valley
to 1850

Beaufort Sea

LaPierre House, 1845?

Ft. McPherson, 1840

Ft. Good Hope
1823–1826

Peel R.

Arctic Red R.

Ft. Good Hope
1804–1823

Ft. Good Hope, 1826–1836

Ft. Good Hope, 1836

MACKENZIE

Great
Bear
Lake

Old Fort

Great Bear Lake Ft.
1799–1815

Carcajou R.

Ft. Norman, 1851–

Ft. Norman
1823–1851

Ft. Castor, 1804?

Ft. Norman, 1804?–1825

RIVER

Ft. Alexander, 1804?

Lac La Martre, 1789

N. Nahani R.

S. Nahani R.

Rocky Mountain Ft., 1800–1803

Frances Lake
1842–1851

Flat R.

Ft. Simpson, 1822–

The Forks, 1803–1822

Livingston's Fort, 1796–1799

Great
Slave
Lake

Liard R.

Old Fort

Old Fort

Ft. Halkett
1832?–1875?

Ft. Liard, 1805–

Petitot R.

Ft. George
1804?

Dease Lake
1837–1839

Ft. Nelson
1807?–1813

0   300
⌞___⌞___⌞___⌞___⌟ km

Source: after J. K. Stager

intense competition, since rivalry for fur trade tended to multiply establishments. Old forts were located at the following places: Fort George at the outlet of Great Slave Lake; one on the left bank of the Mackenzie below Mills Lake; another opposite the old Livingston Fort; old Fort Castor on Old Fort Point of the Mackenzie, downstream from Keele River; and another at the mouth of Carcajou River. It is not possible to say which company built these establishments; in any case, none of them survived very long.

During the period of rivalry, three posts were founded that survive to the present day – Fort Simpson, Fort Norman [Tulita], and Fort Good Hope. In 1803 a post was built at the "Forks" at the confluence of Liard and Mackenzie rivers. Although it is uncertain, it has been supposed that the Forks was a post of the North West Company and may have superceded Rocky Mountain Fort. The next year, 1804, saw the opening of Fort Good Hope on the left bank of the Mackenzie River at Blue Fish Creek (Hare Indian River) – probably a North West Company venture. Wentzel's map, 1822, shows a post, Fort Norman [Tulita], on the right bank about opposite Redstone River. The Robinsons give a date for Fort Norman [Tulita] as before 1810, and it is suggested here that if it were a post of the North West Company, then Fort Norman [Tulita] was probably established about 1804; otherwise, the two new places, the Forks and Good Hope, would have been separated by more than 450 miles [720 km] – a situation that was never tolerated after that time.

Immediately after the North West Company incorporated the X.Y. Company, they built Fort Liard, in 1805, on the right bank of the Liard River, where the Petitot River joins the main stream. Fort Nelson, likely named for England's naval hero, was founded about this time. The usual date for the fort is about 1800. Lord Nelson was then becoming famous in England, and no doubt news of his exploits filtered out to the Northwest. The Battle of the Nile, 1798, might have been reason enough, but Nelson was certainly renowned after Trafalgar, in 1805. It is suggested that Fort Nelson was built about 1807, after the news of Trafalgar could have reached the Mackenzie region. Moreover, it seems sensible that Fort Nelson should have been built after Fort Liard. The old site of Fort Nelson was probably at the confluence of the Fort Nelson and Liard rivers rather than upstream where it now stands. Remarks made in the Hudson's Bay Company Reports on Districts, 1829, refer to Old Fort Nelson at Liard River.

There followed a period of growing competition with the Hudson's Bay Company, which was moving inland from the Bay to take over the trade rightfully theirs under the Charter. Generally, the Athabasca country was not directly affected by these developments although the Hudson's Bay Company succeeded in having a few posts in the region. They did not reach the Mackenzie Valley. Yet the operations in the marginal far Northwest were jeopardized by the new rivalry elsewhere, In 1810 rabbits, a substantial food source in the district, failed to appear, and many skins intended for trade were eaten, further reducing the economic successes of the Mackenzie department. In 1812 the war with the United States threatened supply lines from Montreal, and in that same winter Indians attacked and destroyed Fort Nelson, and massacred the inhabitants. The post was abandoned, probably in retaliation as much as in fear of repetition of the incident. These events compounded the difficulties of continuing business in the Mackenzie region and the Northwesters were forced to withdraw their operations from the entire department in the summer of 1815. However, they returned in 1818 to forestall a takeover of this country by the Hudson's Bay Company, and traders reoccupied The Forks, Fort Liard, Fort Norman [Tulita], and Fort Good Hope.

Outside the Mackenzie River department, the strong competition with the Hudson's Bay Company increasingly put pressure on the Northwesters and demonstrated the inefficiency of the long overland connections with Montreal, compared with the shorter route via Hudson Bay. This was early recognized by Sir Alexander Mackenzie, and he spent considerable effort and money to obtain access for his company through the Bay. Unfortunately, he died about a year before the two great trading concerns united under the name of the Hudson's Bay Company on March 26, 1821. The agreement of union brought monopoly control of the fur trade in all of northern North America, and changes were introduced which would tighten trade practices and increase the returns.

The governor of the new company, Sir George Simpson, saw clearly that progress for the Mackenzie department lay in two directions: first, the reorganization of the present operations, and second, exploration of neighbouring unknown lands. The location of the Forks post was strategically sound, and it continued to function as the administrative centre and distributing point for other Mackenzie and Liard River forts. In 1822, under the direction of W.F. Wentzel, the Company built a new post at the Forks on the island in the Mackenzie River at about the present site, downstream from the old buildings. Soon it was known as Fort Simpson, after the governor. The next summer, 1823, Fort Good Hope was moved downstream on the left bank to the mouth of Trading River, at the request of the Loucheux [Kutchin] people to shorten their upstream travel. It was hoped, also, that Eskimo [Inuit] trade might be encouraged from this location. In order to accommodate the Hare Indians, who traded at the old Fort Good Hope, Fort Norman [Tulita] was moved downstream in the same summer to a place on the west side of the river thirty miles [50 km] above the mouth of Great Bear River. This was just below Old Fort Point, where Fort Castor once stood. Somehow the support of the Loucheux [Kutchin] Indians for Good Hope failed to reach expectations, and no Eskimos [Inuit] visited the post. Moreover, Captain John Franklin, returning from the coast in 1825, confirmed the notion of the traders that the Eskimos [Inuit] were altogether hostile to the whites, and besides, there were no furs worth bothering about on the tundra. Fort Good Hope was moved back to its original site at Hare Indian River in 1825 or 1826, and was possibly built on Manitou Island in the Mackenzie, instead of on the left bank where it had been until 1823. There the post remained until the spring flood in 1836 washed away the entire establishment, which was rebuilt on the right bank at its present site. In addition to the relocating of forts, the Company sought economy by inaugurating the use of "boats" in 1823, presumably similar to boats they employed on the Saskatchewan. Simpson also ordered that the trade tariff be reduced, and more uniform prices were offered to the Indians from post to post. There were recurring but unsuccessful attempts to reduce and eliminate Indian debts.

Governor Simpson's directives on exploration were aimed at opening the country of the Upper Liard River and making contact with the little-known Nahanni or Mountain Indians on the west side of the Mackenzie Valley. In March 1823, A.R. McLeod left the Forks on a month-long journey seeking the Nahannis, but without success. Later, in June, J.M. McLeod set out for the South Nahanni River, explored overland to a tributary of the Flat River, and returned after making contact with Natives near the La Biche Mountains. The next year he performed essentially the same journey and succeeded in bringing back Natives to Fort Simpson. Murdock McPherson sent his Indians from Fort Liard up the Liard River to trade, and they probably reached the Liard Plain. McPherson himself travelled one hundred miles [160 km] upstream in 1824. In 1829 there was another post, on the Liard River, called Fort Halkett. Although there are reports describing the early site, they are difficult to interpret in terms of present place names. In any case it was moved, about 1832, to the mouth of Smith River to please the Nahanni Indians.

In the summer of 1831, J.M. McLeod undertook a commission to explore the Liard River and cross the divide to the Pacific streams, at the mouths of which Russian traders were known to operate. He followed up the Liard to its forks, then along the south branch – the Dease River – and over the divide to the Stikine River, mistakenly called the Pelly. The journey was reported to be five hundred miles [800 km] above the forks. With the notion of removing Fort Halkett to the doorstep of the Russian trade area, McLeod returned in 1834, travelled slightly farther but left his mission incomplete. A.R. McLeod (Jr) and Robert Campbell attempted to start a post at Dease Lake in 1837, and the next summer Campbell travelled to the Stikine River, where he was menaced by the Chilcat Indians because Campbell threatened their role as middleman traders for the Russians. He remained at Dease Lake in the winter of 1838–39, but was forced to abandon the post because the local game supply was inadequate. Later, Campbell explored north from Fort Halkett to Frances Lake and over the divide to Pelly River, from which he returned with news of a rich beaver country, which much delighted Governor Simpson.

Exploration activity was also initiated in the northern part of the department. John Bell, on advice from the Indians, abandoned his original plan to explore Arctic Red River, and in the summer of 1839 followed the Peel River upstream from its mouth. His account is difficult to follow, but he did trace the river for at least one hundred miles [160 km] into the country of the High Peel Plateau; in the following year the Company established Peel River Post (Fort McPherson). In 1842 Bell found the Rat River route across the Richardson Mountains to Bell River of the Yukon drainage. La Pierre House, on Rat River, was probably built a year or so later. Bell was unable to continue these explorations until 1846, when he travelled through the pass to the Porcupine River and went on to reach the Yukon. The next year Alexander Murray, by way of the same route, built Fort Yucon at the junction of the Porcupine and Yukon rivers, where he was greeted enthusiastically by Natives in the heart of the Russian trade territory.

Meanwhile, Campbell continued exploration in the upper Liard Country but the scarcity of game in the early 1840s delayed his progress. Finally, in 1843, he left Frances Lake post, built the previous year, and spent two years examining the upper Yukon drainage, establishing several posts. His energies in this direction soon were the undoing of Campbell's enthusiasm for the country he discovered, because in 1851, as he was exploring a stream he named Lewes River, he came upon Fort Yucon and Trader Murray. Campbell returned via La Pierre House and Fort McPherson. This latter more economic route won out over the long and difficult way via the Liard, and Frances Lake and other posts established by Campbell were abandoned.

By 1850 the main outlines of the Mackenzie drainage and its association with neighbouring river systems was reasonably well known to the Company, and there seemed little reason to incur further expense to embroider with additional detail the knowledge that was already sufficient for the profitable prosecution of the trade.

From the early vision of Peter Pond until the firm concepts and controls of Governor Simpson, the Mackenzie basin had passed from exploration to violent competition that lasted in varying degrees until 1821. In the next thirty years, by trial and error, the fur trade adjusted its organization to the resources of the region and settled into a pattern of posts that today remains essentially the same.

20 The Sponsors of Canadian Arctic Exploration, 1844–1859: The Franklin Search and Rae's Surveys

JOHN E. CASWELL

"On Tuesday the 27th April, 1824, we dined at the Franklins', which we considered a mark of great attention as it was to meet Captn Parry," wrote Jane Griffin, a pretty and witty young Londoner. "Captn. Parry was in the room when we arrived – he is a tall large, fine looking man, of commanding appearance, but possessing nothing of the fine gentleman ... his figure is rather slouching, his face full & round, his hair dark & rather curling. Captain G.F. Lyon was the next object of interest – he is a young man about 30, of good height, & gentlemanly-looking – he has large soft grey eyes, heavy eyelids & good teeth, & is altogether very pleasing." Captain Frederick W. Beechey she found to be a silent, "prim looking little man." At the head of the table was the Second Secretary of the Admiralty, John Barrow, now approaching the midpoint of his four decades of service in that post – "he is said to be humourous and obstinate & exhibited both propensities."

Jane was evidently the belle of the occasion. "I was handed down into the Dining room by Captn Parry, but on letting go his arm as we entered the room, I was desired by Captn Franklin to sit by him.' Looking across the table, Jane's sister Fanny remarked to Captain Parry that Captain Franklin had a fine head. "Yes, indeed," Parry agreed, "it is the finest I know. Inside, as well as out." He held Franklin in the highest respect and confidence.

SOURCE: The Beaver, Outfit 300 (Winter 1969): 45–53. Reprinted by permission of the author and publisher.

Some weeks later Jane and Fanny dined at the D'Israelis. Captain Franklin was among the guests and paid special attention to the Griffin sisters. "As soon as Captn Franklin saw Fanny & me, he gave us each an arm, & seemed to have us under his protection the greater part of the evening, which surely must have made us objects of envy." Dr John Richardson, who had been Franklin's lieutenant on the Mackenzie River in 1819–22, was there with his wife. Of him Jane wrote, he "looks like a Scotchman as he is – he has broad & high cheek bones, a widish mouth, grey eyes & brown hair – upon the whole rather plain, but the countenance thoughtful, mild & pleasing."

Shortly afterward, Franklin and Richardson set out on their expedition of 1825–6. Two years after his return, Franklin, now a widower, was married to Jane Griffin. The next year he received his knighthood, and Jane became Lady Franklin. During the 1840s and 1850s she was to play a leading role in the sponsorship of Arctic expeditions.

After a Mediterranean command, Sir John was appointed lieutenant-governor of Van Diemen's Land (Tasmania). There his colonial secretary intrigued against him, Lady Franklin's presumed influence in the government being one of the chief charges brought against Franklin. He was summarily dismissed by the Colonial Office and returned home in 1843 smarting under the treatment he had received.

On the Franklins return to London they found another Arctic expedition under discussion. In 1838 James Clark Ross had recommended to the Council of the Royal Geographical Society a renewed attempt to find a northwest passage by Parry's old route through Lancaster Sound. Southwest of Cape Walker there was a great void on the charts. Ross felt that a route could be found through this area with Banks Land protecting the vessels from the icefields of the Beaufort Sea. Said Ross, the vessels were available; many experienced officers were eager for the enterprise; and Russia, now outfitting an expedition, might yet forestall Britain in discovering the Northwest Passage. The council "at once and unanimously" agreed to forward the proposal to HM Government.

The Admiralty, instead, sent Ross on his famous magnetic survey of the Antarctic regions. He returned to England only in September 1843. About that time Beechey proposed a voyage towards the North Pole using vessels equipped with the new screw propellers. Others, including Barrow, Parry, Beaufort, and Sabine, countered with Ross's plan of 1838. In December 1844 Barrow proposed it formally to the Admiralty, which referred the plan to the Royal society for a recommendation. The Royal Society gave it a rather perfunctory blessing.

The plan was now submitted to an Admiralty Committee composed of Franklin, Parry, and James Clark Ross. They endorsed the plan and added various suggestions, including Beechey's recommendation of steam engines. The Admiralty gave its endorsement and forwarded the proposal to the Prime Minister, Sir Robert Peel, who granted final approval.

Next came the problem of selecting a commander from a group of distinguished, albeit aging, Arctic officers. James Clark Ross was approached but refused, for on his recent marriage he had agreed to undertake no more Arctic expeditions. Franklin, however, was looking for an active duty assignment on which he might redeem his reputation.

Lady Franklin's pride had perhaps been more deeply wounded by events in Van Diemen's Land than had her husband's. When he had accepted the post she had written him,

It is possible on your return from your present service that you may have two or three years to wait before you succeed in getting another ship, but why not strive during that interval to resume your chieftainship in your own peculiar department, and if you did so, and came back as usual with an increase of credit and of fame, surely a ship when you liked to ask for it would be the least, and a natural reward for your services.

To Sir James Ross, Lady Franklin now wrote:

I dread exceedingly the effect on his mind of being without honorable & immediate employment & it is this which enables me to support the idea of parting with him on a service of difficulty and danger better than I otherwise should.

As Franklin was fifty-nine when he set forth on his tragic venture, one may well speculate as to whether this statement was not a projection of her own motives rather than a reflection of her husband's. Suffice it to say, Franklin was appointed by the Admiralty to command the expedition.

The expedition was made ready and sailed on May 19, 1845. With an experienced if elderly senior officer, emergency steam power, supplies aboard that could be rationed to cover four years, and guns to extend the rations with game, success seemed assured. The only dissenting voice was that of Dr Richard King, who had been with Back on the Coppermine. He reportedly told Barrow that he was sending Franklin "to form the nucleus of an iceberg." King pointed out how easily ships were trapped by Arctic ice and felt that a supporting overland expedition was needed.

HMS *Erebus* and *Terror* with 168 officers and men aboard were last seen headed toward Lancaster sound by the whaler *Prince of Wales* on July 26, 1845. In 1854 some relics and stories were obtained from

Eskimos [Inuit]; in 1859 some skeletons and relics were found, with a single document. Since then still more bones and identifiable remains have been discovered, but debate still continues as to the circumstances surrounding their fate.[1]

The impulse for the expedition had clearly originated within the tight little circle of experienced Arctic naval officers that clustered around Sir John Barrow and constituted a force within the Royal Geographical Society (founded in 1830). Barrow was a charter member and was at the time on the Councils of both the RGS and the Royal Society, as well as being a senior civil servant at the Admiralty. From this clique Captain John Ross and Dr Richard King were conspicuously excluded.

In the 1840s commercial interest in China may well have been the primary factor in the Government's approval of the expedition, but this is speculative. There were good reasons for a naval, rather than a land expedition. Franklin and Back had run into grave difficulties in their expeditions on the Coppermine and Great Fish [Back] rivers. Although Dease and Simpson had been more successful in their survey of the coast, the ability of ships to find a feasible channel was the true test of a Passage. Further, a ship favoured with open water and a fair breeze could sail as far in two hours as an overland expedition could go in a day, and a vessel could remain in the field three or four years.

The Hudson's Bay Company Sends Out Dr Rae

While naval officers at the Geographical Society and Admiralty were planning for a new assault on the Northwest Passage, the Hudson's Bay Company was making its own arrangements to finish the survey of the Arctic coast so nearly completed in the preceding decade by Dease and Simpson.

The initiative the Company now displayed was in contrast to its laggard attitude in the eighteenth century. Possibly one factor in the situation was that rivals might seek to enter the field if others found a sea passage to the north. A desire to avoid requests to succour ill-manned government expeditions may well have played a part. And a very general interest in exploration, whether of the Arctic, Africa, or Asia, created a favourable climate of opinion for such enterprises.

The leader chosen for the expedition was Dr John Rae, an Edinburgh medical graduate who had served the Company around Moose

---

1   Recent investigation of expedition members' remains have revealed that lead poisoning from improperly sealed tinned food was directly responsible for several deaths and in various ways contributed to the ultimate deaths of all. – ed.

Factory since 1833. On May 11, 1844 the Governor of Rupert's Land, George Simpson, wrote to Rae:

An idea has entered my mind that you are one of the fittest men in the country to conduct an Expedition for the purpose of completing the Survey of the Northern Coast that remains untraced, say between the Straits of the Fury & Hecla [between Baffin Island and Melville Peninsula] & the Gulf of Boothia from which Dease & Simpson returned.

In July Simpson wrote in a tone of urgency, "The Govr. & Com: you are aware, are very anxious that the discovery of the Northern Shores of this continent … should be completed by the Hudsons Bay Company …" Preparations should be made for the 1845 season, and Rae should go at once to the Red River Settlement to acquire what knowledge of astronomy and natural history he could in preparation for his journey.

Events conspired to delay Rae's departure. A year later, in July 1845, Simpson transmitted to Rae a resolution of the Northern Council, setting 1846 for the expedition. Finally, on June 15, 1846 Simpson wrote out Rae's final orders. Rae was to go north from Churchill, and survey what was believed to be a deep bay (Committee Bay, the southern part of the Gulf of Boothia) until he joined Dease's and Simpson's survey. Simpson closed with a word of inspiration:

The eyes of all, who take an interest in the subject, are fixed on the Hudsons Bay Company; from us the world expects the final settlement of the question that has occupied the attention of our country for two hundred years; and your safe and triumphant return, which may God in his mercy grant, will, I trust, speedily compensate the Hudsons Bay Company for its repeated sacrifices and its protracted anxieties.

Rae and his party left Churchill on July 5, 1846 in two small boats. From the head of Repulse Bay they hauled and rowed one boat across the isthmus, only to find ice conditions too bad for exploration of the coast. Returning to Repulse Bay, the party built of stones a fourteen-foot by twenty-foot [4-m by 6-m] house for the winter. A hundred and sixty-two caribou were shot which with partridges and salmon stocked the larder.

Regaining the west coast of Melville Peninsula in April 1847, Rae first travelled northwest. He confirmed Captain John Ross's belief that Boothia was a peninsula but failed to go far enough north to discover Bellot Strait between Boothia and Somerset Island. Although falling short of linking his survey with that of Dease and Simpson, he had demonstrated that no passage existed south of the 68th parallel. Returning to Repulse Bay, Rae set out again a week later with five men, on foot, to trace the west shore of Melville Peninsula and linked his survey with that of Parry a little south of Fury and Hecla Strait. Rae then returned to Hudson Bay and sailed to England. A year later he was back on the trail under the command of Dr John Richardson, seeking to bring aid to the Franklin Expedition.

The Franklin Search Expeditions, 1848–51

When Franklin sailed for the Arctic in 1845, the only provision for his relief was the Admiralty's request that the Hudson's Bay Company be prepared to furnish emergency supplies and equipment for Franklin's party at their northernmost posts. Rather than sending foodstuffs, the Company ordered fishing tackle forwarded to stations on the Mackenzie River and Great Slave Lake. The Admiralty further asked that the Eskimos [Inuit] be alerted to watch for Franklin's party in the vicinity of the Mackenzie and Coppermine rivers.

By the autumn of 1846 Franklin's expedition had been out for two summers and no word had come of its reaching the Pacific Ocean. Beechey, King, and others raised their voices proposing relief expeditions. Old Captain John Ross told the Admiralty he had promised Franklin to go to his relief if nothing were heard of the expedition by the following January. The Admiralty queried other friends of Franklin and decided that he had not expected a rescue expedition led by Ross or anyone else. Ross's proposal was rejected.

1847 SEASON

In late February 1847 the Admiralty asked for the opinions of Parry, Sir James Ross, Colonel Sabine, and Sir John Richardson. Parry and Richardson agreed that there was no immediate cause for alarm but made two suggestions: first, that a reward be offered to whalers for assistance to or information from Franklin; second, that an expedition be sent to the Arctic coast and adjacent islands, using the Hudson's Bay Company facilities.

Richardson, in his response, foresaw the possibility of the Franklin ships having been crushed southwest of Cape Walker and the crews making for the continent. He did not believe that the Hudson's Bay Company had facilities for making inquiries along the north coast, so he recommended that four light, strong boats be built and sent to the Mackenzie in the summer of 1847 so that a search could be begun in 1848. When consulted regarding Richardson's proposal, Sir George

Simpson replied, "After very attentive examination ... I cannot suggest any amendments to Sir John Richardson's plan." Richardson, although about 65, volunteered his services as leader and was accepted.

Ross wrote, saying that Franklin and Crozier had considered his uncle's proposal absurd. He felt ships should be sent out on Franklin's route and extra supplies should be sent to Hudson's Bay Company's northern posts, as Franklin's crews would certainly make for them if the ships were unable to proceed.

Sabine, for one, felt that supplies for Franklin should be sent to both Bering Strait and Baffin Bay, although he believed Bering Strait to be "the most sure point for falling in with him."

Of Lady Franklin's activities in 1847 on behalf of her husband we know little, for she seems to have kept no journal of it. As she spent the months from April to August in Italy, only in the autumn could she have begun seeking advice and assistance from her many friends and acquaintances.

1848 SEASON

By 1848 Lady Franklin was busy trying to stimulate action and get the ponderous wheels of government turning. From that time on she spent her own and Sir John's money unstintingly, sought donations, procured vessels and able masters to sail them, spurred the Admiralty, and kept on after every reasonable hope had died, in an effort to learn what had happened to her husband and his gallant crews and to establish his discovery of the Northwest Passage. As she had urged him to take such a command, she may well have felt guilty of sending him to his doom, and thus sought to bridle the demon that drove her on.

The Admiralty failed to offer rewards to the whaling fleet for information or assistance to Franklin during the 1847 season. Only in March 1848 did the Admiralty request the commissioners of customs to advise whaling captains that a hundred guineas and upwards would be paid for accurate information regarding the Franklin Expedition.

In late 1847 the Admiralty had drawn up a three-pronged plan combining the recommendations of Ross, Sabine, and Richardson. Sir James himself took HM Ships *Enterprise* and *Investigator* to Lancaster Sound on Franklin's track. His second-in-command was to search the waters southwest of Cape Walker while Ross sailed to Melville Island. Lieutenant Thomas E.L. Moore was sent in the tiny HMS *Plover* to Bering Strait with instructions to put himself under the command of Captain Henry Kellett in HMS *Herald*. Richardson set forth with Rae, descended the Mackenzie, left caches of food along the coast, and returned via the Coppermine.

On November 1, 1847 Lord Auckland had written Lady Franklin telling her of the plans for the 1848 season. Later she visited Sir John

Barrow, who seemed mortified that the expedition that he had helped to plan had not returned. To her surprise, although he was still Second Secretary, he did not seem aware of the decisions reported in Auckland's letter. No longer was Barrow a moving force in Arctic matters. Shortly after that he retired and died a few months later.

Richardson, Kellett, and Moore carried out their portions of the plan substantially as given. Ross got no farther than Prince Regent Inlet where he was frozen in for the winter, and no party was sent to explore the critical area southwest of Cape Walker. Richardson and Ross were both back in London in November 1849 with nothing to report.

In 1848 Lady Franklin found that there was no plan for exploring Wellington Channel, Franklin's first alternative route. She sought vainly to have Ross's support vessel *North Star* ordered to touch at the headlands of the Channel to look for messages. She asked the Admiralty if she could not be supplied with two dockyard lighters at the least, so that she might send them out at her own expense. And within a few days of time that the Admiralty offered a hundred pounds and upward for word or assistance to Franklin's vessels, Lady Franklin offered a thousand pounds to any whaling ship which should "depart so far from the usual fishing grounds as to explore Prince Rupert Inlet, Admiralty Inlet, Jones Sound or Smith Sound, provided such ships, finding the above expedition in distress, shall communicate with, and afford it effectual relief." A second thousand pounds was offered for extraordinary effort and, if necessary, conveying the party home.

1849 SEASON

The decision whether or not to continue exploration on the Arctic during 1849 was in Richardson's hands. Only one boat was in good order, and supplies were scant. Rae had proved himself vigorous, able, and a good hunter. Richardson wisely left the field to the younger man and betook himself home. In the spring Rae descended the Coppermine, but ice conditions foiled his attempt to cross Dolphin and Union Strait. Returning south, he wintered at Fort Simpson on the Mackenzie, where he was joined by Lieutenant William J.S. Pullen with a boat party from the *Plover*.

In early 1949 the Admiralty raised to twenty thousand pounds the reward for bringing relief to Franklin. Finding that the reward had been announced too late for the British whalers, Lady Franklin wrote to the President of the United States asking that American whalers be notified of the reward. Later Lady Franklin appealed for direct aid from the United States Government. Henry Grinnell, a whale oil merchant and president of the American Geographical and Statistical Society, stepped forward with the offer of two small schooners for which the Navy Department provided crews and supplies during the summer of 1850.

1850 SEASON

By 1850 Franklin had been in the field for five years, and even the most obtuse could realize that the chance of rescuing the expedition was becoming slim indeed. Public opinion was by now deeply stirred; a wide variety of suggestions poured into the newspapers and the Admiralty; the Royal Society urged the wisdom of a search. Within the Admiralty the Hydrographer, Rear Admiral Sir Francis Beaufort, and the new Second Secretary, Captain W.A.B. Hamilton, were called on for reports. Opinions as to the course to be pursued were obtained from Parry, Back, Beechey, Richardson, and Sabine.

A new three-pronged approach was planned. Rae should continue his explorations from the northern coast of the continent. While the *Plover* remained on the Arctic coast, Captain Richard Collinson in HMS *Enterprise* and Captain Robert McClure in HMS *Investigator* were dispatched as promptly as possible to Bering Strait so that they might sail northeastward into the archipelago.

After Collinson and McClure had set forth on their mission the Admiralty put Captain Horatio Austin in charge of a squadron of two sailing vessels, HMS *Assistance* and *Resolute*, and two small steamers which could work close inshore, HMS *Pioneer* and *Intrepid*.

Lady Franklin, meanwhile, had sought and obtained the services of an experienced whaling captain, William Penny. Failing to secure a whaler for him privately, Lady Franklin turned to Government, and got two sailing vessels, the *Lady Franklin* and the *Sophia*. Penny was instructed to go first to Jones Sound, if possible continuing thence to Wellington Channel and Melville Sound. These instructions did not satisfy Lady Franklin. Although previously concerned to search Wellington Channel, Lady Franklin now realized that Ross had *not* followed the first route prescribed in Franklin's orders and that the official 1850 expedition had been directed to beat around the crucial area between King William's Land and Boothia. If Franklin's crews had abandoned their ships, she felt that they would have crossed this area on their way to Fury Beach and Prince Regent Inlet rather than struggling back to Cape Walker or making for the Hudson's Bay Company posts, "a course," said she, "of which Sir John must well know all the dangers and difficulties."

To assure that a vessel would be designated specifically to search the Boothia area, Lady Franklin contributed £2,100 and raised £1,600 by public subscription to obtain the *Prince Albert*, a "small but fine vessel ... a very swift sailer well strengthened." The command was given to Captain Charles C. Forsyth, whom the Franklins had known in Van

Diemen's Land. She directed Forsyth to enter Prince Regent Inlet, cross North Somerset at its narrowest point, and then continue southward from the point Ross had reached.

The late summer of 1850 found seven British and two American vessels in Lancaster Sound. At the entrance to Wellington Channel camp litters and graves were discovered, evidence that Franklin had wintered there in 1845–46. Diligent search failed to reveal any messages giving information regarding Franklin's progress or his future intentions. Forsyth, kept from entering Prince Regent Inlet, was sent back to England with word of the discovery. In mid-October the channel froze over, trapping the eight ships.

1851 SEASON

The summer of 1851 found the American schooners *Advance* and *Rescue* under Lieutenant Edwin J. De Haven borne south in pack ice through Baffin Bay; by the time they were released it was too late to regain Lancaster Sound. The British vessels remained in Lancaster Sound and continued the search through the summer.

On the other side of the archipelago, McClure, against orders, had parted company with his commanding officer and made a dash for Bering Strait in late 1850. Turning northeast, he entered Prince of Wales Strait between Banks and Victoria Islands, where he was frozen in for the winter. In 1851 he extricated himself and at great peril sailed along the west side of Banks Island, turned east, and found refuge in a harbour he named the Bay of God's Mercy. Meanwhile, Collinson entered Prince of Wales Strait, spent the winter of 1851–52 there, and eventually escaped, reaching England in late 1854. McClure's vessel was hopelessly entrapped. His crew was extricated by walking across most of the archipelago and joining Captain Belcher's forces.

Although eight vessels were somewhere in the Arctic in the spring of 1851, Lady Franklin found an additional £1,000 to send the *Prince Albert* out again. A sturdy, part-Indian Canadian, William Kennedy, was captain, with Joseph René Bellot, a youthful French naval officer, as second in command. Entering Prince Regent Inlet, Kennedy wintered there, exploring afoot in the spring of 1852. One significant discovery was made: Bellot Strait. Kennedy, like his predecessors, had failed to push southward far enough to discover the fate of the expedition.

Still another area remained to be searched: the northern coast of Siberia. Lieutenant Bedford Clapperton Pim volunteered to go by land. At the request of the president of the Royal Geographical Society, the Prime Minister furnished £200, Lady Franklin found £300, and Pim set off to Saint Petersburg. Unable to obtain permission for the search, Pim returned to London and refunded Lady Franklin's money.

A second effort to reach Siberia, by screw steamer, was proposed by Captain D. Beatson. With Lady Franklin's assistance, Beatson selected a schooner, the *Isabel*, had 24 h.p. engines installed, and set out around the Horn. Beatson's plan miscarried and the yacht was later turned over to Commander Edward A. Inglefield who used it to search the northern and western shores of Baffin Bay.

The Last Admiralty Search, 1852–54

For the 1852 season Captain Austin was replaced by Captain Sir Edward Belcher, who had passed through Bering Strait with Beechey in 1825–26. He was given Austin's two sailing vessels and two small steamers. Captain Kellett was second in command, while Pullen, now a captain, was given the supply vessel *North Star*.

Kellett was given the *Resolute* and the *Intrepid* to take to Melville Island as a base for sledging parties. Belcher, with the *Assistance* and *Pioneer*, ascended Wellington Channel but found no trace of Franklin's vessels.

On May 30, 1853 a single vessel, the *Advance*, left New York, bound for Smith Sound. Her commander, Dr Elisha Kent Kane, had been with De Haven in 1850–51. He now proposed to explore an area not otherwise covered. Henry Grinnell again was the principal sponsor; funds were raised by private subscription, and the US Government provided some equipment.

Belcher was unable to free his ships in 1853. The next year, when the ice broke up, he stubbornly ignored the advice of experienced ice masters and took the leeward east side of the channel. The ice bore down on him, and he was trapped again. Meanwhile Belcher had already sent word to Kellett to abandon his icebound vessels and, with McClure's crew, join him at Wellington Channel. There, at Beechey Island, the five crews were put aboard other vessels and reached England safely late in 1854.

In January 1854, while these expeditions were still in the field, the Admiralty had notified Lady Franklin that the names of the Franklin Expedition's officers and men would be listed as dead on March 31 in order to clear accounts. On February 24 she responded in a long and vigorous letter, dramatizing her belief in their survival by exchanging her long-worn widow's black for bright greens and pinks.

Lady Franklin advanced four reasons to the Admiralty for believing in her husband's survival: (1) no sign of a wreck or disaster had been found; (2) no search had been made of the area north of the line marked by North Devon, Cornwallis, and Melville Islands; (3) many, including McClure's men, had survived four years in the Arctic and could well have lasted longer; (4) the provisions Franklin was known to have could have been stretched to seven years, and hunting might have provided more. The conclusions of Captain Ommanney and Lieutenant Osborn of Austin's expedition that Franklin had not passed by Cape Walker and down Peel Sound, indeed that there was no navigable channel, Lady Franklin now accepted. But there was evidence that Franklin's men had explored the east coast of Wellington Channel. Now Sir Edward Belcher "had verified the existence of the open sea to the north and north-west of Wellington Channel in the summer of 1853." Commander McClintock had been dispatched to the northwest of Melville Island, and yet, with Belcher and McClintock still in the field, the Admiralty had seen fit to pronounce Franklin and his crews dead, for the sake of clearing accounts!

My Lords, I cannot but feel that there will be a stain on the page of the Naval Annals of England when these two events, the discovery of the Northwest Passage [by McClure, of which word had already reached England], and the abandonment of Franklin and his companions, are recorded in indissoluble association.

The Admiralty, however, was not to be moved, and McClure and his crew in due course received the award for completing the Northwest Passage by discovering two routes from the North American mainland to Melville Island.[2]

Dr Rae's Explorations, 1850–54

While Austin's expedition was under preparation in the winter of 1849–50, Lieutenant Pullen and his boat crew were wintering with Dr Rae at Fort Simpson. Richardson's reaching home in November 1849 apparently stimulated the Hudson's Bay Company to ask for Governor Simpson's advice on continuing the search; shortly after that, the Admiralty approached the Company with the same question. Simpson, unaware that Pullen had reached the Mackenzie, suggested a search of the coast west of the Mackenzie, adding that the task "had better be

---

[2] Members of the Franklin expedition, in fact, had discovered the Northwest Passage before McClure, but none survived the expedition. The first ship traverse of the Northwest Passage was from east to west in 1903 by the Norwegian Roald Amundsen in his small *Gjoa*. The first west to east passage was by the RCMP schooner *St Roch* in 1940–42, which also made the first traverse east to west in a single year in 1944. The first west to east traverse in one year was by the Canadian icebreaker HMCS *Labrador* in 1954. In September 1969 the American oil tanker *Manhattan*, with the assistance of the Canadian icebreaker *John A. Macdonald*, successfully traversed the Passage. – ed.

left entirely to the Company, under the management of Dr Rae, who would do the work at a very moderate outlay."

With the expedition authorized by the Admiralty, Governor Simpson wrote to Rae instructing him to search west along the coast, and northeastward toward Cape Walker. Simpson's letter caught up with Rae and Pullen on June 25. As there were supplies enough for only one party, Pullen's crew descended the Mackenzie and sought vainly to cross to Banks Land, returning to Fort Simpson for the winter of 1850–51.

Ordered to continue the search, Rae set up a food cache at Kendall River (tributary of the Coppermine). In the spring he succeeded in crossing Dolphin and Union Strait to Wollaston Land. Turning eastward, he demonstrated the continuity of the coasts designated Wollaston Land and Victoria Land. He next examined the western shore of Victoria Island and discovered Prince Albert Sound. All unwittingly, he had joined his survey to that of one of McClure's parties, which had been on the opposite side of the Sound only ten days earlier. Returning to the mainland, Rae followed the continental shore to Cape Alexander, where he crossed Dease Strait and explored the southeast coast of Victoria Island.

Rae returned to England in the late spring of 1852. The Royal Geographical Society bestowed the Founder's Gold Medal on him, at the motion of Sir George Back. Weary as Rae may have been, on May 1, 1852 he wrote the Secretary of the Hudson's Bay Company outlining a "plan for the completion of the survey of the northern shores of America, a small portion of which, along the west coast of Boothia is all that remains unexamined."

With his plan approved, Rae ascended Chesterfield Inlet in the summer of 1853 and found a river that seemed to lead towards Back's Great Fish River. As it never came within seventy miles [110 km] of the Back River, Rae was forced to return to Chesterfield Inlet. In the spring of 1854, Rae started overland from Repulse Bay, crossed the isthmus, and followed the west shore toward Boothia.

Rae's party soon met Eskimos [Inuit] who surprised them by showing pieces of silverware and other items. This band said they had not seen any whites, but the Eskimos [Inuit] from whom they had obtained the relics had reported that some "forty white men were seen travelling in company southward over the ice, and dragging a boat and sledges with them" along the western shore of King William Island. Later, corpses and graves were discovered by the Eskimos [Inuit] at the mouth of Back River, and five corpses on a nearby island. Some men had survived until the coming of spring, for shots had been heard, and goose bones and feathers were found. Rae purchased over fifty relics from the Eskimos [Inuit], including crested and initialled silverware and Franklin's star of the Order of Hanover. Having continued the survey of the west coast of Boothia as far as the weather permitted, Rae turned back to Hudson Bay and carried his report on the Franklin Expedition to England. He had not been aware that the Admiralty had offered a £10,000 reward for information regarding Franklin's fate. He now laid claim to it and eventually received £8,000, the balance being distributed among his men.

After having spent several years on the search for Franklin, Rae made his great discovery on a surveying expedition for the Hudson's Bay Company which was organized without any serious hope of finding Franklin.

Lady Franklin Keeps On, 1855–59

Evidently more was to be learnt about Franklin's party at the mouth of Back's Great Fish [Back] River. It was important to try to recover any surviving papers. The Admiralty arranged for another Hudson's Bay man, James Anderson, to lead a party down river in 1855. Unfortunately, no interpreter was available and Anderson was unable to accomplish much.

Elisha Kent Kane and his party were rescued during the same summer after having explored beyond Smith Sound to the 81st parallel.

Many questions remained to be answered about the fate of Franklin and his crews, but the Government had lost interest in further exploration. Not so, Lady Franklin! She wanted to know the circumstances of her husband's death but seemed even keener to claim for him discovery of the Northwest Passage prior to McClure.

Lady Franklin may well have drafted a memorial to the Prime Minster, Viscount Palmerston, which her good friend Sir Roderick Murchison circulated to leading scientists in London on June 5 and 6, 1856. The memorial requested that an expedition be sent to search the area indicated by Rae's discoveries.

Palmerston replied that he wished to carry out the desires of the distinguished signers, but "official authorities" had expressed themselves strongly "that after so many failures the Government were no longer justified in sending out more brave men to encounter fresh dangers in a cause which was viewed as hopeless."

In early 1857 Lady Franklin approached the Admiralty through Joseph Napier, asking for a single ship with stores; she also wrote directly to Palmerston asking for the assignment of a naval officer, only to be turned down. Lady Franklin drafted an address to the Admiralty, asking for the use of the *Resolute*, which she sent to Henry Grinnell in

New York for signatures. The London *Times* relented its opposition and argued for the loan of the *Resolute*. All was in vain.

Anticipating the Admiralty's refusal, Lady Franklin was already negotiating for the steam yacht *Fox*. She offered Captain Frederick Leopold McClintock the command and through Prince Albert obtained leave for him from the Admiralty. Lady Franklin contributed some £4,000 by dipping into capital; £3,000 was raised by public subscription; and the Admiralty was generous with stores and equipment. Lady Franklin's only instruction to McClintock was a letter stating three objectives: rescue of possible survivors, recovery of records, and establishment of Franklin's discovery of the Northwest Passage.

McClintock was trapped in Melville Bay in the summer of 1857. Returning to Greenland, he renewed his stock of provisions and set forth again. As Peel Sound was blocked he entered Prince Regent Inlet and sought to pass through Bellot Strait. Having become expert at sledging under Austin and Kellett in 1851, he spent the winter of 1858–59 in planning carefully for spring sledge parties. On the first major sledging party McClintock charted 120 miles [190 km] of new coast. This "completed the discovery of the coast-line of continental America ..."

After having established caches, McClintock and Lieutenant William Hobson set out for King William's Land. McClintock took the eastern shore, leaving to the younger man the honour of searching the western shore where relics were most likely to be found. There Hobson discovered a memorable note that told of Franklin's death on June 11, 1847, and of the crews' abandoning ship and heading for the Back's Fish [Back] River. A boat, skeletons, and a number of mementoes were found.

With the return of McClintock to England, an era in the exploration of the Arctic had closed. A number of alternative routes through the archipelago had been discovered, although a century would be required for mankind to develop the ships and submarines to exploit them. The northern limits of the North American continent had been established, and a large part of the Canadian Arctic Archipelago charted. Much of the credit for this achievement must go to Lady Franklin for her single-minded persistence. Recognition must be given to the British Government, particularly the Admiralty, for the very considerable expenditures made. In the United States, Henry Grinnell's and Elisha Kent Kane's activities produced significant achievements. The honourable role of the Hudson's Bay Company in undertaking surveys of the Arctic on its own and in efficiently executing the requests of the Admiralty contributed substantially to the total accomplishment.

21   Albert Peter Low

F.J. ALCOCK

The discovery of immense deposits of iron ore in the heart of the great peninsula lying between Hudson Bay on the west and the Gulf of St Lawrence and the Atlantic Ocean on the east, and their recent development have drawn attention again to the explorer whose name is inseparably connected with this region. The traverses of Albert Peter Low by canoe, dog team, and on foot in various directions across it and his work by boat along its coasts furnished the first information concerning much of it. The region politically belongs to two provinces of Canada, the eastern coastal portion, Labrador proper, to Newfoundland and the remaining and much larger part to Quebec. To Low the whole was the Labrador peninsula, over half a million square miles [1.3 million km²] in extent, at the time he began work the largest unexplored area in Canada.

Low was born in Montreal on May 24, 1861, and in 1882 graduated with honours in Applied Science from McGill University in that city. Shortly afterwards, on July 1, he joined the staff of the Geological Survey under Alfred R.C. Selwyn. His first independent assignment was in Gaspé Peninsula in 1883, where he had served during the two preceding field seasons as an assistant to R.W. Ells. His principal traverse in this connection was up the Ste Anne River from the St Lawrence, through the Shickshock Mountains to Lake Ste Anne, across a low divide to one of the headwater branches of the Little Cascapedia River and down that stream to Chaleur Bay. In 1923, while

SOURCE: *The Canadian Geographical Journal* 48, no. 4 (April 1954): 160–3. Reprinted by permission of the author and publisher.

working in central Gaspé, the writer found near Lake Ste Anne what he believed to be a souvenir of that traverse. This was a rusted Geological Survey micrometer disk that had apparently been lost by Low forty years earlier.

The next two years saw the first of Low's Labrador investigations. Starting from the St Lawrence in 1884, he ascended the Bersimis or Betsiamites River and the upper part of the Peribonca. Leaving the canoes on the latter stream, the party started on foot for the Hudson's Bay Company post on Lake Mistassini on November 27 dragging their outfit on toboggans. They arrived on December 23, the last ten days part of the long and difficult tramp on snowshoes having been made on very short rations and with the thermometer ranging to forty degrees below zero [–40°C].

An incident in connection with this project illustrates the character and hardihood of Low. The work was being done in conjunction with a provincial party. During the winter, certain disagreements arose regarding how the work should be continued. In order to establish who was really in charge, Low on February 2 started for Ottawa which he reached on March 2. Three weeks later he received his instructions and with a letter stating that he was in full charge of the work he started back rejoining the others at Mistassini on April 29. In the field season that followed Lake Mistassini, which has a length of over one hundred miles [160 km], was surveyed and later in the summer the Rupert River was descended to James Bay.

For the next six years Low was engaged in a variety of exploratory and geological work in other fields. In 1886 he traversed a belt of country between Lake Winnipeg and Hudson Bay by ascending the Berens River from the former, crossing the divide, and descending the Severn River to the Bay returning by the old York Boat route from York Factory up the Hayes River to Nelson House at the head of Lake Winnipeg. In 1887 he descended the Missinaibi River of northern Ontario to map the islands of James Bay. Then for several years he carried out geological mapping chiefly in Portneuf and Montmorency counties northwest of the St Lawrence.

In 1892 Low returned to the Labrador peninsula and the explorations he carried out from then until 1899 are the ones for which he is best known. Starting from Lake St John he crossed to Lake Mistassini and thence by the upper Rupert waters to the Eastmain and down the latter to James Bay. In the following year he once more started at Lake St John and proceeded again by the same route to the Eastmain. This time he ascended that stream. Crossing the divide to waters flowing north he descended the Kaniapiskau and Koksoak Rivers to Fort Chimo [Kuujjuaq] at the foot of Ungava Bay. He had planned to pass

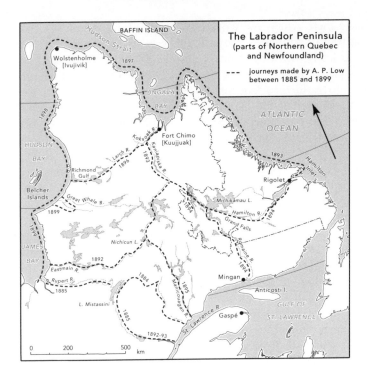

the winter there but learned that during the previous winter the herds
of barren ground caribou, upon which the Indians and Eskimos [Inuit]
of the region depended almost wholly for their food and clothing, had
failed to make their usual migration southward to the region, and that
as a result upward of one hundred and fifty persons had died of starvation. He therefore took passage on the Hudson's Bay Company's
steamship *Erik* to Rigolet on Hamilton Inlet on the east coast.

The work of the 1894 season began in March, the party of six pro-
ceeding up the Hamilton on snowshoes hauling on toboggans their
outfit and supplies sufficient to last them for five months. This necessitated at least three loads and often four for everyone so that the same
ground was passed over five or seven times. On May 20 when they had
reached a point near Grand Falls they were halted by the break-up of
winter and from there on travel was continued by canoe. The summer
was spent mapping a number of streams and lakes in the headwater
region of the Hamilton. One of these lakes was Michikamau, upwards
of seventy miles [115 km] long and at one place twenty-five miles [40
km] wide, the largest body of water, after Mistassini, in the whole
peninsula. In the autumn the party descended the Romaine and St John
Rivers to the mouth of the latter stream near Mingan on the Gulf of St

Lawrence. In these two years' work Low travelled 5,460 miles [8,700 km]: by canoe 2,960 miles [4,700 km], on vessel along the coast 1,000 miles [1,600 km], with dog teams 500 miles [800 km], and on foot 1,000 miles [1,600 km]. It was during these two years also that he outlined a belt of stratified rocks extending in a north-northwest direction from the upper part of the Hamilton waters across to the Koksoak. Low reported that these rocks contain a quantity of valuable iron ore so great that, in the exposures seen by him, it was estimated in millions of tons.

In 1885 Low ascended the Manicougan River which flows south joining the St Lawrence about 220 miles [350 km] below Quebec City. The river was followed for 200 miles [320 km] and then in order to reach Summit Lake, the source of the stream, a parallel route 125 miles [200 km] long, marked by many long portages was followed. From there another route was followed westward to Lake Nichicun to tie up with the 1893 traverse. The route followed in returning was down the Manicougan including the portion that had been bypassed in the ascent. This included a narrow gorge fifty miles [80 km] long where it was impossible to make portages and where rapids were continuous. In a particularly dangerous part a canoe was upset and one of the Indian canoe-men was drowned.

In 1896, starting from Missanabie on the Canadian Pacific Railway, Low descended the Moose river to James Bay. A government fishing boat was fitted up and in it the party sailed 450 miles [725 km] along the east coast of Hudson Bay to Richmond Gulf. Leaving their boat, Low and his party proceeded inland by canoe, crossed the height of land to northeastward-flowing waters and descended the Stillwater and Larch rivers to the Koksoak and on to Fort Chimo [Kuujjuaq]. Here a passage was taken on the *Erik* to Rigolet on Hamilton Inlet, where a transfer was made to a schooner bound for Quebec. Ottawa was reached on October 10.

In 1897 the northern coast of the peninsula along Ungava Bay and the south side of Hudson Strait was explored and mapped by boat, and in the following year a similar examination was made of the east coast of Hudson Bay from Cape Wolstenholme south to Great Whale River [Kuujjuarapik], where he remained for the following winter. Early in February 1899, preparations began for a renewal of work. Two exploration surveys were made on foot – one eastward into the barren interior and the other up the Great Whale River. In the following spring the examination of the east coast of Hudson Bay was continued and the work carried south to the foot of James Bay. An important phase of the work was the investigation of the iron-bearing beds of the Nastapoka islands.

Though perhaps best noted for these Labrador traverses, Low's name is also an important one in connection with Arctic exploration. In 1903 he was appointed to the command of the "Government Expedition to Hudson Bay and Northward." The ship chosen was the *Neptune*, the largest and most powerful of the Newfoundland sealing fleet. It had a register of 465 tons and a cargo capacity of 800 tons. Its wooden sides where a contact with ice was expected were nearly eighteen inches [45 cm] thick and its bow was reinforced with iron plates. With a total company of forty-three, the *Neptune* left Halifax on August 22, passed along the Labrador coast, making certain stops to study the geology, and then on to Cumberland Gulf on the east side of Baffin Island. Returning south to Hudson Strait an examination was made of parts of the coast and of certain islands, particularly Southampton at the northwest corner of Hudson Bay, and then Fullerton Inlet on the mainland west of Southampton Island was chosen as the site for winter quarters. The ship was roofed in and banked all round with snow, making it dry and comfortable. In the months of April and May, traverses on foot were made to explore the adjacent region. It was July, however, before the *Neptune* could sail north. It passed through Hudson Strait and up along the east side of Ellesmere Island. Turning south the expedition entered Lancaster Sound studying the south shore of Devon Island and landing at Beachey Island, where Sir John Franklin with the crews of the *Erebus* and *Terror* passed his last winter in harbour. The return journey was down Baffin Bay and through Hudson Strait again to Fullerton and thence back to Halifax, which was reached on October 12 after a voyage which had lasted fourteen months. During it, the *Neptune* had steamed 10,000 miles [16,000 km]: 9,100 [14,600 km] in open water and 900 [1,450 km] through ice. This voyage had accomplished many things. Over 1,100 miles [1,800 km] of previously unsurveyed coast had been charted, numerous astronomical observations and soundings made, a large amount of rock specimens and fossils collected, studies and photographs taken of the Eskimos [Inuit], and collections of northern birds, mammals, fish, etc. brought back. In a very interesting volume, *The Voyage of the Neptune*, Low has furnished details regarding the expedition.

In 1906 Low became Director of the Geological Survey, and in the following year when a new government department, that of Mines, was organized to include the Geological Survey and other divisions that became known as the Mines Branch, Low became its first Deputy Minister. He occupied that position until 1913, when, due to a very severe sickness which incapacitated him, he took his superannuation. he died at his home in Ottawa on October 9, 1942.

22 The North in Canadian History

W.L. MORTON

I put forward the thesis that Canada is different. It is not another
United States; still less is it a United States that failed to come off. It is
a different country, one profoundly different and one different from the
beginning. And what makes the difference is the North, the fact that
Canada is a northern country, with a northern economy, a northern
way of life, and a northern destiny.

I believe it is terribly important that we Canadians should recognize
this fact, both in order to keep our own notion of ourselves clear and
distinct, and also in order to understand our own history. Certainly to
approach Canadian history with this thesis is to alter one's view of that
history considerably. For example, one then notes that the lands which
are now Canada were discovered, not as a result of the voyages of
Christopher Columbus, but centuries earlier as a result of the Norse
settlement of Iceland and Greenland. The Columban voyages meant little to Canada; and Canada, unlike Latin America, unlike the United
States, would have happened without them. Now that we may be reasonably sure, as a result of the work of Stefansson and Icelandic scholars, that there was no gap between the last Greenland voyages and the
first modern voyages, we can see that Canada is the result, not of this
early modern attempt to sail westward to Asia, but of the extensions
across the North Atlantic of what might be called the "maritime frontier" of the northern peoples of Europe.

SOURCE: North 7, no. 1 (Jan.-Feb. 1960): 26–9. Reprinted by permission of the author and
publisher.

The Norsemen were Forerunners

For if one looks at the way the Norsemen who settled the Faroes, the Orkneys, Iceland, and Greenland lived, one sees that they lived by a combination of fishing, farming, and hunting which was very like the way early Canadians lived – and live – in the Atlantic Provinces and Quebec. The great voyages of discovery, such as Cabot's and Cartier's, and the colonization of Canada itself by Champlain and the Hundred Associates were actually government ventures imposed on the existing private enterprise of the fisherman and fur trader, right down to La Vérendrye's entrance to the prairie west.

And, quite recognizably, the national economy of Canada to this day continues the same basic features of the northern economy. Like the Norse farmer with his narrow shelf of ploughland in the fjord, this country has in its south, in the Annapolis, St Lawrence, Saskatchewan, and Fraser valleys, about enough farmland, broadly speaking, to grow food to feed itself and to exploit the Precambrian Shield, the Arctic, and the mountains of British Columbia and the Yukon. The one exception is the short-grass plains of Saskatchewan and Alberta which produce our wheat surplus. And the great money-making enterprises now, as over most of our history, are the big extractive industries of the north, timber and pulp, the precious and the base metals, and hydroelectric power. These are the modern equivalents of the fisheries and the fur trade of our early colonial days. As compared with the United States, Canada had little western frontier, a scant agricultural frontier; it possessed from the time the Huron Indians grew corn by Lake Simcoe, agricultural bases to feed the men bringing the furs, timber, and minerals out of the northern frontier. The Canadian frontier is a northern frontier and it is an extension overseas of the northern frontier and northern economy of the north lands of Europe. It is no wonder Scots and Icelanders have played so large a part in Canadian life. They were, as northern people, at home here from the first. The softer rest of us have had to learn the hard way how to live and prosper in a hard land.

It is because I see Canadian history in these terms – and I must say that I do not mean them to be thought anti-American terms – that I think the present interest in our north and farther north is well based historically. The national task of Canada has been to create the resources and institutions by which the simple northern economy of the past could be developed on a national scale to an American standard of living. This we have been doing.

Snowshoes to Tractor Trains

The task is, however, one of enormous difficulty and risk. Canada is probably the toughest country in the world to develop, after Antarctica and Tibet. Siberia by contrast is easy. The development depends on special techniques – the snowshoe, the dogsled, the tractor train, bush flying. The economic risks are of the greatest, for the resources are scattered, the costs are high, the selling prices are world prices and often unstable. Hence the peculiar Canadian blend of government and private enterprise, a delicate and sometimes a difficult thing to work, is ever present, and I would suppose would be more and more present as the frontier goes on into the Arctic archipelago.

Finally, there is the human factor. A northern country needs a northern people, people prepared to live in the North, on the terms the North imposes. Either that, or we must install in the North the essential furnishings of life here in the south, central heating, refrigeration, schools, hospitals, and – dare I mention it? – television. I believe myself that we shall always have northern people, but that we shall have to reinforce them with southerners and that the southerners will have to have the material conditions of life which prevail down here on the United States border.

A Northern People

But the matter of great concern to me, and I believe to us all, is that we should see our country as a distinctive and integrated nation, not as a second class United States, not as United States failed, but a different enterprise to be played on the forest-bound and rocky northern half of the continent as part of the manifold drama of human history. I would not for the moment be political, but I must say it is important that Canadians should have, if not a vision, at least a notion, or conceit, of themselves as a northern people, that with Stefansson we should dare to call the Arctic friendly, that we should believe that we are called, not to people a last best West, but in the dour fashion of our northern forebears, to make something of a grim, but a challenging, a tough but a rewarding proposition, our north.

CHANGE AMONG THE ABORIGINAL PEOPLES/FIRST NATIONS

23 Changing Settlement Patterns amongst the Mackenzie Eskimos [Inuit] of the Canadian Northwestern Arctic

M.R. HARGRAVE

Following 1576, when Martin Frobisher encountered Eskimos [Inuit] on Baffin Island, intermittent contact was kept up by explorers and nineteenth-century American and Scottish whalers until the establishment of Hudson's Bay Company posts in the present century. Yet Bisset's sketch of present conditions in the life of the Melville Peninsula Eskimos [Inuit], indicates that in spite of the superimposition of trapping, the proximity of DEW Line radar stations, the introduction of firearms and trade goods; in spite of schools, medical, and other facilities the pattern of life and settlement in the 1960s is not so radically different to the traditional one. A relatively long navigation season and wide approaches to Baffin and Hudson Bays and to Foxe Basin, and proximity to markets and home ports permitted many of the whalers to complete their hunt and return to civilization in the same season. The pattern of Eskimo [Inuit] settlement was so scattered that what depredation by undisciplined whaling crews did occur had little effect on the population in general. The eastern and central Eskimos [Inuit] were also shielded from southern influences by a broad buffer zone of barren lands.

A very different situation has prevailed in the Canadian Western Arctic where more recent and more intensive contact has profoundly influenced Eskimo [Inuit] society. Different geographic conditions have contributed to the destruction not only of traditional society but also

SOURCE: *The Albertan Geographer*, no. 2 (1965–66): 25–30. Reprinted by permission of the author and publisher.

of almost the entire original population. Firstly, the Beaufort Sea had large numbers of bowhead whales, but the configuration of the "peninsula" of Alaska, together with the long distance from markets and bases, made it imperative that the commercial whalers winter in the Arctic. Secondly, a different settlement pattern obtained from that of the eastern Eskimos [Inuit] – the concentration of the Mackenzie Eskimos [Inuit] in large villages – facilitated cultural contact and ensured rapid transmission through the entire group. Thirdly, the Mackenzie River fostered the entry of southern goods and influences. Lastly, the proximity of the tree line and the presence of a rich muskrat resource in the wooded Mackenzie Delta encouraged white settlement, and offered the Eskimos [Inuit] an alternative economy to that of the coast.

The nineteenth-century explorers were generally unanimous in their admiration of the Mackenzie Eskimos [Inuit] as a robust and self-reliant group. Lieutenant Pullen, RN, during his search for Sir John Franklin in 1850, had this comment to make on the Cape Bathurst villagers:

I cannot help saying here, that I never saw a finer body of men, nay women too, than I found the natives here, and I have often thought what a glorious expedition it would be to introduce the blessed Gospel among them.[1]

In contrast, though some of the same attributes noted by the explorers are still discernible today, serious social and economic problems are evident, and what may be called a *"metis* society" is developing.

This paper endeavours to sketch the traditional life and settlement patterns of the Mackenzie Eskimo [Inuit] and contrast it with that of today. In accounting for changes which have taken place, the writer pretends no significant original research: rather he aims to point to the need for research which might shed light on causes which have led to the present distributions and problems.

Traditional Settlement Patterns and Economy

From the journals and writings of the explorers, especially of Sir John Franklin (1826), Sir John Richardson (1826 and 1848), and of Oblate Father E. Petitot (1865), together with later information gathered by Vilhjalmur Stefansson during his travels in the Western Arctic (1906–18), it is possible to reconstruct the pattern of traditional life before modern influences disturbed the equilibrium.

It is commonly thought that the Eskimos [Inuit] of Arctic Canada were a nomadic people whose environment compelled them to disperse

[1]   W.J.S. Pullen, "Pullen in Search of Franklin," *Beaver*, Outfit 278 (1947): 25.

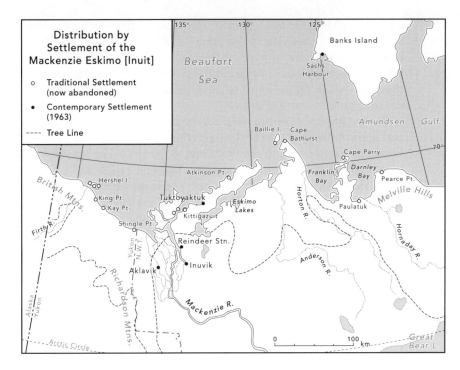

in small groups. This was not so in the case of the Mackenzie Eskimos [Inuit]. Franklin and Richardson report meeting numerous groups along the coast from Herschel Island to Cape Bathurst, but settlement appears to have been concentrated at five locations: (a) Kittigazuit, at the mouth of the easterly channel of the Mackenzie Delta; (b) Atkinson Point, almost seventy miles east of the Delta; (c) Baillie Island and adjoining Cape Bathurst; (d) at three locations along the southerly coast of Herschel Island; and (e) at the mouth of the Anderson River (see map above). Several sub-groups of the Mackenzie Eskimos [Inuit] have been recognized but older Natives refer to the whole group as the Tariumiut (the sea-dwellers).

Richardson reports that in 1848 about two hundred kayaks propelled by generally younger mature males (constituting probably the "home-defense force") surrounded his boats at Kittigazuit. Franklin's and Richardson's journals indicate that the average family consisted of no more than four or five individuals; therefore if one able-bodied male from each family met Richardson it would appear that the population of Kittigazuit was between eight hundred and one thousand individuals. There were also villages at Shingle, Kay, and King Points, and three villages on Herschel Island, all to the west of the Mackenzie Delta. Stefansson estimates the population of Atkinson Point and Baillie Island-Cape Bathurst at five hundred each, and suggests a total population of four thousand. Petitot estimates the total population at two thousand which figure seems to be more in accordance with the journals of Franklin and Richardson.

The traditional economy was largely based on the resources of the sea. In the immediate area of the Delta the sea is shallow, silty, and only faintly brackish owing to the discharge of the river. The sea off Atkinson Point, Herschel Island, and Cape Bathurst is less influenced by river water, and is deeper and clearer. The large bowhead whale and seal were hunted at the eastern and western extremities of the region, which together with fish (whitefish, anadromous herring, and Arctic char) caught with nets of caribou sinew, were the basis of life. The more numerous Kittigazuit people, and the whaling village of Shingle Point subsisted on fish (mainly whitefish and herring) and on the smaller white whales (beluga) which migrate from the west to enter the shallow estuaries of the lower Delta in July and August to bear their young in the warmer Mackenzie waters. The Anderson River Eskimos [Inuit] primarily depended on a rich fishery resource, sealing (many large bearded seals, or "ngruk," can be seen on the ice basking in the early summer sun near the mouth of the river), and caribou hunting. The journals of Franklin and Richardson make almost daily reference to caribou along the coast, and this was an important auxiliary resource, particularly for its value in furnishing clothing and sinew for sewing and fish nets. Of lesser importance were wildfowl, ptarmigan, and the collection of berries on the tundra.

Trade was carried on with the Eskimos [Inuit] of North Alaska, exchanging furs for Russian-made knives, and with the Copper Eskimos [Inuit] of Coronation Gulf but appears to have terminated prior to 1850, probably due to the development of trade with the neighbouring Indians. Petitot refers to the commencement of trade in 1849 with the Hare Indians of the Fort [Good] Hope area, via the Anderson River, and with the Loucheux [Kutchin] Indians of the Fort McPherson area, via the Mackenzie River. At that time the Indians became the middlemen between the Hudson's Bay Company post and the coast-dwellers. In return for furs the Indians introduced firearms, tobacco, beads, and basic hardware items such as files, matches, cooking utensils, knives, and traps.

Yet in spite of the rudimentary beginning of trade in furs the pattern of settlement and economy remained basically unchanged. The continued dependence on the bowhead and beluga whales necessitated community cooperation in large villages.

Present Settlement Patterns and Economy

The "Eskimo Identification Disc List," District W–3, 1963 edition,[2] listing all Eskimos [Inuit] to January, 1963, enumerated some 1,560 residents in the region. The composition of the group, the settlement pattern, and economy were radically different, however, than those of a hundred years ago.

Of this number some four hundred[3] lived in Tuktoyaktuk, the terminus of the Mackenzie River transportation system, and trans-shipment point for all water freight for Distant Early Warning radar stations and settlements in the Western and Central Arctic as far east as Spence Bay. [Taloyoak]. At Sachs Harbour, Banks Island, the centre of what is probably the Arctic's richest white-fox trapping area, there lived eighty-five Eskimos [Inuit]. A similar number lived at Cape Parry, in which area numerous caribou are still hunted, and where the rare muskoxen seem to be regaining a foothold.[4] Fifty lived at Reindeer station, the headquarters of the Canadian government reindeer herd. About forty were living at DEW Line stations where the family heads were employed. The remainder, about one thousand, lived in the wooded southerly half of the Mackenzie Delta, mainly within the settlements of Aklavik (total population, Eskimo [Inuit], Indian, and White, in 1965 estimated at six to seven hundred), the former regional "capital" and centre of muskrat trapping, and Inuvik (total population estimated at 2,200 to 2,300 in 1965),[5] the modern community completed in 1960 to serve as an administrative centre for many thousands of square miles of the northwestern Mackenzie District.[6] Clairemont states that, in 1961, 106 "bush" Eskimos [Inuit] lived in log cabin camps close to Aklavik.

2   Compiled by the R.C.M. Police, the northern representatives of the Registrar of Vital Statistics, Northwest Territories.

3   Author's calculations are estimates. The figures for Tuktoyaktuk, Sachs Harbour, and Cape Parry are accurate. The Eskimo [Inuit] population is mobile, and small fluctuations are a constant feature.

4   Since hunting was first banned and is now closely regulated, the numbers of muskoxen have increased significantly, and they have recolonized their historic range. In 1997 their total numbers were estimated to be just under 140,000 – 120,000 of which are in the Arctic islands. The largest numbers of these are 65,000 on Banks Island. – ed.

5   In 2001 the population of Aklavik was 632, and that of Inuvik was 2,894. – ed.

6   At the time, the NWT were divided into three districts for administrative purposes: 1) Mackenzie, north of 60°N and west to the Yukon border; 2) Keewatin, north of 60°N and including the mainland east of the Mackenzie District to Hudson Bay, excluding the Boothia and Melville Peninsulas; and 3) Franklin, including the Arctic Archipelago and the Boothia and Melville Peninsulas. – ed.

A hunting economy based on caribou and seal, supplemented by white fox trapping, and the sale of a few polar bearskins, prevails at Cape Parry. At Sachs Harbour the economy is based almost wholly on the white fox, which is trapped in large numbers by an elite group of aggressive Eskimo [Inuit] trappers. At Reindeer Station the population is supported entirely by salaried reindeer herders. Wage-labour, divided about equally between permanent and casual, maintains the Eskimo [Inuit] population at Inuvik, and is also the predominant source of income at Tuktoyaktuk and Aklavik. In Aklavik, of twenty-nine settlement Eskimos [Inuit] between the ages of sixteen and twenty-nine only two earned over $400 by trapping in the 1960–61 season, and what in the early 1950s was referred to as the "million dollar business" (referring to muskrat trapping and shooting which has for many years accounted for 95 per cent of the wooded delta's fur catch) was in 1961 referred to as "$100,000 activity." Most of the more than fifty trappers at Tuktoyaktuk set their traps along the sea coast between Atkinson Point and the outer edge of the Delta islands, and even in the best years trapping income is inadequate owing to the proliferation of short traplines. With the availability of some wage-labour, trapping as a way of life appears to have been rejected not only by many younger men but also by some of the older Eskimos [Inuit], especially in Inuvik, Aklavik, and Tuktoyaktuk.

The Delta and surrounding waters are rich in anadromous herring and whitefish, and throughout there is heavy dependence on subsistence fishing, not only for the table but also for the numerous dog teams.

The traditional occupation of hunting the beluga (white whale) is still carried on in July and early August at Shingle Point, Kendall Island, and Whitefish Station by a few Delta Eskimos [Inuit], and from Tuktoyaktuk, but the hunt is no longer a communal pursuit but an individual chase by a few old and small schooners and river craft purchased during the rich trapping years of the 1920s, by small motor craft and even outboard canoes. The beluga is trailed by following the slight swell it makes in the shallow water, shot when it surfaces, and harpooned before it sinks.

Of the almost 1,600 Eskimos [Inuit] in 1963 probably less than 20 per cent could be traced back to the original Tariumiut – and that only by virtue of generous infusions of white and immigrant Alaskan Eskimo [Inuit] blood. With the death in 1964 of an old Tuktoyaktuk man, who was able to recall hunting with bow and arrow as a boy on Herschel Island prior to the arrival of the whales in 1889, it is likely that no pure-blood Tariumiut remains.[7] The other significant fact with

---

7 Present-day Inuit in the region are identified as Inuvialuit. – ed.

regard to the present Eskimo [Inuit] population is its alarming fecundity. Clairemont has noted that roughly 50 per cent of the population of Aklavik was under fifteen years of age in 1961, compared to a corresponding figure for Canada of 34 per cent. Ferguson reported 55.4 per cent below the age of fifteen in Tuktoyaktuk in 1957, and in the same year a natural increase rate of 5.59 per cent as compared with 2.02 per cent for Canada in 1955. The writer calculated a natural increase of approximately 6 per cent in 1962. Truly the term "population explosion" is valid here!

The Changing Pattern of Life

A number of factors can be recognized as having affected or influenced the changes which have taken place since the traditional stage of Mackenzie Eskimo [Inuit] life. Some are obvious and important: others are tenuous and their importance is less evident, or not as yet measured. Certainly three factors can be seen as crucial, and chronologically it is possible to divide the history of change into three periods: the commercial whaling period, 1889 to 1912; the trapping period, 1912 to 1954; and the modern period, 1954 to the present.

THE COMMERCIAL WHALING PERIOD, 1889–1912

The New England whalers, who had penetrated the Bering Sea by 1850, slowly increased the range of their operations until in 1889 seven steam whalers reached Herschel Island. From that year until 1910 huge profits were made from the harvesting of baleen, used in the manufacture of ladies' corsets and horse whips. By 1912 substitutes and the development of the automobile made whaling in the Beaufort Sea unprofitable. In 1894–95 fifteen whaling ships wintered at Herschel Island, Bodfish reports that as many as six hundred men wintered there some seasons. Ships also wintered at Baillie Island and in southern Franklin Bay.

The contact with the whalers was drastic in its effects. Eskimos [Inuit] flocked to Herschel Island from the Delta and Nunatamiut (land dwellers), Eskimo [Inuit] caribou hunters from interior Alaska immigrated into Canada to enter the service of the whalers as professional hunters. Bodfish indicates that during two winters, 1897–99, his ship, the *Beluga*, received 47,000 pounds of venison, and that during the whaling period the professional hunters killed all the muskoxen "within an area of one hundred and fifty miles [240 km]."[8] The material culture of the Mackenzie Eskimos [Inuit] was to all effects des-

---

8 H.H. Bodfish, *Chasing the Bowhead Whale* (Cambridge, Mass.: Harvard University Press, 1936), 186.

troyed: the whale boat replaced the kayak and umiak almost overnight. The bowhead whale resource was decimated and the Eskimos [Inuit] were debauched, exposed to diseases to which they had no immunity. "The R.C.M.P. census of 1911 showed only forty descendants of the local people, although there were also in the country considerably over a hundred immigrants, and since this census of two years ago six of the forty have died and three have gone permanently insane."[9]

THE TRAPPING PERIOD, 1912–1954

The year 1912 marked the visit of the last whaling ship and the establishment of the first Hudson's Bay Company post close to the present site of Aklavik. The high prices for white fox during World War I and during the 1920s led to fierce competition between free traders and the Hudson's Bay Company, and the Western Arctic proliferated with trading posts. The competition was encouraged when Eskimos [Inuit] purchased schooners and whaleboats from the proceeds of their trapping and were able to choose the best posts at which to buy or sell. In 1924 the Eskimo [Inuit] fleet at Aklavik totalled thirty-nine schooners, nineteen with auxiliary power, and twenty-eight whaleboats. These were valued at $128,000 and had been bought within the previous five years. The price of white fox skins soared up to as high as $70.00 in 1928, and a good silver fox could be exchanged for a whaleboat. In the Mackenzie Delta muskrat prices rose from $0.50 in 1914 to an average of $1.31 for the period 1921 to 1929. The price plunged to $0.31 in the early depression years but by 1935 had risen again to $1.00.

The expansion of the fur trade after 1912 brought about the dispersion of the Eskimo [Inuit] population along the coast as far east as Pearce Point, and few settlements had more than three or four families. The competition between the free traders and the Hudson's Bay Company ended with the purchase by the latter of some of the free traders' posts and by declining prices when other traders, having reaped rich profits, left the country. High muskrat prices in the Delta during the 1920s attracted new immigrants from nearby Alaska, and white trappers from the south. The rising price in 1935, following the earlier decline, brought a new wave of Alaskan immigration: reference to the W–3 Eskimo Identification Disc List indicates that no less than twenty-nine present family heads born in Alaska arrived in the Delta area between 1935 and 1946. The population of Aklavik and immediate area in 1931 was 411, composed of 180 Indians, 140 Eskimos

---

9 V. Stefansson, "The Distribution of Human and Animal Life in Western Arctic America," *Geogr. Journal* 41 (1913): 453.

[Inuit], and 91 whites. By 1958 it had risen to 1,500, comprising 384 whites, 242 Indians, and 883 Eskimos [Inuit].

After 1939, and the abandonment of the Hudson's Bay Company posts at Herschel and Baillie Islands the small populations there moved to Tuktoyaktuk where a store and transportation centre had been set up, and where there existed limited opportunity for wage employment during the busy summer months. Two small groups remained east of Tuktoyaktuk at Stanton, near the mouth of the Anderson River, and at Paulatuk, in Darnley Bay east of Cape Parry. The Eskimo [Inuit] had lost his independence: the location of trading stores governed his distribution. In the absence of commercial posts the Roman Catholic missions at Paulatuk and Stanton engaged in trading until 1954 when their two functions were found to be incompatible. The Stanton residents, discouraged by low fur prices, moved to Tuktoyaktuk while the Hudson's Bay Company was prevailed upon to re-open a store at Lettie Harbour to cater to the needs of the Cape Parry area Eskimos [Inuit]. The possibilities of sales to DEW Line Eskimo [Inuit] and other employees together with the attractions of casual employment resulted in the moving of the store to Cape Parry in 1959, close to the large radar station there.

By 1954 the present settlement pattern had largely emerged. Only Sachs Harbour remained a viable trapping community. The low prices for fur in the late 1940s and 1950s were accompanied by higher food and equipment prices. The muskrat population remained high, but apart from cyclical fluctuations in the numbers of white fox there appears to have been an actual downward trend in the numbers of these animals along the mainland coast. Three reasons have been proposed. Firstly, at the end of the whaling period many whole carcasses were strewn along the storm beaches, and, the rate of decay being very slow in the Arctic, the carcasses were for many years a source of food for the foxes – and the foxes for the trapper. Secondly, the introduction of white fox trapping in Banks Island in the late 1920s and the high catches which prevailed there since may have reduced the number of animals crossing to the mainland after a westerly migration through the Arctic Islands. Thirdly, the radar stations along the coast have been accused of disturbing the ecology of the white fox. It would appear that at least the first two reasons have some validity.

THE MODERN PERIOD, 1954–

Trapping continues to be an important feature of the Eskimo [Inuit] economy in Aklavik and Tuktoyaktuk, but for some years wage employment has surpassed trapping as the major economic activity. The decline in trapping coincided with an intensification of the Cold

War between Russia and the West, an awareness of the need for hemispheric defence against possible transpolar air attack, and a national awareness of the social and economic needs of the indigenous people of the North. A surge of economic activity started in 1955 when the construction phase of the DEW Line began, followed by the construction of government health, educational, administrative, and other facilities throughout the North. It was decided that a modern town was to be built in the Delta when the site of Aklavik proved subject to flooding and unsuitable for further expansion. The construction of Inuvik in the late 1950s created many casual and permanent jobs as did the construction and manning of the DEW Line stations.

There is evidence of a levelling off in construction activity in the North. In the Mackenzie Delta area some Eskimos [Inuit] are finding employment and developing new skills, but opportunities are limited, and there is a high rate of unemployment and under-employment.[10]

In the absence of research information other more tenuous factors may be best treated by asking questions rather than by hazarding answers. What has been the influence of the missions, medical facilities, schools, public assistance, and other community facilities on population movement and settlement patterns? Can the decline of trapping be explained in purely economic terms or are other factors involved?

Certainly the structure and functions of northern settlements has changed radically during the last few years, and new institutions – good in themselves – militate against traditional pursuits and the development of a stable economy. The problem is recognized and alternative pursuits have been fostered by the federal government. As long ago as 1935 reindeer herding was introduced to the Mackenzie Delta Eskimos [Inuit] and since 1960 there have been fishing and whaling projects, and small fur-garment shops (each employing about twelve Eskimo [Inuit] women and making fur parkas and sportswear) have been set up in Aklavik and Tuktoyaktuk. With the exception of the latter, success has been limited. While the long range aim is assimilation there remain short term problems referred to above which are perhaps equally important.

Conclusion

As opposed to the Canadian Eastern Arctic, the influence of the first sustained white contact was violent to the extent that it virtually

---

[10] Since this article was written, the increased educational qualifications of many younger Aboriginals and the 1982 signing of the Inuvialuit Final Agreement on comprehensive land claims have provided greater economic opportunities for the Inuvialuit. – ed.

destroyed a numerous and self-reliant group. Under the influence of the southern material culture and the vagaries of the fur industry – itself governed by changing fashions and cyclical fluctuations of fur-bearing animals – settlement patterns in the Western Arctic have "revolved" from the large traditional villages based on communal hunting of sea mammals, to dispersed and mobile "camps," and back again to the present nuclear settlements. There is, however, a basic difference. Under the physical and social conditions prevailing, the traditional village was an economically viable unit, and with an apparently low birth rate[11] the Eskimo [Inuit] lived in harmony with his environment. On the other hand, the present large settlements are not economically viable. The rejection of dependence on the traditional diet, the factors which militate against dispersion and more thorough harvesting of resources in areas not presently utilized, the restricted employment opportunities, and last but not least, the fecundity of the population, together comprise a serious problem.[12]

The changing patterns of settlement and the concomitant social and economic changes engendered are not unique to the Mackenzie Delta. Similar patterns have developed, are developing, or will develop elsewhere. The process is merely more developed there than in most of Arctic Canada (again, Labrador must be excepted). It would appear then that the area is an excellent research laboratory for the social scientist – a laboratory in which the anthropologist, sociologist, and the geographer can pursue not only academic studies but also pragmatic solutions. To reconstruct the initial, and report on the present stage is not enough; the process too is important. The principle of uniformitarianism, that "the present is the key to the past," has long been held in geology. Similarly in human geography, the past (the dynamic past) can be the key to the present – and of assistance in planning for the future.

---

11   Older informants were questioned closely with regard to the possibility of high infant mortality or infanticide but no evidence was adduced for either as the governing factor.

12   In an introductory paragraph it was suggested that the Mackenzie River was a cultural corridor. It was then a factor in the breakdown of the Mackenzie Eskimo [Inuit] society. It may well be that the Mackenzie system will change its role to that of a highway of progress. Whether the movement will be south or north is as yet too soon for conjecture.

24 Changing Patterns of Indian Trapping in the Canadian Subarctic[1]

JAMES W. VAN STONE

The importance of trapping to the Indians of the Canadian Arctic and Subarctic is a matter of historical fact, and the changes brought about in the traditional Indian way of life by the introduction and development of a trapping-trading economy have been well documented for many tribes throughout Canada by historians and anthropologists. The author's field work at Snowdrift [Lutselk'e], a Chipewyan community at the eastern end of Great Slave Lake in the Northwest Territories, provided information about recent acculturative factors affecting the trapping pattern. The purpose of this paper is to show that these factors are not peculiar to Snowdrift [Lutselk'e] alone, but are widespread and appear to be altering the significance of trapping in the present-day economy of peoples throughout the eastern and western Subarctic.

I

The community of Snowdrift [Lutselk'e] is located on the southeastern shore of Great Slave Lake in a region that is entirely within the area of

[1] The field research on which this paper is based was supported by the Northern Co-ordination and Research Centre, Canada Department of Northern Affairs and National Resources during the summers of 1960 and 1961 and for one month in the winter of 1961. The author is grateful to Mr Victor F. Valentine of the Research Centre for his assistance and encouragement. The map was drawn by Dr Edward S. Rogers. The author is also indebted to Drs Rogers, F.W. Voget, Ronald Cohen, and T.F.S. McFeat, for many helpful suggestions during preparation of this project.

SOURCE: *Arctic* 16, no. 3 (Sept. 1963): 159–74. Reprinted by permission of the author and publisher.

Precambrian rocks. The eastern end of the lake has an extremely intricate shoreline with large numbers of bays and innumerable islands. The country around the village is characterized by wooded, rolling hills from five hundred [150 m] to one thousand feet [300 m] above sea-level; many lakes of various sizes dot the area; the vegetation and fauna are essentially Subarctic in character.

Snowdrift [Lutselk'e] with a population of approximately one hundred and fifty persons in 1961,[2] is a very recent village whose physical existence in its present form goes back no more than ten years. However, the area has been a focal point for residents of the surrounding region since 1925 when the Hudson's Bay Company established a post at the site of the present community. Prior to that time the population at the eastern end of Great Slave Lake consisted of an unknown number of Chipewyan families who hunted, fished, and trapped throughout the area and moved about the country as single families or groups of families. Most of these families traded at Fort Resolution, a long-established post on the southwestern shore of the lake, and considered that community to be their trading centre. When the Snowdrift [Lutselk'e] post was established many of these families, together with some who had traded at posts to the south and southwest on Lake Athabasca, shifted their centre of activity to the new post. Factors responsible for the recent concentration of a permanent population at Snowdrift [Lutselk'e] are not, however, specifically connected to the fur trade. They include (1) the increase in government services that have reduced reliance on income derived from trapping, (2) the recent establishment of a federal school in the village, (3) improved housing, and (4) wage-employment.

The yearly round of subsistence activities at Snowdrift [Lutselk'e] includes fall fishing near the village, winter and spring trappings, a certain amount of hunting throughout the year for moose and caribou, and some wage-employment during the summer. Winter and spring trapping keeps men away from the village for varying periods of time, usually not more than two weeks at a time, and the late summer and early fall run of caribou usually takes many families to the extreme northeastern end of the lake for a period of about two weeks to one month during late August and early September. Opportunities for wage-employment in the form of commercial fishing and tourist guiding are of growing importance, and there is also a government sponsored road-way clearing project that takes many young men away from the village during January or February.

Trapping begins officially on the first of November when the season for most fur-bearing animals opens. The area that has been used by

---

[2] In 2001 its population was 248. – ed.

Snowdrift [Lutselk'e] trappers is bounded on the north by Walmsley and Clinton-Colden lakes, on the east by Whitefish Lake area, on the south by the southern Taltson River–Nonacho Lake area, and on the west by the Doubling–Meander Lake region (see map). This is a large area but the majority of trappers have not trapped to the peripheries of the region in recent years. In 1961, for example, only one man trapped north of Narrow Lake in the Doubling Lake–Meander Lake area and only occasionally do trappers extend their activities as far north as Clinton-Colden Lake or as far east as Whitefish Lake. In fact the number of trappers who operate beyond the tree line is very small indeed. During the winter of 1959–60, when fox prices were relatively high and the animals few in the wooded country, only three men trapped in the Barren Grounds. Most men do a large part of their trapping within a radius of from sixty-five to eighty miles [105–130 km] from the village and the Taltson River–Nonacho Lake area is the centre of concentration.

The length of time that the trappers stay away from the village varies considerably. A very few men go out in the late fall, return for Christmas and stay several weeks, then go out again and do not return until late in the spring. However, a much larger number return frequently to trade furs and obtain supplies. Most informants maintained that they seldom stayed out more than two weeks before returning to trade. Thus a great deal of time is spent in travelling to and from the trap lines. A man seldom comes to the village without staying at least a week.

Trappers always have to do a certain amount of hunting because the supplies they take with them from the village usually consist of little more than staples, regardless of the amount of credit they have received from the Hudson's Bay Company manager. The most important hunting during the winter and spring is for caribou. Some men feel very strongly that their dogs should be fed on caribou meat during the winter, especially when they are being used on the trap line; thus, there are trappers who will not set nets while they are on the line, but will spend time hunting instead of trapping. When they are unsuccessful, they must feed their dogs flour or cereal from their own food supply and therefore run out of these staples rapidly and must return to the village. it seems certain that for many trappers, looking for caribou is the most important thing they do on the trap line and always takes precedence over trapping. Some men are even unwilling to leave the village if no caribou have been reported in the area where they trap.

In this connection it is worth commenting on a statement made about thirty years ago by a manager of the Hudson's Bay Company store at Snowdrift [Lutselk'e]. He believed that the success of the Snowdrift [Lutselk'e] Indians as trappers was directly related to the

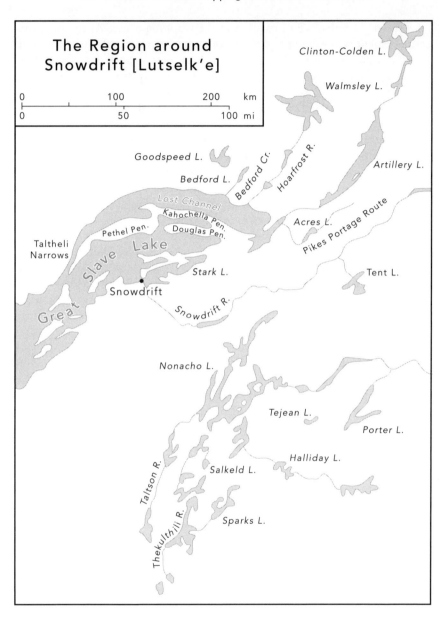

The Region around Snowdrift [Lutselk'e]

| 0 | | 100 | | 200 | km |

| 0 | 50 | | 100 | mi |

Clinton-Colden L.

Walmsley L.

Goodspeed L.

Bedford L.

Bedford Cr.

Hoarfrost R.

Artillery L.

Lost Channel

Kahochella Pen.

Pethel Pen.

Douglas Pen.

Acres L.

Pikes Portage Route

Taltheli Narrows

Slave   Lake

Great

Snowdrift

Stark L.

Tent L.

Snowdrift R.

Nonacho L.

Tejean L.

Porter L.

Taltson R.

Halliday L.

Salkeld L.

Thekulth;li R.

Sparks L.

presence or absence of caribou in the trapping areas. At the time this statement was written into the records, fur prices, particularly for white fox, were very high, but unless caribou were reported in the Barren Grounds, the Indians would not go there to trap foxes. Thus trapping seems, to some extent, to have been incidental to hunting even as late as 1925 or 1930. The same is true today even though caribou meat is no longer quite as essential a food item as it was thirty years ago.

The existence of clearly defined family trapping territories among the Chipewyan has been the subject of some dispute. At least one early observer believed that these territories did exist (Seton 1920), while another has denied their existence (Penard 1929). It must be admitted that the evidence collected at Snowdrift [Lutselk'e] does not support the theory of family trapping territories. Most informants stated that they had learned to trap in one of the areas trapped by their fathers and afterwards often continued trapping in that area. However, few trappers could be found who had trapped more than five years in any single area. Two or three years was the more usual length of time, depending on how successful the trapping was in a particular locality. Some trappers had trapped at one time or another throughout most of the area within a one-hundred-mile [160-km] radius from the village. During the winter of 1961, one trapper relocated his trap line in the middle of the season because he felt that the snow was too deep in the region where he had originally set his traps. Consistent opposition to trap line registration on the part of the villagers is a further indication that they desire freedom to determine for themselves where the trapping is best.

There can be no doubt that the total area trapped by Snowdrift [Lutselk'e] residents has been shrinking steadily in recent years, particularly since the Indians began to live permanently around the trading post. The factors primarily responsible for this situation have already been mentioned. However, it is worth discussing them in more detail, particularly as they relate to the changing trapping pattern.

The payment of family allowances on a monthly basis provides an additional form of income that can be obtained at regular intervals. This has meant that trappers, even when they normally trap without their families, are reluctant to be away from the post when there is a cheque to be cashed. Other government services such as rations and old age pensions have a similar effect and the income from these sources reduces the reliance on income derived from trapping.

The recent establishment of a federal school in the village must be regarded as a stabilizing feature whose effects will mainly be felt in the future. However, even in the winter of 1961, when the school was only in its second year of operation, many families were beginning to realize and accept the fact that their mobility during the trapping season would be reduced. Several families who had previously spent most of the winter in the bush on their trap lines have been forced to change their pattern of operation. Now only the heads of these families go out on the line and the tendency is for them to return to the village at frequent intervals. It should be emphasized that it is the most vigorous, and therefore usually the most successful, trappers who take their families with them on the trap line; it is the activities of these men, usually middle-aged with large families, that have been most affected by the construction of the school.

Improved housing should probably be considered as being closely associated with population stability since the people appear to be interested in improving houses that they are going to live in all or most of the year. The Canadian government has encouraged this by instituting home improvement and home building programs. Thus it is possible to note a greater contrast between the difficult, uncomfortable life on the trap line and the comfortable, gregarious life in the village. It is probably also safe to say that the discomforts of a bush trapper's camp are more difficult to tolerate when the trapper is alone or with another man rather than with his family. It is little wonder, then, that the trappers tend to become easily discouraged and desire to return frequently to the comforts of their homes in the village.

Wage-employment is, as yet, not of much importance to the people of Snowdrift [Lutselk'e] during the winter months. However, in recent years the government has introduced a roadway-clearing project at the northwestern end of Great Slave Lake. This project, which involves Indians from every community in the Great Slave Lake area, takes many men away from Snowdrift [Lutselk'e] during January or February. Government policies for the Indians in the Great Slave Lake area include developing alternative means of livelihood to trapping and the roadway-clearing project is thought of as a step in that direction. So far, the Indians of Snowdrift [Lutselk'e] have only participated in the project after Christmas when the main trapping period has ended. Thus its direct effect on trapping has not been great. An indirect effect, however, must be taken into consideration. During December 1961, the author noted that many trappers were not particularly worried about their lack of success and adopted a definitely desultory attitude toward the whole trapping procedure. The feeling was that since such a comparatively large amount of money was to be earned clearing roadways in January, it was not necessary to worry about the number of mink or marten trapped in November and December. Thus we must think of this government project, as well as others more extensive that are likely to follow in the years to come, as being another factor that greatly influences the Snowdrift [Lutselk'e] trapping pattern.

If one is prepared to say that the total area trapped by Snowdrift [Lutselk'e] men is shrinking, then it is equally certain that the region around the village is not being trapped as efficiently or effectively as it might be. At first glance there appears to be a contradiction here. If the Indians no longer trap to the peripheries of their areas because of their desire to visit the village frequently, why is it that they do not trap more intensively than formerly in the area close to the village? The answer to this is that the factors that are responsible for the shrinking of the total Snowdrift [Lutselk'e] trapping area also operate to reduce trapping effectiveness and, in short, to reduce the number of hours spent by the men on their trap lines.

Keeping these factors in mind, it is not difficult to understand why trapping has become increasingly unpopular with the Snowdrift [Lutselk'e] Indians. Nevertheless, one would suspect that enthusiasm for the activity might have remained relatively high if definite and pre-dictable benefits continued to be derived from it. This, however, has not been so. Not only is life on the trap line very hard and carried out under extreme climatic conditions, but the rewards are unpredictable. It is this last point that is of particular importance. As Honigmann[3] has pointed out for the Attawapiskat Cree, "the enthusiasm which is an important motivating factor in the Indian hunter and trapper is keen-est when his efforts promise to bring him not a few but many animals." This is also true for the Indians of Snowdrift [Lutselk'e] and this enthu-siasm is also observable when the prices paid for furs are high. Honigmann further points out that in some societies relative scarcity of resources means that people will work harder. At Attawapiskat, and also at Snowdrift [Lutselk'e], the reverse is true. It is the author's opin-ion that at Snowdrift [Lutselk'e] it is the extreme fluctuation of fur prices as much as anything else that has had a discouraging effect upon Indians' interest in trapping as a means of making a living. This fact, along with others previously mentioned, has simply meant that the men, with few exceptions, are spending less time on their trap lines regardless of how far these lines are from the village. During the main trapping period, which extends from about the first of November until Christmas, most of the village trappers are on their lines for no longer than four weeks. In recent years there has been only a moderate amount of trapping after Christmas. The value of mink and marten pelts drops rapidly after the first of the year and only white-fox pelts continue to be in prime condition. Thus, largely because of the unwill-ingness of most Snowdrift [Lutselk'e] trappers to go into the Barren

---

3 J.J. Honigmann, "Folkways in a muskeg community. An anthropological report on the Attiwa-piskat Indians," (Ottawa: Dept. of Northern Affairs a. Natl. Res., NCRC–62–I, 1962): 89–90.

Grounds to trap white foxes, there is relatively little trapping activity until muskrat and beaver trapping begins in the spring.

The material on contemporary trapping presented in the preceding pages leaves the distinct impression that this activity is a changing aspect of the Indian economy and that, particularly in recent years, this change has been rapid, extensive, and influenced by factors that are characteristic of Indian acculturation throughout much of Canada, particularly among Subarctic peoples in the east and west. It therefore seems probable that the changing pattern of trapping at Snowdrift [Lutselk'e] can be duplicated in other parts of Subarctic Canada. Whether and to what extent this actually is so will now be examined.

Although neglected for many years by anthropologists, both the eastern and western Subarctic of Canada have recently received more attention, particularly with regard to problems of social and cultural change. Much of this work is still unpublished, and, unfortunately, even the available reports do not deal with trapping in such a way as to make them comparable with the information collected at Snowdrift [Lutselk'e]. This is, of course, not the fault of the investigators but stems from the particular emphasis of the research designs. Nevertheless, enough information does exist so that some statements can be made about similarities in trapping patterns, particularly with regard to the problem of area utilization as it was described for the Snowdrift [Lutselk'e] Chipewyan.

Turning first to other studies in the western Subarctic, it is found that Helm, in describing trapping activities of the people at "Lynx Point," a Slave Indian community on the Mackenzie River near Fort Simpson, notes that the time spent on the trap lines varies considerably. Most of the men have short lines and seldom spend more than one or two nights on the trail. Those with longer lines may set up tents and spend as much as two weeks in the bush. Only one old man in the community has a winter trapping camp in the bush and stays away from the community as long as two months at a time. Helm mentions that in spite of the definite economic advantages of such an arrangement, most "Lynx Point" trappers would prefer more frequent visits to the community because they do not like the social isolation of trapping-camp life. They prefer the comforts of their own homes where children can be more easily tended, and they like to be near the store. Although not specifically stated by Helm, one gets the definite impression that there are as many factors that draw the trappers of "Lynx Point" closer to the community as in Snowdrift [Lutselk'e] and that it is only the older trappers who think in terms of long periods of intensive trapping with infrequent visits to the post.

A similar situation can be said to exist among the Dogrib Indians of Lac la Martre [Wha Ti]. Here, according to Helm and Lurie, a trap line is set only an overnight distance from the village, and even an extended trapping tour undertaken by several men together seldom lasts more than two weeks. The authors also mention that an important factor affecting the fur trade at Lac la Martre [Wha Ti] is the increasing access to other sources of income, "both in terms of providing less arduous income activities and of cutting down on time-energy remaining for trapping."[4]

In the eastern Subarctic the picture is somewhat less clear and is complicated by the Indians' trapping in many areas in government apportioned and registered trapping territories. This means, for one thing, that a trapper may construct a log cabin at a convenient point on his trap line and generally will not hesitate to improve his line, both by using it better and by providing greater comfort. A Snowdrift [Lutselk'e] trapper, on the other hand, seldom traps more than four or five years in the same area and for that reason hesitates to make elaborate preparations or improvements on any one trap line. Thus the trapper in the eastern Subarctic, if he is working a registered trap line, presumably finds that there is less contrast between life in the bush and in the village, at least as far as physical comfort is concerned. His cabin and other improvements are also likely to commit him as far as intensity of trapping and time away from the village is concerned. Therefore, he would be less likely to abandon all or part of a line because of distance from the community or fluctuations in fur prices. Thus the trapper in the eastern Subarctic would seem to be less susceptible to some of the influences that are affecting trapping areas in parts of the west.

Although the situation appears, then, to be more complex in the east than in the west, it is nevertheless possible to isolate the factors under discussion. At Northwest River in the Melville Lake region of Labrador, for example, there was during the early 1950s a growing tendency for the Montagnais Indians to spend more time near the trading post. McGee indicates that with greater dependence on the Hudson's Bay Company store, the summer camp has increasingly become a base of operations from which the Indians move out and to which they return frequently. Wives and children are usually left at the camp, which discourages the trappers from staying away for long periods of time. In emphasizing the importance of pensions and relief payments as part of the total picture of changing trapping patterns in the Melville Lake region, McGee further points out that more Indians now have a basis

4 J. Helm, "The Lynx Point people: the dynamics of a northern Athapaskan band," (Ottawa: Nat. Mus. Can., Bull. 176, 1961), 40.

for credit at the Hudson's Bay Company store that is not connected with the trading of fur. It is possible for a family to subsist more or less entirely on relief allotments and other forms of unearned income. "Because of the availability of welfare money, and because of the resident missionary with a church and school, the number of individuals who do not go to trapping grounds has grown rapidly to encompass at least 25 per cent of the (Northwest River) band."[5] It should be emphasized that those families living permanently in the village have given up trapping entirely and are living on welfare payments and some wage-employment during the summer. Trapping areas close to the Northwest River have been taken over by whites so that it is impossible to operate a trap line from the village. Since the trends described by McGee were already well established at the time of his field work in 1951 and 1952, it is probable that they have continued and become more significant during the past ten years.

Moving west into northern Ontario, we find that at Attawapiskat many inland trappers leave the village shortly after the beginning of the new year and do not return until Easter. However, some remain in the community until the end of February. A large number of trap lines are located relatively close to the coast and the trappers who work these lines leave their families in the village and, consequently, return frequently during the trapping season. Men who trap beaver on Akimiski Island also leave their families at the post and make a number of trips to the island by sled. These men remain away from the village for about ten days at a time and then return, each visit to the post lasting about one week. Thus one receives the distinct impression of diminishing use of trapping areas and the gradual development of a more sedentary community life. Honigmann's field work was done in 1947 and 1948, and it should be emphasized that although trends that point toward changes similar to those documented for the west are observable, traditional trapping patterns were still of considerable importance. Thus we learn that in the two seasons, 1944–45 and 1945–46, 43 per cent of the listed Attawapiskat trappers operated within sixty miles [100 km] of the post and that these men earned 36 per cent of the trapping income for those years. On the other hand, 57 per cent of the trappers operated at a distance greater than sixty miles [100 km] from the village and earned 64 per cent of the income. It is obvious from these figures that a greater proportion of earnings are achieved by those men who travel farther from the post into the area where fur-bearing animals are more plentiful. It is possible too, in the light of previous statements, that these trappers work harder because of the

5 J.T. McGee, "Cultural stability and change among the Montagnais Indians of the Lake Melville region of Labrador," (Catholic University Am. Anthrop. Ser. no. 19, 1961), 57.

greater potential rewards awaiting their efforts. At any rate, it will be seen that there were still many factors that in 1948 encouraged Attawapiskat trappers to maintain a trapping routine involving long absences from the community.

In describing field work among the northern Ojibwa at Pekangekum in northwestern Ontario, Dunning has little detailed information to offer concerning changes in trapping over time, but he does document some of those factors of economic and social change that have been seen to influence trapping at Snowdrift [Lutselk'e]. Thus we learn that government buildings, a school, a nursing station, and houses have been constructed on the reserve within the past ten years. This, together with a steady increase in government subsidy of the economy until it constituted more than 41 per cent of the total income in 1955, has resulted in a change in the pattern of residence in summer "from domestic units spread widely over the trapping territories to a cluster of population at each of two centres."[6] The author points out that this concentration of population "could not have occurred under the condition of the traditional hunting and trapping economy," and we can assume that such a fundamental demographic change has had a definite effect on the pattern of trapping, very likely in the direction that has been documented here for other communities.

At Winisk, less than two hundred miles [320 km] northwest of Attawapiskat on Hudson Baby, Liebow and Trudeau say that the construction of a radar base between 1955 and 1957 offered the Cree Indians of the area alternative ways of making a living. Prior to the construction of the base each family group would leave for its own trap line area in late September, return for a week at Christmas or Easter or both, to sell furs and obtain food supplies, and then go back to the trap line until late May or June. Thus there would only be a little more than two months during the summer when all families were in the village. The authors also say that some families had only a few miles to travel to their trap lines and it seems safe to assume that they visited the post more frequently. With the construction of the radar base, the pattern outlined above changed almost immediately to nearly complete dependence on wage labour and year-round residence in what became a community.

It certainly would be difficult to find a more dramatic example of the preference that Subarctic Indians have for wage labour over trapping as a means of making a living. The reasons for this preference, documented in some detail for the Winisk Cree by Liebow and Trudeau and for the Snowdrift [Lutselk'e] Chipewyan by the present author, are of

---

6 R.W. Dunning, "Some implications of economic change in northern Ojibwa social structure," *Can. J. Econ. Pol. Sci.* 24 (1958): 565.

no concern in this context. What is of particular interest is the great change in trapping patterns, actually the almost total disappearance of trapping, in response to the sudden introduction of an alternative means of making a living. What is seen here is the rapid culmination of a process that is going on also in other parts of the eastern and western Subarctic.

In a paper discussing changing settlement patterns among the Cree-Ojibwa of northern Ontario, Rogers gives the clearest statement of the series of trends that have already been noted for Snowdrift [Lutselk'e] and other areas. He finds that there has been an increasing tendency for larger groupings of people to come together and remain more sedentary for a longer period of the year than formerly. This change to larger settlements and a more sedentary existence can be mainly attributed to the impact of Euro-Canadian culture in much the same way as can be documented for other parts of northern Canada. With regard to trapping, Rogers documents certain changes in the yearly cycle at Round Lake, Ontario, as follows:

With the advent of fall the men make ready for the coming season of trapping. When the time arrives, primarily during October, nearly half the population of the village departs for their winter camps, located as a rule within the boundaries of their trapping territories where they will reside until just before Christmas. During this period some of the men will occasionally return to the village to sell their furs and obtain more supplies. Some trappers take their families with them, others leave them in the village. Those men who exploit the area in the immediate vicinity of the settlement do not establish winter camps in the bush but rather operate directly from their homes in the village.[7]

This pattern continues with the return of those trappers, who have been in winter camps, to the village for Christmas. After Christmas, trapping is again resumed but some men do not return to their bush camps; rather they "borrow" a territory, or, more likely, secure permission to trap in a territory, near the village.

It appears, then, that at least half the Round Lake trappers trap near the village. Some have done so since the post was first established, but for others this form of trapping represents a break with the old tradition of winter trapping camps. In his Round Lake study, Rogers states that in spite of the growing tendency for trappers to exploit areas close to the village, there has been no noticeable lowering of the yield per trapper. It is probable, however, that this will occur in the future. What Rogers does not state, and what would be of particular interest in this

7 E.S. Rogers, "Changing residence patterns among the Cree-Ojibwa of northern Ontario," *Southwestern J. Anthrop.* 19 (1963): 75–6.

connection, is just how frequently those trappers who do stay in winter camps before Christmas return to the village to sell furs and obtain supplies. Another point of interest would be whether the number of trappers who return to the winter camps after the holiday festivities is declining or remaining approximately the same. One would suspect that the number of men staying in the village after Christmas is growing.

At the risk of reading more into Rogers' information than it actually contains, it could be said that among the Round Lake Ojibwa the trapping pattern is undergoing changes similar to those noted for the Snowdrift [Lutselk'e] and other communities in the eastern and western Subarctic. Namely, there is a general decline in the efficiency with which the total available trapping area is used. Less time is being spent on the trap lines and more either in the community or going to and from the community. These are general statements that seem to apply, to a greater or lesser degree, wherever fur-bearing animals are being trapped today in Subarctic Canada.

It would, of course, be naive and over simplistic to suggest that the factors responsible for this situation are everywhere exactly the same or are taking place at the same rate. However, it is probable that lack of uniform documentation over the entire area prevents, to some extent, the easy recognition of those similarities and parallel developments that do exist. One factor that complicates this situation and about which little has been said here is time. Even such information as has been presented indicates that changes in trapping patterns have not taken place at an equal rate. Thus it is the opinion of one experienced field worker that in many areas, particularly in some parts of the eastern Subarctic, the Indians trap more today than they did a hundred years ago. This may be true in the west as well since the proliferation of trading posts and the availability of consumer goods on a large scale, both of which have encouraged trapping, have occurred during the past thirty or forty years. Similarly, although it has been possible to make general statements about the decline in trapping effectiveness and land use, which apply over the entire area of Subarctic Canada, it is undoubtedly true that statements of this kind must be tempered by the recognition that the concentration of Indian populations in permanent communities has sometimes resulted in greater use of the trapping area near the community as the trappers have increasingly withdrawn from the peripheries. This is not true at Snowdrift [Lutselk'e] but it is at Round Lake, and there are indications that it may also be true of other communities in the eastern Subarctic.

All this appears to suggest that in spite of conditions that are working to the detriment of continued interest in trapping, there are still a number of important factors that tend to insure that it will be a long time before this means of livelihood entirely disappears. Undoubtedly the most important of these, at least in the western Subarctic, is that trapping remains the only source of income besides uncertain wage-labour, relief, and welfare. It is unlikely that very many Indian groups will desire, or be permitted, to live entirely on unearned income, and it has been only under unusual circumstances, such as at Winisk, that wage-employment opportunities in any part of the Subarctic have existed in sufficient abundance and permanence to enable the Indians to make a complete change in their means of subsistence. Even if job opportunities were to increase considerably, it is by no means certain that trapping would rapidly disappear. There exist, in many areas, factors that encourage a continued interest in trapping. In the east, the logging and mining industries, as well as tourist guiding, have tended to keep the Indians in the bush and helped to maintain their interest in and close association with the environment they know thoroughly. Commercial fishing and the growing importance of tourist guiding has had, and is likely to continue to have, a similar effect in the west. It must also be remembered that trapping has been important in the eastern Subarctic for more than three hundred years and in the west for nearly as long. Time is an important factor in consolidating adaptive forms of culture and it required perhaps a dozen or more generations for patterns of socialization, family holdings, and marriage preferences based on trapping to be built up and strongly consolidated in both the east and the west. Indeed it is the strength of these developed patterns that is likely to determine the reaction of any one group of Indians to the introduction of the various factors mentioned above that have the potential to inhibit trapping.

In this connection one point in particular must be examined; that is the presence or absence among Subarctic Indians of normative pressures to hunt and trap. Although information on this point is not as complete as it might be, the author feels that this is not a significant factor among Indians in the western Subarctic. At Snowdrift [Lutselk'e], for example, the men seemed to take very little pride in their trapping skill, and the author never heard comments about the pleasures or compensations of bush living. In the Mackenzie–Great Slave Lake area the field worker will hear a great deal about how the Indian is most happy and content when he is in the bush on his trap line, but these comments always seem to originate with local whites and never with Indians. At Snowdrift [Lutselk'e] all informants maintained that they would give up trapping at once if an opportunity for steady wage-employment presented itself. It has been previously noted that the "Lynx Point" trappers prefer the comforts and conveniences of the village to the social isolation of the trapping camp.

The extent to which this is also true in the east can only be surmised since, with one exception, the authors cited in this paper have not discussed the matter. However, it seems clear that there were definitely more strongly developed normative pressures to hunt and trap than have been documented for the west. For example, in speaking of the more or less permanent Indian population living in the vicinity of Seven Islands [Sept Îles] on the coast of the St Lawrence, Speck notes that for the period of his field work, between 1915 and 1925, these people could claim no prestige through their close contact with whites and assimilation of white values. Both social and financial prestige lay with the interior hunters and trappers. The coast people are spoken of as being glad to give up their precarious employment and restriction of freedom, should the opportunity arise, for "the adventure and possible greater profit of furs of the big woods." With reference to general aspects of Montagnais-Naskapi economy, Lips makes a similar statement. The Naskapi have always been hunters and trappers and they wish always to remain so. McGee, writing about the people in the Northwest River region of Labrador is more explicit concerning the compensation of bush living as opposed to the more comfortable life in the village, and his statements have added significance because he is dealing with the more or less contemporary scene. First of all, trappers are close to good caribou territory and are thus in a position to satisfy their need for meat. Secondly, old people who might have difficulty getting on alone at Northwest River have relatives who will keep them supplied with meat and firewood on the trapping grounds; in other words, it is not so easy to ignore old people in a trapper's camp. Finally, and perhaps most important of all, in the bush, for young and old alike, there are no non-Indian sanctions on conduct to worry about.

It cannot be denied, therefore, that factors exist throughout the Subarctic, although perhaps most notably in the east, that work to maintain interest in and dependence on trapping as a means of livelihood. These are essentially conservative factors and they arise from a tendency to associate trapping with the old Indian way of life that is right and good and as a reaction against the insecurities and uncertainties of the newly developing community life and increased contact with whites. It seems, nevertheless, to be true that in most if not all parts of the Subarctic, factors favouring the persistence of trapping as a major economic undertaking are in rapid decline. That is why the trends and changes described in this paper have, for the most part, become significant within the past fifteen years, or even much more recently than that.

By way of summary and conclusion, then, it can be said that in most Subarctic communities there are, at the present time, factors operative which tend to reduce the reliance on income derived from trapping. These factors, which for the most part involve additional sources of income, have also been effective in developing tendencies toward sedentary community life. The Indians are turning increasingly toward wage employment and dependence on various forms of government assistance as they attempt to achieve a higher standard of living. It is not surprising, therefore, that both Indians and white administrators are agreed that one of the most important problems facing the people of Subarctic communities today is the need to achieve financial stability and to be free from the uncertainties that are characteristic of an economy based on trapping.

25a Frequency of Traditional Food Use by Three Yukon First Nations Living in Four Communities

ELEANOR E. WEIN AND MILTON M.R. FREEMAN

This study documented the frequency of use of traditional food species among 122 adults from three Yukon First Nations. The informants resided in four communities: Haines Junction, Old Crow, Teslin, and Whitehorse. Food patterns were examined in two ways: (1) estimated frequency of household use of traditional food species over a one-year period, and (2) frequency of traditional foods in four daily diet recalls of men and women, collected once per season. On average, Yukon Indian households used traditional foods over 400 times annually. Moose was consumed on average 95 times yearly, caribou 71, chinook salmon 22, Labrador tea 20, cranberries and crowberries each 14, and blueberries 11 times yearly. According to household estimates, traditional foods were consumed almost as often in Whitehorse as in Haines Junction. Teslin surpassed both these, while Old Crow had the highest frequency. Daily diets of adult individuals indicated that traditional foods were consumed on average 1.14 times per day. Traditional foods were reported twice daily in Old Crow diets, once daily in each of Teslin and Haines Junction, and 0.5 times daily in Whitehorse diets. Measured by frequency of use, traditional foods – especially moose, caribou, and salmon - remain extremely important in contemporary diets of these Yukon Indian people.

SOURCE: *Arctic* 48, no. 2 (June 1995): 161–71 (abstract). Reprinted by permission of the publisher.

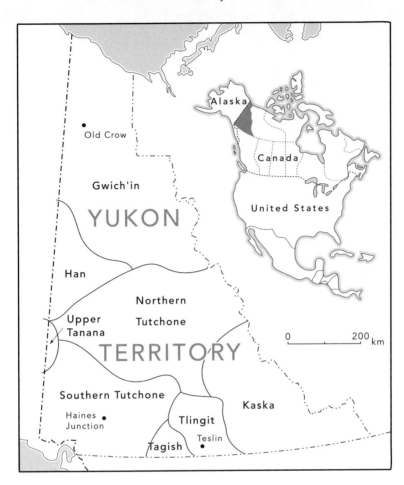

Study communities, shown within the Aboriginal language and culture areas of the Yukon (modified with permission from a map kindly supplied by the Aboriginal Language Services, Executive Council Office, Government of Yukon).

25b Native Subsistence Fisheries: A Synthesis of Harvest Studies in Canada

FIKRET BERKES

Subsistence fisheries, as distinct from commercial and recreational, exist throughout much of the Canadian North and satisfy local needs for fish protein. These fisheries have been investigated quantitatively only since the 1970s. Many of these studies are in the "grey literature"; methods of study and reporting are not standardized, and interpretation of data is often problematic. Nevertheless, some generalizations can be offered from a preliminary survey of harvest study data from ninety-three communities and from ten regional studies representing Labrador, Quebec, Ontario, Manitoba, Saskatchewan, British Columbia, and the Northwest Territories. The data indicate a wide range of harvest values, clustering at about 60 kg of whole fish per capita per year. If these data are representative, there is a significant subsistence fishery sector important for the local economies of hundreds of communities. Most of these fisheries are not being reported in fishery statistics, nor are they being monitored and assessed.

SOURCE: *Arctic* 43, no. 1 (March 1990): 35–42 (abstract). Reprinted by permission of the publisher.

26 This Land Is Our Life

JIM BOURQUE

The Northwest Territories today offers a way of life that is no longer possible to achieve in other more densely populated areas of North America and Europe.

People in Canada's North still live in harmony with nature. We are not divorced from nature to the same degree as those who live in the larger population centres.

Aboriginal people, who are in the majority, insist that any industrial development takes place in a manner that does not interfere or destroy wildlife habits or traditional relationships with the land.

As a result, the Northwest Territories is one of a very few populated places in the world that still reflects a natural balance between man and nature – a balance that its Inuit, Inuvialuit, Dene, and Metis residents have respected from earliest times. The relationship is spiritual. It is on the land that the Aboriginal people gain a sense of achievement and identity from their traditional economy of hunting, trapping, and fishing. The land and the people are one. Without this relationship, we would not have survived.

To former Government Leader Richard Nerysoo, a Loucheux [Kutchin] Dene born in a tent on the shores of the Peel River, it can be explained in these terms:

SOURCE: Jim Bourque, *This Land Is Our Life* (Yellowknife Dept. of Renewable Resources, Government of the Northwest Territories, 1986), 5, 7. Reprinted by permission of the publisher.

It is very clear to me that it is an important and special thing to be an aboriginal person. It means being able to understand and live with this world in a very special way. It means living with the land, with the animals and with the birds and fish as though they were your brothers and sisters. It means saying that the land is an old friend, an old friend that your father knew and your grandfather knew ... indeed, a friend that your people have always known.

We see the land as much, much more than others see it. Land is not money. To the aboriginal person, land is life. Without our land, and the way of life it has always provided, we can no longer exist as a people. If the relationship is destroyed, we too are destroyed.

"The land has been, and is, our life and it is our industry, providing us with shelter, food and income similar to the industries down south supporting the white people." – Inuit Hunter David Nasolgaluak, Sachs Harbour, NWT.

In today's world, reality dictates the need for a decent financial return on seal skins, furs, and skin and fur manufactured products in order for Aboriginal peoples to subsist and in order for us to provide for our families on a continuing basis.

Aboriginal people in the North live in small communities scattered throughout the Subarctic and Arctic regions of Canada and no longer live nomadic lives following the seasonal habitats of the various species of wildlife. As a result, it costs more money to hunt and trap and to provide a traditional subsistence.

Furs. Money. Subsistence. The process is cyclical. But, the anti-trapping and anti-sealing lobbies have gotten out of hand and have broken that cycle.

Faced with the lowest fur prices in ten years, aggravated by spiralling inflation and world-wide recession, Aboriginal hunters and trappers can no longer afford the equipment and provisions required to harvest the land and the Arctic ocean. In many cases, our only alternative is welfare and that can be pretty demeaning to anyone whose identity, lifestyle, and way of life is being pulled out from beneath his feet by an emotionally-based campaign directed at his livelihood.

Instead of being able to afford a regular diet of nutritionally-sound wild meats and fish, we are being forced into diets of more expensive canned goods and other products imported from the southern North American marketplace. These products are purchased with public tax dollars obtained from welfare payments and unemployment insurance benefits.

It's not a nice picture. It's nothing to be proud of.

"Without seals," according to Inuk hunter Louis Tapardjuk, "our life is not full."

27 Bewildered Hunters in the Twentieth Century

ABE OKPIK

This problem came into being within the last fifteen years, and still continues to grow, and will be in existence for some time to come yet.

The situation, as it existed in time past, was based on the economic resources of the land, which furnished some needs in most communities throughout the north. Trapping was a big factor because it provided income and power to buy equipment such as rifles, boots, outboard motors, tents, toboggans, etc. Hunting comes into this category as well, for if you had a good hunting ground you can picture yourself getting more fur. Fishing was another resource in large parts of the north and still is.

Each individual with his family was the bread winner for his family and sometimes for other unfortunate ones. Sometimes two or three families agreed to stay together to hunt and trap in the same area providing that the surrounding area was economically sound. The people did not dwell in the settlement simply because, in times past, the resources did not allow it. So, people in the north never lived together in one place at all times, because the situation did not allow it. The location of homes was not a problem, because it was the general trend, and accepted that the house they built would be no more than a log house or sod house. To the east, a tent insulated with moss was used. Sometimes the house or tent was no larger than eight by ten feet [2.4–3 m]. Some coastal people who live mainly near the sea use ice houses, or ice frame with a tent inside: a method found mostly in the central

SOURCE: North 13, no. 4 (July-August 1966): 48–50, Reprinted by permission of the author and publisher.

Arctic. Still, in a large part of the north, for groups comprised of movers-from-place-to-place, an igloo is most used.

All over the Arctic, this type of housing was pretty well accepted for the place of dwelling did not matter too much in space, structure, or shape as long as it was a place to keep the body and soul warm and together. In the summer the families moved into a tent. This housing among the people did not change until the last decade when the employment programs started developing in the north.

The population of each community came into being from the surrounding areas, and the houses grew around the settlement without too much observance by anyone. Instead of being apart and distant neighbours, the people of the area now found themselves cuddled up in and around the settlement. Any type of structure or material was accepted. Although the intention was to stay only for a short while, the people who found jobs began to stay around, bringing more and more of their relatives and people into the village. Some schools were set up in order to teach the children, thus bringing more population into the community. A small settlement which often consisted of a missionary, a Hudson's Bay Store, an RCMP building, and a few houses found itself developed into a larger community where a school, a hostel, or nursing station and administrators, power plants, and welfare housing, and other people entered into the picture. Thus a settlement became a larger community within the last few short years.

I will try to outline the problems involved among the northern communities. Mentioned earlier, the people congregated around the settlement in order to get jobs and an opportunity to work for a short while without any real intentions of staying there. Sometimes a hunter became a community garbage collector, and sometimes a camp drudge became an assistant-to-the-administrator though often he did not understand the circumstances involved where a person hanging around town could be a valuable helper to the administrator but in time the hunter sometimes gets the idea of the reasons. Then the hunter and family who occasionally visited the settlement began to come to the settlement more frequently, sometimes even for a short time to work. They would camp in the same place they normally visited for the first few times, then their visits became longer and the children started attending school part time. Eventually the families move into the settlement without being too conscious about what they were really moving into. The hunter could no longer take extended trips to make a successful hunt as he used to.

The usual camping visit to the settlement was now permanent. The tent he sets up is renovated into a temporary house. The family is stuck in the settlement by this trend. The people who have lived in the area are now living in settlement to either work or have their children go to school. This accumulation turned the village into a larger community without any planning involved. The planner came later, and this is where concern for poverty started to be seen in the community.

The community grew without too much pressure from the authorities or planner as to how the shacks or houses should be kept, or what type of structure the homes should be. Once an individual hunter who still thinks as a hunter is absorbed, even without his awareness of it, he loses control of his hunting equipment: his dogs, boats, canoes, traps which were at one time his vital possessions for living off the country. He gives this a thought and soon the whole thing flows over: "I'll be all right." So therefore his beginnings are left in the mist, and this new life starts eruption. At first he is without bewilderment but then his first wages are earned, and more and more necessities are required in order to settle the mind. Now the ideal life is a mystery to him. In his own opinion the whole dream will be over when he gets out on the land again. Then the expectation of this dream is not fulfilled; it materializes into the facts of a new way of life.

This gives him no point of return, and he loses all livelihood since his life possessions are now gone. As time carries on, he is watching for the opportunity to go somewhere. But where? He finally decides to get work elsewhere, wherever a job was found available. But his does not solve anything because when he leaves his family for an indefinite time he worries about them. This is where his stability is really tested by those strange things that lead from one thing to another. To keep the family together he decides to leave his job and make an attempt to go back to the family and the old environment. But this seldom brings anything new. In fact, he now finds it harder to survive on the land than he used to. He finds himself more or less alone, caught in a situation between two great gaps. His mind is bewildered and confused. He pictures the both sides. The outsiders came here to give him an opportunity to change to a new and conformed way of living. But he believes that he cannot really change his ways. His children are learning the other side of conforming since they have been going to school. The hunter realizes that he is too old to take any schooling. For the sake of his children he must try to make some adjustment, but how? First he hears that his cousin was arrested for drinking in a public place and given a light sentence in jail. Then he hears of another cousin or relative who was involved in stealing some money. He then gets a shock by hearing of one of his nieces who got pregnant while going to school. Perhaps by someone in school or some white man who works with the people who are building the houses in town. He then may hear a rumour that a cousin and his friend were arrested by the police for fighting and resisting arrest, and were given a sentence by a local Justice of the Peace. This time each was given three months. To him this seems odd – once the policeman was a friend, but now what has he become towards the people? During these incidents his mind is restless. Once or twice he tries hard to start again on the land, but his children are not happy out there. No one in the family cares to stay out any length of time. So he goes back to the community to try and make enough money to start over on the land. This time when he goes back it seems like all his relatives were really glad to see him. They will come and eat with him bringing their share, just the way it used to be. Pretty soon the country food which he brought back to town with him is gone. He goes on the hunt alone but he cannot go far because the equipment and dogs are not as ready and good. Disappointed, he goes back to town and goes over to see the trader he used to know to see if he can be given a grubstake to start for a better trap line. But even the trader has little to offer. He is busy, selling to others who have more money to spend. He could not convince him to give him a grubstake. He wonders what happened to the storekeeper who used to help him with anything he wanted, especially nets, shells. But now the trader does not even trust him. Still he thinks in his mind that everything will turn out better. He goes and sees some friends he knows who can perhaps help him. But they too have changed their attitude and it seems to him they don't care.

He manages to get a little money here and there. Soon another addition to the family arrives. His housing condition is the same. He knows this used to be okay in the bush and there is nothing to be ashamed of as his was better than others. He has nothing to worry about although his home is of tar paper, cardboard, and a double tent.

He has been working on part-time jobs off and on but he cannot understand where all his money has been disappearing to.

He then decides to have a drink with friends he knew to ease his mind and talk with them about some things which have been bothering him. This has no effect either. Finally, on the last effort, he goes and has a serious talk with the administrator. But the administrator has very little time to spend with him as he is also very busy. He finally ends up talking to the Welfare Officer. The officer is prepared to listen so he tells him all the things that have been bothering him and why. He tells him about what is happening to his family. The closely-knit group of relatives is falling apart, their words do not have a meaning anymore.

Within a few short years these good people he used to know seem to have deteriorated to a lowly thinking type of people. They do not seem to care. Something else has taken over their responsibility. He does not think of it as bad or poor, he thinks of it as a broken machine which cannot be repaired because it is completely useless and helpless.

Even the Missionary who used to be helpful does not come and visit anymore. It seems more and more teachers and other people are arriving from outside and they are all supposed to be helping the Native. Now they say his house is not good as it has no washing facilities, no flush toilet. He wonders, "what do they mean by that? What do they want me to do about it. It used to be all right. Anyway, I am not going to do anything about it until maybe when someone else starts getting new houses."

This is the beginning of poverty – when a person is bewildered and has no way to improve his ways in a completely new environment.

This example in this writing is what has happened to the hunter when he is caught between two cultures: one he knows, and one he guesses at.

This is a beginning and the end of an era which has confused the once leader, hunter, provider, and honest man. He is caught between this new influence and the old way of living.

This is what is happening to many people of the north, especially the people who were born before 1945 and who are not going to change. This group is the core who refuses to change.

What is their future?

28 The Great White Hope

ED STRUZIK

On a warm, sunny August day in 1975, Justice Thomas Berger landed in the Dene Indian village of Fort Good Hope in the Northwest Territories. He was conducting the now famous inquiry into two gas pipeline proposals: one would link northern Alberta to the Mackenzie delta, the other would add a trunk line across the northern Yukon to Alaska's Prudhoe Bay.

That day chief Frank T'Seleie greeted Berger warmly at the dirt airstrip but hours later at the hearing, he vowed to put his life on the line if the pipeline was built. "It is for the unborn child, Mr Berger, that my nation will stop the pipeline. It is so that this unborn child can know the freedom of this land that I am willing to lay down my life."

Turning to Bob Blair, president of Foothills Pipelines which was behind the all-Canadian line, T'Seleie declared: "You are the twentieth century General Custer. You have come to destroy the Dene Nation. You are coming with your troops to slaughter us and steal land that is rightfully ours. You are coming to destroy a people that have a history of thirty thousand years. Why? For twenty years of gas?"

When the Arctic Gas consortium first proposed the $7 billion pipeline in 1974, no one expected this kind of response from the Aboriginal community. In fact, governments on both sides of the border had counted on immediate approval and a quick start on construction. Even the value on the right of way had been worked out – $3.10 per

SOURCE: *Nature Canada* 24, no. 3 (summer 1995): 29–34. Reprinted by permission of the author.

acre [0.4 ha]. In return for just under $8,000 a year, private interests would undertake the biggest project of its kind in Canada – one that would employ as many as 7,500 people.

Federal officials didn't even expect to recover any expenses for studies or land claim settlements from the pipeline companies, who would have the option of renewing after five years. According to Jean Chrétien, who was minister of Indian Affairs and Northern Development at the time, Berger's mandate as commissioner of inquiry was "to find a way to build a pipeline, not to stop the pipeline."

After 238 days of testimony, 1,700 witnesses, and over 40,000 pages of transcripts, Berger had other ideas. Not only did he recommend against the Arctic Gas proposal, he called for a moratorium on development in the Mackenzie Valley for ten years. Most importantly, he laid down a blueprint for the protection of the North's environment and Aboriginal culture. It included establishment of a national park in the northern Yukon, a whale and several bird sanctuaries in Mackenzie Bay and the Mackenzie Valley, settlement of Native land claims, and preservation of Native culture and renewable resources.

Following publication of his report, Berger emerged as a national hero. *Northern Homeland, Northern Frontier*, the first volume of his two-volume report, turned out to be the best-selling document ever published by the Canadian government. One southern-based pundit called its author "the therapist of the North," another the "tundra-striding Solomon, adjudicating between cruel technological forces and harassed Native people on Canada's last frontier." Peter Gzowski wrote in *Canadian Magazine* that Berger "is one of the great Canadians of our time – a man of fathomless integrity who ... refuses to be badgered or shaken ... and who has guts, wisdom, and charm."

Twenty years after the launching of the Mackenzie Valley inquiry, Berger's legacy continues. A national park has been established in the northern Yukon to border the Alaska Wildlife Refuge in Alaska; the Inuvialuit of the western Arctic have settled their land claim with Ottawa; and most of the other Aboriginal groups have come to an agreement in principle on theirs.[1]

Meanwhile self-government is also on the way. By the turn of the century, the Northwest Territories will be divided in two, with the Inuit governing in the east, and the Dene, Metis, Inuvialuit, and non-Native peoples controlling the west.

T'Seleie, who went on to represent his constituents in the Northwest Territories legislature, says "The Berger inquiry opened a lot of eyes in

---

[1] Most comprehensive land claims of the First Nations in the Northwest Territories (and Yukon) have been settled since this was written. In April 1999, Nunavut, the new Arctic Territory, came into existence, carved out of the previous Northwest Territories. – ed.

the North and the south. It was an education for us all. It helped us prepare for the development which we all assumed was inevitable."

Yet, on the eve of another huge development in the North – the building of one or more diamond mines in the central Arctic – many northerners are taking a second look at the Mackenzie Valley pipeline inquiry. This time, the verdict coming out in print is that Berger steered Aboriginal people in the wrong direction and sidetracked the kind of economic development that northerners really needed.

Some people think that Berger is a saint and he will go down in history as the greatest man who saved the North, says Cece McCauley, a newspaper columnist and former chief of the Inuvik band. "Saved the North from what? We don't need to be saved. We need our dreams fulfilled; the good life; cheaper food; easy access to the outside world."

"After twenty years, look at the mess of our social fabric now," she adds. "The fools who gave us the kiss of death on development couldn't see past their noses. As for the 'legacy of bitterness' it would have left, they left a legacy of bitterness anyway. For the past twenty years, the Mackenzie region has remained isolated. The multinationals moved out. We are overrun by jobless people. Too many people depend on welfare. Booze is the one big consolation for some. Youth are paying for all of this inactivity and the absence of growth in development."

McCauley's view may be shared by the northern business community but most Aboriginal leaders do not support it. Yet it is difficult to avoid the fact that many Native leaders who embraced Berger twenty years ago, are now supporting the kind of development that he and his inquiry helped stop.

Nellie Cournoyea, the Inuvialuit leader of the Northwest Territories government, is a big booster of a $650 million road running through the heart of the great caribou migrations. She approves of the route from Yellowknife to the central Arctic in spite of concerns from the wildlife department she once headed. The Dogrib people north of Yellowknife also support development, specifically a proposal for one or more diamond mines within their claims region. They're even contemplating damming a river or two to provide hydroelectric power to the operations.

Joe Rabasca, the grand chief of the Dogrib, makes no excuse for the shift in thought. "The time has come when Aboriginal people exploit development opportunities. Too often in the past we've been left behind," he says.

If a new ethic toward development has invaded the hearts and souls of Aboriginal leaders in the North, the impact on the environment will be profound. It will almost certainly slow or put a halt to the establish-ment of some national parks, bird sanctuaries, and national wildlife refuges, and threaten other sensitive areas important to grizzly and polar bears, wolves, caribou, muskoxen, and marine mammals. In the long run, the pristine North as we now know it may be carved up by mines, oil and gas rigs, pipelines, roads, quarries, sewage lagoons, and garbage dumps.

That, of course, may be a worst case scenario. But Chris O'Brien, the head of Ecology North in Yellowknife, is concerned. He says that in the twenty years he has lived and worked in the North, he has never experienced the kind of pro-development, anti-conservation sentiment that is now being voiced at nearly every level of society.

"I used to think that this was the one part of the world where things would be done differently, but now I'm not so sure," says O'Brien. "People are no longer thinking about the future, about the preserva-tion of the land and waters they love. They're talking about jobs. And it seems like they don't care what form that employment comes in."

The North has long been fertile ground for this kind of frontier men-tality, and southern business leaders have had little trouble rallying Chambers of Commerce and Mines. For the boosters – the miners, oil and gas men, and small businessmen who benefit from them – the North represents the kind of unbridled economic opportunity that has been tied up by red tape or exhausted in the southern provinces. But few, if any, of them anticipated that the Aboriginal community would become an ally in their cause so soon.

Then again, no one expected traditional renewable resource activities such as sealing and trapping to experience the kind of catastrophic col-lapse that has occurred over the past twenty years. The seal industry in North America is all but dead, and trapping is alive only because it's on life-support. As a consequence, government assistance has be-come the new economy in most northern Aboriginal communities.

Currently, the federal government spends $28,000 a year, or two-and-a-half times the national average, on each person in the North. Combine that with a birth rate that is twice the national average, and an unemployment rate that is likely to reach 40 per cent in three years, and one can see why Aboriginal people are entertaining industrial options they wouldn't have dreamed of considering two decades ago.

Don Gamble, an engineer and former director of the Canadian Arctic Resources Committee, is sympathetic towards the economic plight of northerners. But he says it is a gross misreading of history to lay the blame on Berger and his inquiry. Gamble was working on behalf of the Department of Indian Affairs and Northern Development when Berger headed north. He switched to the inquiry side part way through.

"People forget that it was the National Energy Board, the official government regulatory agency, that recommended against the Arctic Gas proposal just two months after Berger delivered his report," he says. He adds that the Arctic Gas project "had a lot of technical problems. It also required huge government subsidies. Combine that with the fact that Native people and national sentiment was against the pipeline at the time, and you can see why the federal government was glad to get out."

Gamble thinks Berger's critics should be reminded that the kind of environmental review process we take for granted today was not in existence twenty years ago. "Berger gave a wide variety of interests, most notably Native people in the North, the chance to say what they thought of the pipeline and industrial development." But he confesses that "none of us who worked with him back then thought he was going to take them so literally. The consensus among us, in fact, was that he should add another chapter to his report – one that would outline minimum requirements if a pipeline were to go ahead."

"We thought we were being realistic," Gamble says. "We thought acceptance of his recommendation would be improbable otherwise. But Berger said 'no.' He said it's what the people told him and he wasn't going to give the government a way out."

Kevin McNamee, coordinator of the Canadian Nature Federation's wildlands program, isn't so much worried about the changing philosophy of northern Aboriginal people towards industrial development as he is about the federal government's failure to follow up on conservation initiatives. He points out that the International Biological program, the Northern Conservation Task Force, and the Green Plan all had a vision for the North. But he believes that neither the previous Conservative government, nor the current Liberal one, has acted on them in a meaningful way.

"It is true that some northern Native leaders are supporting industrial economic development, and there is reason to be concerned about that," says McNamee. "But at the same time, many of them continue to support the establishment of national parks and wilderness areas. The problem is that our government does not.

"The federal government doesn't think twice about expediting approval for a huge development project in the North, but it will take twenty years to approve a national park. There's always some official who can find a reason not to go ahead with a conservation initiative. These people seem to be winning out with the Liberal government today, and that's what the people of Canada should be worried about."

Cindy Gilday was a school teacher in the Dogrib community of Rae-Edzo when Berger and his inquiry came to town twenty years ago. As

a Slavey from Fort Franklin [Deline], she did not think it was her place
to state her case on environmental and cultural concerns at that time.
But she has been doing just that ever since.

Over the past decade, Gilday has been instrumental in spearheading
Indigenous Survival International, an international organization dedicated to preserving Native culture. She has co-chaired the International
Union on the Conservation of Nature's task force for indigenous people, and overseen the Northwest Territories government's strategy to
preserve and promote the trapping industry. She has also been named
Northwest Territories Woman of the Year, awarded a national Aboriginal achievement honour, and sat on a federal review panel which will
look at Broken Hill Proprietary Ltd.'s proposal to build a $500-million
diamond mine in the central Arctic north of Yellowknife (see *Nature
Canada*, fall 1994).

Gilday is mindful of the pro-development pressures in the North, but
she isn't buying any of the negative talk about Berger nor accepting the
predictions that big business is going to win the day. "The legacy of
Berger is that nothing is done without the consent of the people," she
says. "[He] legitimized the concerns of the ordinary person in the community – the person who has known nothing other than a hunting or
trapping life. I buy into that. I also buy into the idea that the hunting
and trapping life is not dead. It must always be an option for Native
people."

That said, Gilday is not dismissing the idea of industrial development on Aboriginal lands either. "My view is that if it's going to happen, if it is an option that the people want and can live with, then they
should know what they're getting into. And they should know what is
the best available technology that will minimize the impacts.

"I don't want economic desperation to be the driving force in any
future development decision. I don't want to see what we've seen on
southern reservations where Aboriginal people have gone about clear-
cutting forests or allowing nuclear wastes to be dumped on their land.
You can say I'm optimistic about the future, but cautiously so. The
bottom line is that we need to give our people choices."

Berger himself is guarded about making comments on where the
North is today, and what the future holds for it. Thinking back to
twenty years ago, however, he says he never imagined that the inquiry
would have such an impact – or that it would still be stirring up controversy today.

He recalls his recent trip to India as deputy chairman of a World
Bank commission looking into a huge dam proposal. "We travelled to
many rural villages to speak to people just as we had during the
Mackenzie Valley inquiry, and there were some who attended the hear-

ings who had knowledge about what happened in the Canadian North back twenty years ago. It was the same in Bombay. So, good, bad, or not, it did seem to have an impact," he said.

With respect to his own recommendations, Berger is particularly proud that the government decided to establish a national park in the northern Yukon. He is also heartened by the recent progress in the settlement of land claims, although he had predicted the process could be completed in the decade after his report was tabled.

What concerns him still "is the interpretation by some that I was trying to turn back the clock to another time in nature, and that I was denigrating the accomplishments of industrial society. That was not my intent. My report was intended to reflect the choices that would determine the kind of country Canadians wanted in the North at the time. I still think the choices for the future have to be made by the people themselves."

Berger concedes that choices have become a little confused over the years as national attention on the North comes in and out of focus. But he is confident that Canada can avoid the kind of mistakes that the Americans made when they embarked on the exploitation of their last western frontier and virtually wiped it out.

"The North [is] a special place that has something to do with our idea of ourselves as a country. There is a northern dimension to our psyche," says Berger. He believes the environmental and Aboriginal issues the North represents are here to stay, and he's hopeful that in the end "we will make the right choices."

29 Northern Aboriginal Toponymy

RANDOLPH FREEMAN

As we celebrate one hundred years of official 'standardization' of Canadian toponyms, this would be an appropriate time to reflect upon the state of toponyms used by the Aboriginal people of Canada. Few among us would dispute the belief that the ancestors of North American Aboriginal people crossed the Bering Land Bridge at least 12,000 years ago. It would also be difficult to dispute the likelihood that these 'First Nations' were not unlike modern cultures in their need to label the geographical features they encountered. Whether or not any of these very early traditional names survive through to today would be impossible to say, though some tantalizing evidence exists that some have. This evidence comes to us in the form of accurately descriptive names for ancient archaeological sites, or in traditional names made up of words no longer used in the Aboriginal language.

Within Aboriginal language groups, traditional toponyms tend to form cohesive systems that can only be fully appreciated when viewed in their entirety. Over the past one hundred years, Canadian naming authorities have given official recognition to only small parts of these once extensive systems and those Aboriginal toponyms that have been approved have been transformed, deformed, mangled, and anglicized or francisized, prior to absorption into the Canadian names system. The resulting toponyms can no longer be considered Aboriginal, only as having been derived from Aboriginal sources. Most Canadians are unaware that names such as Canada, Manitoba, Saskatchewan, Ottawa, etc., are, at best, poor renditions of the original Aboriginal names.

SOURCE: *Canoma* 23, no. 1 (July/juillet 1997): 53–4. Reprinted by permission of the author.

These names now form a solid part of the Canadian cultural mosaic and it is unlikely that the general public would approve of any substantive orthographic changes to names of this type.

The situation in northern Canada is considerably different. Aboriginal name systems remain relatively intact and are still in use by a large, but quickly diminishing, portion of the population. In many parts of the North, these systems are breaking down as names, assigned to many geographical features by non-Aboriginal people, are becoming accepted and used by the younger generations of Aboriginal northerners. While some would suggest that this is a perfectly natural process in the evolution of toponyms, the NWT Geographic Names Program was created specifically to counteract this natural process! The creation of the Geographic Names Program in 1985 came about as a direct result of the alarm felt by many Dene, Metis, and Inuit Elders that their traditional toponyms would be lost forever, if not given official recognition. The mandate of the Program was clear from the beginning; it was to be an active part of a much broader effort to slow, and perhaps reverse, the loss of Aboriginal culture and language.

While the Geographic Names Program has all the usual responsibilities of a provincial-type naming authority, the largest portion of its resources are expended on the recording, preservation, and dissemination of traditional Aboriginal names for geographical features, populated places, and formerly populated places. The NWT Geographic Names Program is actively involved in providing financial, technical, and training support to community-based research projects oriented towards the gathering of traditional Aboriginal toponyms. To date, the Program has participated in more than seventy community-based projects, many directly funded by the GNWT. These projects have recorded more than 15,000 traditional Aboriginal names for creeks, mountains, lakes, islands, archaeological sites, etc. The development of easier public access to the NWT Geographic Names Data Base, which currently contains information on more than 22,000 official, historic, and proposed geographical names, also continues as a priority of the Program. Much of this information will soon become available on the Internet.

Communities undertaking projects that gather traditional land-based knowledge are encouraged to develop the means by which research data can be used in the community. This will ensure that valuable cultural information is not lost with the passing of Elders and that the trend for younger generations to use non-traditional toponyms is reversed. Projects that receive funding from the Geographic Names Research Contribution Program are required to submit copies of maps, data forms, cassettes, video tapes, etc., to the NWT Geographic Names Program. All other toponymic research projects are encouraged to submit the results of their research. This information is entered into the Geographic Names Data Base and subsequently use to support the official recognition of traditional geographical names.[1]

Many of the Aboriginal people of the NWT know and use traditional names for geographic features but as each year passes, fewer and fewer of these remain in common use among the general population. The opportunity to record and preserve these toponyms, and associated valuable culture information, is quickly passing. Traditional toponyms are closely associated with traditional land use; the generations of Aboriginal people having intimate knowledge of the land will soon be gone. Similar conditions exist in many areas of southern Canada; therefore; a concerted and coordinated effort, by all levels of government in Canada, including Aboriginal, must be made to record and 'standardize' traditional Aboriginal toponyms, so that this facet of Canadian culture and history is not lost.

---

[1] Numerous examples of Aboriginal place names replacing previous non-Aboriginal toponyms now are to be found in the North, e.g. Frobisher Bay to Iqualuit, Snowdrift to Lutselk'e, Eskimo Point to Arviat, Fort Norman to Tulita, Fort George to Chisasibi, Port Harrison to Inoucdjouac, etc. – ed.

ECONOMIC RESOURCES

30 Resource Development

ROBERT M. BONE

The northern economy has always been based on the exploitation of its natural wealth. In pre-contact times, the major resource was wildlife; after contact with Europeans, it was fur-bearing animals; and now it is a wider mix of resources, including energy, forests, and minerals. During the 1980s, the annual value of resource production from the North was in the billions of dollars. In 1989 the Northwest Territories accounted for $1.2 billion worth of resource exports (GNWT 1990: 109). A conservative estimate of the total value of all northern resource production in 1990 would be $30 billion.

In the 1950s, the resource economy offered a fresh economic approach for northern Canada. Increased demand for energy and raw materials, and federal financial support for development not only created a new economic environment in the North but also expanded the national, provincial, and territorial economies. The North soon became an important staple hinterland serving Canada, the United States, and other industrialized nations. The new economic landscape consists of mines, oil wells, pulp mills, and hydroelectric power stations, all connected to southern markets by a modern transportation network. The construction of these industrial projects and the building of a complementary infrastructure consisting of towns, roads, and a wide range of public facilities generated a demand for goods and equipment produced in southern factories; created job opportunities in both northern and southern Canada; and increased the volume of Canadian exports,

SOURCE: Robert M. Bone, in *The Geography of the Canadian North: Issues and Challenges* (Toronto: Oxford University Press, 1992): 99–103. Copyright © Robert M. Bone 1992. Reprinted by permission of the author and Oxford University Press Canada.

thereby improving Canada's balance of trade. Both levels of government have benefited from increased tax revenues generated by the resource industries and the associated service industries, as well as from their employees' personal income taxes.

The Demand for Northern Resources

At first, increased demand for resources came mainly from the United States. In the early post–World War II years, American industrialists were unable to satisfy their need for raw materials from within their borders, and began to look at securing these resources from foreign countries. Canada's North was an attractive area for three reasons. It contained the desired resources, relatively close to factories in the United States; Canada was perceived as politically stable and therefore a safe place for American investments; and Canadian resources were viewed by American firms and governments as falling within a North American Trading bloc. In 1947 American iron and steel interests announced plans for the first large-scale, privately funded resource project in northern Canada. This plan called for the exploitation of the vast iron deposits in northern Quebec and Labrador and the building of a rail-sea transportation system capable of supplying iron ore to American steel plants. By the early 1950s this massive iron mining and transportation system was completed. In more recent years, other foreign countries have turned to the resources of northern Canada. Japanese companies, for example, have leased enormous tracts of timber in northern Alberta for their pulp and paper plants. Canadian companies have also been drawn to the North, particularly to smaller projects requiring less capital and with a relatively short pay-off. The high risk of large-scale projects is real. The closure of the Cyprus Anvil mine at Faro,[1] Yukon and the bankruptcy of the leading oil exploration firm in the Beaufort Sea, Dome Petroleum Company, represent the consequences of such risk. Offsetting the possibilities of economic failure, and drawing companies to the North, are the prospects of large profits.

Accessibility and Resource Development

In the past, resource development in the northern frontier was often stalled by the absence of a modern transportation system. The cost of building the necessary transportation link to the south was usually

---

[1]   The Faro mine has been the fifth largest zinc mine in the world and the largest private-sector employer in Yukon. Since its start-up in the late 1960s, however, it has been closed several times and reopened several times, reflecting a fluctuating world demand and the financial problems of various owner companies. In 1999 it was placed in receivership. – ed.

beyond the financial capacity of the developer. Over the years this handicap has been reduced, but the basic problem remains for much of the North and certainly for the Arctic.

The high cost of construction projects in the North is due to four main factors. These are the physical terrain, particularly the Canadian Shield and the presence of muskeg and permafrost; the short construction season; the need to assemble men and equipment in remote areas; and the long distance from southern supplies. Because of the short construction period in the summer, and the lower labour productivity associated with winter construction, costs are much higher in the North than in the rest of the nation. If the workforce, equipment, and supplies must be assembled in the south and then transported to a remote northern location, the costs are extraordinarily high. The presence of permafrost at the construction site calls for special – and costly – construction methods. For the same reasons, maintenance of rail lines, pipelines, and highways is much more costly than in southern Canada.

Because of these high costs, resource development which has minimal transportation needs often occurs first. Historically, the more valuable resources are exploited first, e.g., gold and silver. When refined, these highly valuable minerals can be shipped to market by small aircraft. In such cases, there is no need for an expensive all-season highway or railway. On the other hand, base metals like lead and zinc remain bulky products even after milling the ore. They require rail, road, or sea transportation. The degree to which various mineral deposits are deemed exploitable is largely a function of the value of that commodity per unit of weight, that is, the higher the per-unit value of the mineral, the greater the distance it can be shipped.

Both provincial and federal governments have tried to overcome this distance barrier to resource development by building modern transportation routes into the North. In the 1920s the British Columbia government tried to stimulate resource development in its northern territory by building the Pacific Great Eastern Railway (PGE). In 1952, some thirty years later, the PGE was extended to Prince George and North Vancouver, thereby providing an effective link from the main port on the west coast to the regional capital of the northern interior of British Columbia. At the same time, the PGE reached further north to Fort St John and further east to the Dawson Creek and the Peace River country. At Dawson Creek, the PGE established a link with the Northern Alberta Railway. During the 1960s resource development of northern British Columbia began in earnest and the PGE both assisted and benefited from the resource boom in forestry and energy.[2] This

2 In 1971 the railway was extended another 400 km (250 miles) north to terminate in Fort Nelson, BC. – ed.

railway was renamed British Columbia Railroad in 1972 and BC Rail in 1984.

In spite of the efforts of individual provincial governments, the Canadian North remained a remote and inaccessible place. For resource companies, this remoteness meant that known resources in the Arctic and Subarctic were not economical to develop. From their perspective, the cost of building northern roads or railways to ore bodies was too great, and if the North were to be opened up by the private sector, the federal and provincial governments had to provide suitable access routes.

Under the terms of Confederation, the federal government had no responsibility for building roads in the provinces and, until the 1950s, there was little need for roads in the territories. But circumstances changed and new policies were needed. By the late 1950s the old laissez-faire policy was abandoned, largely because the federal government saw northern development as a way to strengthen the nation. As well, because it had initiated a policy to 'modernize' the North, Ottawa hoped resource development would provide jobs to Native northerners who were now living in settlements. In 1958 there was a major shift in federal government policy. The newly elected Conservative government sought to increase the rate of resource development in the North by using public funds to extend the southern transportation system into the North. In its Roads to Resources program, the federal government agreed to fund half the cost for building roads in northern areas of the provinces, leaving the provinces responsible for the remainder. A similar program was available in the territories, but in this case the federal government covered all the costs. Under these two programs, roads were built to help private companies reach world markets. One example is the road built to connect the asbestos mine at Cassiar in northern British Columbia with the sea port of Stewart.[3] Other highways were constructed to provide major northern centres with a land connection to southern Canada. For example, the Mackenzie Highway from Edmonton to Hay River was extended to Yellowknife. In all, from 1959 to 1970, over 6,000 kilometres of new roads were built at a cost of $145 million (Gilchrist 1988: 1877).

Social and Political Implications

The economic development of the North was closely followed by social and political changes. The federal, provincial, and territorial governments now provide basic education and health services to all commu-

---

3   Asbestos was trucked over the Alaska Highway to Whitehorse and shipped to Skagway by rail for export. The mine closed in 1992, and the community of Cassiar was cleared from the site. – ed.

nities in the North. The cost of building this education and health infrastructure has been considerable and its annual operating cost is high. In the territories and the more remote areas of the provinces, the local tax base is far too small to pay for operating these institutions and most operating funds come from federal and provincial sources. Besides the transfer of public funds into the North, political power has been shifted from Ottawa to the two territories. This devolution of powers has not run its full circle to provincial status but it has come a long way in a short period of time. No such transfer of powers has occurred in the provinces' northern areas.

These developments have exposed northern Natives to a different way of life, one that they are encouraged to join. Much of this encouragement takes the form of education and job training. So far, the participation of Native adults in the work force is low. Though it has increased over the past half-century the percentage of Native adults is far less than the percentage of the non-Native population. The alternative source of income for Native northerners is trapping. With low fur prices and high costs of outfitting a trapper, few northern Native Canadians are able to satisfy all their needs from the land. The Native economy now involves periodic wage employment and transfer payments. Many families cannot obtain sufficient income from the combination of trapping and wage employment, and have become dependent on welfare payments and public housing programs.

From the Native perspective, the cultural cost of participating in this wage economy is high – too high in the eyes of some. Substantial social changes are required, such as living in a resource town, working in a non-Native environment, and participating in a lifestyle practiced by fellow workers. Few Natives have made this change. Most still live in Native communities where there are few jobs but where the social environment is a Native one. The complex cultural and economic issues affecting Native Canadians are more fully discussed in subsequent chapters.

The Nature of Resource Hinterlands

The Canadian North is but one of many resource hinterlands. All hinterlands have a number of common characteristics.

1 World demand for primary resources and energy determines the course of hinterland development.
2 Multinational corporations, with their capital, management skills, and technical knowledge, are the leading force in resource development.

3 The global demand for raw materials and energy is cyclical, following the global business cycle. These cycles are more pronounced in resource hinterlands, leading to a "boom-and-bust" economy.
4 Resource hinterlands in other countries often compete against each other, driving the price of primary products down.

References

Gilchrist, C.W. 1988. "Roads and Highways." *The Canadian Encyclopedia*, 2nd ed. Edmonton: Hurtig.

Government of the Northwest Territories (GNWT), Bureau of Statistics. 1990. *Statistics Quarterly* 12: 4.

31 Forty Years of Northern Non-Renewable Natural Resource Development

W.W. NASSICHUK

Introduction

The Arctic regions of the globe are now, more than ever, important both economically and strategically, and their importance to international relationships will continue to grow. The Soviet Arctic, in particular, is well endowed with non-renewable resources such as nickel, cobalt, platinum, copper, gold, tin, iron, and diamonds, in addition to which oil, gas, and coal are being exploited. In Spitsbergen coal is currently being mined and petroleum exploration is under way. A variety of metals and oil and gas are abundant and being exploited in Alaska. In Arctic Canada, exploration for and development of resources have increased steadily since World War II, and it seems clear that the importance of northern mineral and energy resources will continue to increase in the future.

The purpose of this report is to provide a chronological summary of the exploration and development of non-renewable resources (minerals, oil, gas, coal) in northern Canada during the past forty years, with emphasis north of the Arctic Circle. Equally important is the development of renewable resources in northern Canada, including wildlife, fisheries, forest products, and hydroelectric power, but a review of those is beyond the scope of this study.

Heightened concerns for northern Canada's communities and renewable resources during the past twenty to thirty years have led to the

SOURCE: *Arctic* 40, no. 4 (December 1987): 274–84. Reprinted by permission of the publisher.

development of policies and procedures that have necessarily had a direct impact on non-renewable resource exploration and development. Regulatory procedures have been developed to monitor land use and to assess the environmental and social impact of resource development. Happily, northern residents have become increasingly able to influence the regulatory process by participating in public hearings and, indeed, by serving as members on various review panels or boards.

In 1974 Thomas R. Berger was appointed by the Government of Canada to conduct an inquiry into the social, economic, and environmental impact of a gas pipeline possibly followed by an oil pipeline in the Northwest Territories and Yukon. One year earlier a Pipeline Application Group under the leadership of J.G. Fyles was assembled to review the massive data presented by Arctic Gas, the consortium seeking to build the pipeline, and the Fyles report was a basis for formulation of the Berger inquiry. The National Energy Board was also reviewing the Arctic Gas proposal independently and was assisted indirectly by the Fyles report. The Berger inquiry was unique in that previously a major frontier project had never been reviewed through public participation before construction was begun.

The introduction of a new National Energy Program (NEP) for Canada in 1980 resulted in significant changes in oil and gas exploration in Arctic regions. This policy was designed principally to ensure security of supply, increase Canadian ownership of the domestic petroleum industry, and establish a pricing and revenue regime of benefit to both the nation and the industry. Key elements of the policy included establishment of a Petroleum Incentive Program, which provided about 80 per cent of the cost of exploration; requirements of 50 per cent Canadian ownership of production; government ownership of 25 per cent of any discovery; and a new system of land administration under the Canada Oil and Gas Lands Administration (COGLA). Under the land administration system, the petroleum industry entered into a number of exploration agreements. The exploration agreement confers upon the interest-owner the right to explore and the exclusive right to drill in specific areas, and, as well, it describes the conditions of tenure and the work program to be undertaken. Most of the current agreements require the relinquishment of a portion of the lands initially held. The purpose of relinquishment is to focus exploratory activity and to maintain a bank of Crown reserve lands available for disposition. Although geology is the major consideration in the Crown's selection process, any special environmental considerations or other sensitivities that may be identified in a locality are also taken into account.

Historical Background

The Inuit, the Aboriginal people of Arctic Canada, arrived from Siberia more than eight thousand years ago. They were followed by the Vikings, the first Europeans to visit Arctic Canada, by about AD 1000. The Vikings arrived by way of Iceland and Greenland and probably introduced iron implements to the region. As recently as three hundred years ago the Inuit utilized native copper in the Coppermine River area.

Martin Frobisher visited Baffin Island in 1576 and raised the first hopes for mineral wealth in the Arctic of the New World. Frobisher's "blacke stone," amphibolite that he thought contained gold, was collected from Kodlunarn Island on the south side of Baffin Island. Several loads of "ore" from Canada's first "mine" were shipped to England by Frobisher, but the rock proved to be barren of gold.

British and Danish explorers searched for a sea route to the Orient through Hudson Bay and the Arctic Islands during the following two hundred years. Many hardships, and even disasters, were suffered due to the frailty of ships and the inadequacies of equipment, but gradually maps were improved and the nature of the region that contained the "Northwest Passage" become better known. Samuel Hearne in 1771 proved that no passage lay west of Hudson Bay by travelling overland from the mouth of the Churchill River to the Coppermine River and the Arctic Coast. Hearne recovered a single two kilogram piece of native copper, a material used by the Inuit of the Coppermine area and now known to be associated with Proterozoic volcanic rocks from the northern mainland and Victoria Island. Alexander Mackenzie in 1789 travelled the river now named after him and reached the Beaufort Sea. Mackenzie observed yellow, waxy bitumen and coal during his travels.

In 1818 the search for the Northwest Passage was directed to the Arctic Islands, and by 1830 the central islands had been explored as far west as Melville Island. The mainland coast had been mapped from Bering Strait to Boothia Peninsula. In 1845 Sir John Franklin, Royal Navy, set out with two ships to complete the few miles of passage remaining to be discovered. The ships and their entire crews vanished. Naval searches, some on a large scale, were organized, but only in 1859 was evidence at last found to account for the tragic loss of Franklin's party. A great deal of the archipelago was mapped, however, by the "Franklin search" expeditions, and important fossil and mineral collections were signposts to future discovery.

Coal has been used sporadically in northern Canada for more than a hundred years. The Nares expedition to northern Ellesmere Island (1875–76) mined some Tertiary coal for general use, and in the same period whalers used locally mined coal in northwestern Baffin Island.

Coal has been known in the Yukon Territory, particularly near Carmacks, where it was used for mine power generation, and Dawson City, where it fueled river steamboats, since the early part of the twentieth century.

During World War I small quantities of mica, graphite, and garnet were mined on Baffin Island and shipped to England for use in manufacturing. In 1919–20 Canada's first significant oil discovery was made at Norman Wells on the Mackenzie River. During 1928 and 1929 the lead-zinc deposits near Pine Point and the silver-pitchblend deposit near Great Bear Lake were delineated. A short time later, in 1930, the Great Bear Lake deposit was developed into the Eldorado silver mine.

The Great Depression resulted in serious retrenchment in the mineral industry. Silver and gold were not affected as much as other metals because of their high value per unit weight. Exploration on a modest scale was supported by government and industry throughout the early 1930s, and in 1935 the Yellowknife mining district was discovered.

The advent of World War II cause a shift in the emphasis of mineral exploration. Labour and material shortages and the demand for strategic minerals resulted in a decline in gold and silver mining. Silver production ceased at the Eldorado mine in 1940, but the mine reopened to produce uranium in 1942. By late 1944 no gold was being mined in the Northwest Territories.

During the war, the search for strategic minerals resulted in the discovery of tungsten-, tantalum-, and lithium-bearing pegmatites. In 1944 and 1945 the delineation of new ore in the Giant Yellowknife Gold Mine led to a resurgence of interest in precious metals in the North. The end of the war alleviated shortages of men and materials, and gold production was re-established at the Negus, Con, and Thompson-Lundmark mines near Yellowknife between 1945 and 1947.

Exploration for Minerals in the Arctic Mainland

Areal geological mapping in the northern Canadian Shield prior to 1940 had been carried out by traditional means, usually canoe and foot traverses. In 1952 an experimental light helicopter-supported mapping project by the Geological Survey of Canada, Operation Keewatin, attained dramatic success in increasing the pace of 1:250,000 scale mapping in remote areas. Operation Keewatin, and other similar projects that followed, resulted in the virtual completion of reconnaissance mapping of the Canadian Shield.

The mapping operations in the districts of Keewatin [Kivalliq] and Mackenzie resulted in the discovery of occurrences of gold, silver, nickel, copper, lead, zinc, molybdenum, and asbestos. Pegmatites were found

Figure 31.1
Location of selected mines and mineral deposits, northern Canada.

to contain lithium, beryllium, niobium, and tantalum (Wright 1967). The North Rankin Nickel Mine, established as a nickel-copper mine on the northwest coast of Hudson Bay (Figure 31.1) during the 1950s, had by 1962, when it closed, extracted about 8.7 million kg of nickel and 2.3 million kg of copper with a total value of over $8 million.

Several mines in the Yellowknife area in the Northwest Territories, including the Giant Yellowknife Mine and the Con Mine, have been producing gold since the early 1950s. The Negus mine closed down in 1951 but some of its shafts were incorporated in the Con Mine. South of Great Slave Lake, the Pine Point lead-zinc mine was brought into production in 1964. After twenty-three years of production the Pine Point Mine was closed in 1987.

During the 1960s the main emphasis in mineral exploration shifted to base metals and mining activity in the northern mainland centred around the eastern end of Great Bear Lake. Copper, silver, bismuth, lead, and other associated metals were recovered from the Echo Bay, Northrim, and Terra mines (Figure 31.1) and also from the nearby Norex mine. The Echo Bay Mine produced copper and silver from 1964 until 1976, when the old Eldorado silver-radium workings were reopened. From 1976 to 1981, when the mine closed, silver was produced only from the Eldorado property.

Exploration for gold has continued in the northern Barren Lands to the present time. In October 1981 Selco Inc. brought the Cullaton Lake gold mine into production. In April 1982 Echo Bay Mines Limited began production near Contwoyto Lake, and the Lupin gold deposit, which was discovered by Canadian Nickel Limited in 1961 (Figure 31.1).

In the eastern District of Mackenzie, over 40 million tonnes of massive sulphides containing copper, lead, zinc, silver, and in some instances gold have been discovered in approximately thirty deposits. The deposits range in size from 11 million tonnes to less than 39,000 tonnes. None of these has yet been brought into production mainly because of their remoteness.

In the District of Keewatin [Kivalliq], uranium deposits are common in Proterozoic meta-sediments, within fracture zones, and in various extrusive rocks, gneisses, and pegmatites, but none has been developed to the mining stage.

Two large iron deposits containing 23–38 per cent iron are located on the east and west coasts of Melville Peninsula. The western deposit may contain over three billion tonnes of ore and the eastern deposit more than one billion tonnes. A similar large deposit occurs near Mary River in northern Baffin Island (Figure 31.1).

Metallic minerals have been produced in the Northern Cordillera since 1898, when the gold rush in the Dawson area initiated mining operations in the Yukon Territory. Silver, lead, and zinc have been mined continuously at Keno Hill since 1946. Production of the same metals from the Faro (Curragh) Mine at Ross River began in 1969 and continues to the present time.[1] Tungsten was mined at Cantung Mines from the early 1960s until 1986. Other interesting deposits of tungsten, stratabound cupriferous redbeds, stratiform and replacement silver-lead-zinc, iron taconite, and diamond-bearing diatremes also have been discovered. Gold and silver have been known in the vicinity of Mount Skukum in the southern Yukon since the early 1930s, and production from Total Erikson's Mount Skukum Mine began in 1986.

[1] The Keno silver mine shut down in 1989. The Faro mine was the fifth largest zinc mine in the world. It accounted for 40 per cent of Yukon's economic output but closed in late 1996 and was placed in receivership in 1999. – ed.

Petroleum Exploration in the Northern Mainland-
Beaufort Sea Region

Only eight wells were drilled in Arctic Canada between 1920, when the
Norman Wells field was discovered, and 1941. Five were exploration
wells in the vicinity of Norman Wells and three were follow-up discovery wells. In 1941 and later, during World War II, numerous development wells were drilled in the Norman Wells field and some twenty
exploratory wells were drilled in the surrounding area. Up until mid-
1945, 31,800 cubic metres of oil had been transported through the
Canol pipeline to a refinery in Whitehorse. The strategic importance of
the pipeline and refinery ended with the war, and Norman Wells production thereafter supplied only local demand along the Mackenzie
River; up until 1955 total production from Norman Wells was one million cubic metres.

In reviewing petroleum exploration in the Mackenzie Delta-Beaufort
Sea region, Young and Lyatsky (1986) suggested that the availability of
vertical air photographs and new topographic maps, improved helicopters, and newly published geological research had markedly positive effects on petroleum exploration in northern regions of Canada in
the early 1950s. Peel Plateau Exploration Limited initiated exploration
in Arctic regions in 1953, and petroleum exploration north of the Arctic Circle advanced quickly in the following years. In 1956, while four
development wells were being drilled in the Norman Wells field, enthusiasm for expanded exploration in the northern Yukon and adjacent
areas was growing.

In 1958 exploration drilling was initiated a considerable distance
from the Norman Wells area and the Peel Plateau Eagle Plains Y.T.
No. 1 well was drilled in the Eagle Plains Basin. A follow-up well, Western Minerals Chance Y.T. No. 1, drilled in 1959 discovered gas in the
basin. The discovery of hydrocarbons after only two wells were drilled
in previously unexplored sedimentary basin can be considered partly
good luck but also certainly a *coup* in terms of geological exploration.
The next well was spudded in 1962; it was completed as a discovery in
1965. The high level of industrial interested in the oil and gas in northern mainland caused the Geological Survey of Canada to mobilize a
large-scale mapping project, Operation Porcupine, to cover the northern Cordillera and much of the Mackenzie Delta in 1962.

Geophysical surveys were begun by private operators in the Arctic
mainland in 1958, and in 1959 seismic surveys were being conducted
in both the onshore and offshore areas of the Beaufort Sea-Mackenzie
Delta region. During the same period Shell Canada Ltd. and Gulf
Canada Corporation Ltd. filed on large tracts of land onshore in the

Mackenzie Delta. From the early seismic surveys a working hypothesis was developed of a seaward-thickening wedge of Mesozoic sediments up to 10 km thick under the Mackenzie Delta.

Although sixteen wells had been drilled throughout the northern mainland by the early 1960s to reach Paleozoic reservoir targets, none of these wells produced significant evidence for hydrocarbon accumulation. In 1962 the Texcan C. & E. Nicholson G-56 and N-45 wells were drilled on Nicholson Peninsula between Liverpool Bay and Wood Bay to bring petroleum exploration to the shores of the Beaufort Sea. Both of these wells reached Lower Cretaceous beds but were declared dry and abandoned. The Reindeer D-27 well, spudded in Richards Island in the Mackenzie Delta in 1965, confirmed that a thick Tertiary and Mesozoic section existed in the outer part of the Mackenzie Delta.

Exploration continued throughout the Arctic mainland between 1962 and 1968, but of the thirty-eight wells drilled none encountered significant hydrocarbons.

The 1968 discovery of the Prudhoe Bay field in northern Alaska inspired a dramatic increase in the number of seismic surveys and land applications in the Beaufort Sea-Mackenzie Delta region. Six wells were drilled in the delta in 1969 and exploration continued unabated until 1986, when sharply reduced oil prices resulted in a dramatic curtailment of exploration activity.

The first significant discovery of oil in the Mesozoic-Tertiary strata in the Beaufort Sea-Mackenzie Delta area came in 1970 after eleven wells had been drilled. Oil from the Lower Cretaceous strata flowed from the IOE Atkinson Point H-25 well (Figure 31.2). During 1970 and 1971 thirteen more wells were drilled onshore, and hydrocarbons were discovered in the Myogiak J-17 well on Tuktoyaktuk Peninsula.

In 1971 Candel and its partners drilled the East Mackay B-45 well, approximately 45 km south of Norman Wells. Although classified as a suspended well, it may now be of commercial significance because it is close to the newly built Norman Wells pipeline that transports oil to northern Alberta. In the same year Aquitaine and its partners ventured into the Foxe Basin in the Eastern Arctic to drill the Rowley N-14 well. A Paleozoic stratigraphic section of about 500 metres thick was found and the well was abandoned as a dry hole.

The giant Taglu gas field was discovered in the Richards Island area of the Mackenzie Delta in 1971. The following year, the slightly smaller Parsons field was found. The Taglu field contains 68 billion cubic metres of recoverable gas and 3.6 million cubic metres of recoverable condensate. With the completion of the Niglingtak H-30 well in the spring of 1973, forty-three wells had been drilled in the area. The Niglingtak field contains 23 billion cubic metres of recoverable gas and

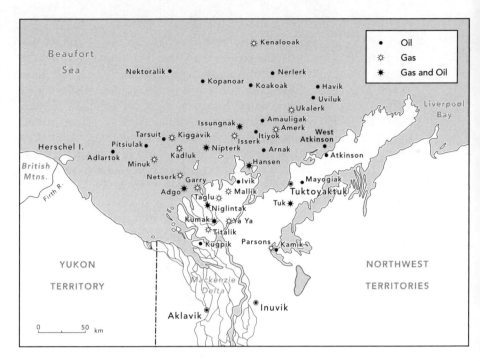

Figure 31.2
Main oil and gas discoveries in the Mackenzie Delta-Beaufort Sea area; after Young and Lyatsky (1986)

3.7 million cubic metres of recoverable oil (Procter et al. 1984). In 1973 the Kugpik well encountered oil in Cretaceous sandstones of the Parsons group and the YaYa South, Reindeer, and Titalik wells discovered gas in the Tertiary Reindeer Formation, each well establishing a new field. That year, also, exploration moved into the shallow offshore areas when artificial islands were built from materials dredged from the seabed in water two to three metres deep to serve as platforms for drilling and support services.

Between 1968 and 1974 approximately ninety-five wells were drilled in the northern mainland without success. A glimmer of hope appeared in 1974 when the Ashland Tedji Lake K-24 recovered gas from Cambrian sandstones 1136 metres beneath the Anderson Plains. Production from Cambrian beds, it must be noted, is unusual in most of North America.

In 1975 twenty-one wells were drilled in the Beaufort-Mackenzie area. Although no new fields were discovered, several of the wells delineated previously encountered oil and gas accumulations.

The introduction of reinforced drill ships to the Beaufort Sea in 1976 marked a new era of Arctic exploration. These drill ships allow wells to be drilled in water depths greater than twenty-five metres during the ice-free summer and autumn seasons. The Kopanoar D-14 and Nektoralik K-59 wells discovered new pools in the deeper, offshore area (Figure 31.2). During the same year gas was discovered in the Netserk F-40 well, which was drilled from a nearshore artificial island, and oil and gas pools were discovered in the onshore Kamik D-40 and Garry P-04 wells; the latter was drilled on a small island adjacent to Richards Island (Figure 31.2). In 1977 the deeper offshore areas were probed from reinforced drill ships and the Ukalerk C-50 well discovered an accumulation of gas.

During the period 1978–81 the Kopanoar M-13 well in the Kopanoar field tested the combined oil flow to that date in the Beaufort-Mackenzie area at 1545 $m^3 \cdot d^{-1}$. The Issungnak O-62 well and the follow-up 20-61 well discovered another giant gas field containing 71 billion cubic metres of recoverable gas and 16 million cubic metres of recoverable oil. The Tarsiut A-25 well and a follow-up well discovered the Tarsiut field, in which 2.4 billion cubic metres of recoverable gas and 24 million cubic metres of recoverable oil are estimated as reserves (Procter et al. 1984). An additional well, the Koakoak O-22, also discovered oil (Figure 31.2).

Another exploration era was initiated in the Beaufort Sea in 1982 when the Tarsiut N-44 well was drilled from an artificial, caissoned island. Such a structure allows year-round drilling in up to twenty-two metres of water. The following year a conical reinforced floating platform, the *Kulluk*, was introduced into the area. Moored by cables to numerous heavy anchors, the platform is designed to withstand pack-ice approaching from any quarter.

In 1982 the West Atkinson L-17 well discovered modest amounts of oil in the nearshore area of the Tuktoyaktuk Peninsula, and the Kenalooak J-94 well in the offshore discovered gas. The giant Nerlerk structure yielded minor amounts of oil from several separate zones.

During 1984 several important discoveries were made in the Beaufort Sea. The Amauligak J-44 well oil and gas discovery is still being delineated, but it now appears to be an oil reservoir with perhaps the largest reserves in the area. The Tuk L-09 well was thought to have discovered one of the largest wet gas reservoirs onshore from the Lower Cretaceous Kamik Formation; in a follow-up well oil was also recovered from the Tertiary clastic strata.

Several important discoveries were made in the Beaufort Sea during 1985 and 1986. These included gas and oil in the Nipterk L-19 well, oil in the Adlartok P-09 well, and oil and gas in both the Havik B-41 and Arnak K-06 wells (Figure 31.2).

OIL AND GAS RESERVES

Approximately 375 wells have been drilled in the Arctic portion of the mainland since 1947. The geological setting and resource potential of this area have recently been summarized by Nassichuk (1983) and by Procter et al. (1984). Reserves in the earlier discovered Norman Wells field have been increased to approximately 95 million cubic metres, and oil is currently being shipped through a 25 cm pipeline to northern Alberta at rates up to 4,700 $m^3 \cdot d^{-1}$.

More than two hundred wells have been drilled in the Beaufort Sea-Mackenzie Delta region. Of these, thirty-six were drilled from artificial islands, thirty-four from drill ships or floating platforms, and the remainder lie in onshore areas. More than forty hydrocarbon accumulations have been discovered, and according to Procter et al. (1984), these accumulations contain more than 286 billion cubic metres of recoverable gas and more than 117 million cubic metres of recoverable oil and condensate. The 1984 Amauligak discovery may increase the recoverable oil reserves by about 112 million cubic metres. The total discovered and undiscovered resources in the Mackenzie Delta-Beaufort Sea area may amount to 2,131 billion cubic metres of gas and 1.35 billion cubic metres of oil (Procter et al. 1984).

The first shipment of Beaufort Sea oil reached the market in 1986, when approximately 50,400 cubic metres of oil from the Amauligak field was delivered by tanker to Japan.[2]

COAL RESOURCES

Coal is widely distributed through the Yukon and District of Mackenzie and ranges in rank from lignite to anthracite. Low rank coal was used to generate steam in river boats at Dawson near the turn of the century. A short time later, in 1923, the Tantalus Butte mine near Carmacks opened and for the next fifty-five years supplied bituminous coal to a variety of base metal mines, the most recent being the lead-zinc mine at Faro. Minor amounts of coal were also delivered to the Faro Mine from deposits near Ross River.

Recent exploration has delineated large reserves in four principal areas, most of which could be used for power generation in the future. The reserves include; about 200 million tonnes of Cretaceous and Tertiary lignite to bituminous coal in the Bonnet Plume Basin; at least 300 million tonnes of subbituminous and bituminous coal of Upper Cretaceous and Carboniferous age in the Fort Liard area; 50 million tonnes of Tertiary subbituminous and lignite coal at Rock River, southeast of

---

[2] In 2000 between 11 and 12 million barrels of oil were produced from the Norman Wells field. Natural gas is produced in the Fort Liard field and flows into the west coast pipeline system in British Columbia. The Ikhil field provides natural gas for the town of Inuvik. – ed.

Yukon; and more than one billion tonnes of Tertiary lignite and subbituminous coal in the Brackett Basin, near Fort Norman [Tulita], and near the Mackenzie River, a potential transportation route.

Exploration in the Arctic Islands

1947-50: ESTABLISHMENT OF WEATHER STATIONS

A most important step in the development and opening of the High Arctic islands was the establishment between 1947 and 1950 of high-latitude weather stations by the United States Air Force and Navy under an international agreement between Canada and the United States. The stations were known as the Joint Arctic Weather Stations. Resolute was established by sea from Boston in 1947 and Eureka the same year by airlift from Thule Air Base in Greenland. Isachsen and Mould Bay were established by airlift from Resolute in the spring of 1948. Alert was established by airlift from Thule in 1950. The stations were operated jointly by the Canadian and US weather services between 1947 and 1972, when they were taken over by Canada. The stations provided a radio communication network, weather information, and all-season airstrips, all of which provided improved access for geological work in this remote part of Canada.[3]

The US Air Force, following shortly on the close of World War II, played a significant role in the advance of our knowledge of the Arctic Islands; during 1946 and 1947 the US Air Force flew seventeen long-range photographic flights from Fairbanks, Alaska, and from Edmonton to cover virtually the entire archipelago. The aircraft carried Canadian observers. During the flights, which covered distances of 4,000–7,500 km, some of the last major geographic discoveries of the Arctic Islands were made. Borden and Cornwallis islands were found to be, in each case, two islands, and Stefansson Island was discovered. The US aeronautical charts that were made following the exercise were the first to succeed the British Admiralty charts of the nineteenth century. Strangely, these historic flights have never been documented in an official Canadian publication.

TRIMETROGEN AERIAL PHOTOGRAPHY: TOPOGRAPHIC MAPS

An important function of the Royal Canadian Air Force (RCAF) since the establishment of this service in 1924 was survey photography. By

---

3 For an overview of the Joint Arctic Weather Stations, see Wonders, William C., "The Joint Arctic Weather Station (JAWS) in the Queen Elizabeth Islands," in Hage, K.D. and Reinelt, E.R. (eds), *Essays on Meteorology and Climatology: In Honour of Richmond W. Longley* ("Monograph 3, Studies in Geography"). Edmonton: Department of Geography, University of Alberta, 1978. 399–418. – ed

1931 this had reached the Arctic mainland. By 1948, using converted Lancaster bombers, the RCAF reached the archipelago, and between 1948 and 1953 trimetrogen photography of the islands was completed. The photographs were used in the production of air navigation charts at a scale of 1:506,880. In 1948 the RCAF was able to add two new large islands to the map of Canada: Prince Charles Island and Air Force Island. This was the last major discovery of new land in Canada – and perhaps in the world. Between 1948 and 1957, the RCAF, using DC-3 aircraft and Lancaster bombers, completed the shoran net of the whole of the archipelago, starting from a fixed point south of Winnipeg.

In 1947 the Canadian government entered into contracts with civilian aircraft companies to obtain vertical aerial photographs of the whole of Canada. Photographic coverage of much of the archipelago was completed during the late 1950s (the entire program took nineteen years). The Surveys and Mapping Branch (Department of Energy, Mines, and Resources) then used the photography and geodetic and other positional surveys to produce topographic maps at 1:250,000 scale.

ADVANCES IN TRANSPORTATION

Early exploration of the Arctic Islands and northern mainland of Canada was carried out by sailing ship and by man-hauled sledge. Dog teams, canoes, and other means of travel were used later, and eventually, as now, field exploration became dependent upon aircraft (see Christie and Kerr, 1981).

The aircraft era of geological exploration of the islands began in 1947, when Y.O. Fortier carried out a geological reconnaissance using an RCAF flying boat. Among the early activities by civilian aircraft in the archipelago was the use of helicopters for close support of field parties of the Geological Survey of Canada. In the summer of 1955, Operation Franklin, which initiated a modern era of scientific study in the High Arctic, was supported by two Sikorsky S-55 helicopters. Eleven geologists and ten assistants of this project successfully studied about 260,000 km^2 of previously little-known terrain in the Queen Elizabeth Islands.

In 1958 the Survey assisted in pioneering the use of light aircraft on oversized, low-pressure tires for landings on unprepared terrain. R. Thorsteinsson and E.T. Tozer, cooperating with W.W. Phipps, the well-known pilot and developer of "big wheels" for aircraft, mapped Melville, Brock, Borden, Mackenzie King, and Prince Patrick islands in one season using a Piper Super Cub (PA-18). The revolutionary wheels with oversized tires were inspired by earlier experiments by the US military on the tundra of northern Alaska. The De Havilland Beaver,

Otter, and Twin Otter eventually were fitted with oversized tires and the special wheels and brakes needed and are still operated in this mode.

The Polar Continental Shelf Project (PCSP) was created by the Government of Canada in 1958 to conduct scientific research over Canada's polar continental shelf and contiguous ocean areas. For thirty years the PCSP has provided logistical support, communications, navigational aids, and air, ground, and sea transport to innumerable government, university, and industrial research activities and has been an important force in asserting Canada's sovereignty in the High Arctic.

The problems of transporting marketable products (e.g., coal, oil, natural gas, metallic minerals) from Arctic regions to more populated areas have been subjects of considerable research for a decade at least. The development of improved, strengthened, and ice-breaking steel ships has enabled seasonal maritime transport to be extended to many parts of the Canadian Archipelago (Franklin 1983; Haglund 1983). Designs for pipelines, liquefied natural gas (LNG) carriers, and other specialized surface ships, and even submarine cargo carriers, are undergoing research and development (Bailey 1983; Jacobsen and Murphy 1983; Wilckens and Freitas 1983; Kaustinen 1983). Clearly, any implementation of these modes of transporting Arctic commodities must await improved economic conditions.

MINERAL EXPLORATION

Mineral exploration of the High Arctic has been extremely limited compared with that on mainland Canada; a summary of that exploration up until 1980 was prepared by Findlay et al. (1981). In contrast to oil and gas exploration in the Arctic Islands, where expenditures since 1957 approximate $1.5 billion, total expenditures for mineral exploration and mine developments in the same period were less than $500 million. The predominance of Phanerozoic sedimentary strata over igneous rocks of all ages in the archipelago naturally favours oil and gas exploration, and the potential value of oil and gas discovered thus far exceeds the value of discovered minerals by a considerable margin. An extremely important result of mineral exploration in the High Arctic has been the development of new communities in support of successful mines. Inuit families work and live at Polaris and Nanisivik, lead-zinc mines on Little Cornwallis and Baffin islands.

NANISIVIK LEAD-ZINC MINE, BAFFIN ISLAND

Sulphide-bearing carbonate rocks containing lead and zinc near Strathcona Sound on the northwest corner of Baffin Island were first noted by members of Captain Bernier's 1910–11 expedition. Reports of mineralization by Blackader (1956) aroused the curiosity of Texas

Gulf Sulfur Company Ltd. geologists. A visit to the area in 1957 confirmed the worth of the occurrences, and claims were staked to cover surface showings about 80 km northeast of the Inuit settlement of Arctic Bay.

Texas Gulf Sulfur conveyed ownership of these claims to a group headed by Frank Agar, and Nanisivik Mines was created to explore the claims. Guided by Graham Farquason and Strathcona Mineral Services Limited, a group was put together under the control of Mineral Resources International, the Government of Canada, Metallgesellschatt Canada, and Biliton B.V., and the mine was brought into production by 1976 (Figure 31.1).

Olsen (1984) postulated that ore-bearing fluids derived from the Arctic Bay Formation reacted with hydrocarbon-bearing fluids and that the sulphides were deposited in already existing cavities. The ore body has a sinuous, flat, lenticular shape more than 3,048 m long with a narrow keel-like ore shoot extending vertically below the main body. The pyrite-rich, galena-sphalerite deposit at Nanisivik is enriched in silver (up to 60 $g \cdot t^{-1}$), and the silver "sweetener" has made the Nanisivik Mine viable. With European smelters committed to long-term contracts, the Nanisivik mine has been one of Canada's highly successful base metal producers. From October 1976 to March 1986, Nanisivik treated 5.5 million tonnes of ore with grades of 9.0–11.0 per cent zinc, 0.80–1.7 per cent lead, and 42–60 $g \cdot t^{-1}$ silver. Total metal production was 600,000 tonnes of zinc, 70,000 tonnes of lead, and 250 tonnes of silver. Annual production is currently 700,000 tonnes of ore at a grade of 9 per cent zinc, 0.8 per cent lead, and 42 $g \cdot t^{-1}$ of silver.

THE POLARIS LEAD-ZINC MINE, LITTLE CORNWALLIS ISLAND

The Polaris lead-zinc mine, located at latitude 75°30'N and longitude 96°20'W, is the most northerly mining operation in the world (Figure 31.1). The discovery of the deposit had a strong element of luck tied to exploration that was not directly related to mineral prospecting. In the late 1950s, when enthusiasm was growing for the energy potential of the High Arctic, the North American oil industry looked north in search for untapped hydrocarbon wealth. In 1959 Bankeno Mines Limited of Toronto, in conjunction with Talent Oil and Gas Limited of Calgary, acquired more than 1.2 million ha of oil and natural gas priority applications in the Arctic and placed mapping crews in the field to conduct geological studies. These applications were granted in anticipation of the government issuing exploration permits once proposed regulations governing prospecting in the Arctic were released.

In 1960 Lionel Singleton and George Wilson of J.C. Sproule and Associates discovered mineralized gossans on the west coast of Little Cornwallis Island. The widespread nature of the sulphide alteration in the area and the presence of galena and sphalerite caused a diversion from oil and gas exploration to base metals exploration.

Under extremely arduous conditions, Bankeno Mines drilled nine holes totalling 206 metres in the fall of 1960. The results indicated the presence of widespread lead-zinc mineralization in dolomitic rocks of Paleozoic age. Bankeno Mines turned to Canada's principal lead-zinc producer, Consolidated Mining and Smelting Limited (Cominco), for assistance in evaluatingand possibly developing the deposit. In spite of early misgivings, Cominco came to an accord with Bankeno Mines and agreed to develop the Little Cornwallis deposit. Cominco undertook a gravity survey over the surface showing and discovered one of the biggest gravity anomalies recorded in the history of Canadian mineral exploration. A several milligal anomaly was delineated, which, upon drilling, turned out to be a massive body of high-grade galena-sphalerite ore.

During the 1970s the main ore body was further delineated by drilling, and thirteen other smaller deposits and showings were identified in the immediate area. Unique to the mining industry, the mill and concentrator were constructed as a unit at Trois Rivières, Quebec, and in 1981 were floated on a barge about 5,600 km by way of Davis Strait and Lancaster Sound to the mine site. Production commenced in 1982, and within months the mine achieved its rated milling capacity of 2,300 t·d^{-1} at grades of 17 per cent zinc and 6 per cent lead.

Since it opened in 1982, the Polaris mine has produced an average of 200,000 tonnes of zinc concentrates and 45,000 tonnes of lead concentrates annually. The entire output of the mine is transported directly to smelters in Europe during a shipping season that extends from August to late October. The Polaris Mine contains over 4 million tonnes of lead-zinc in a single ore-body (nearly twice the amount delineated in the whole Pine Point district).[4]

DIAMONDS AND OTHER MINERALS IN THE ARCTIC ISLANDS

Between 1973 and 1975 Diapros Limited, and later Cominco, investigated some diatremes reported by Blackader and Christie (1963). Using geological and geochemical techniques special to diamond exploration, Diapros successfully located diamond-bearing kimberlite pipes on Somerset Island. Inspired by the Diapros activity, Cominco also acquired prospecting permits on Somerset Island and conducted a rather extensive exploration program for diamonds.

The program resulted in the discovery of a few small diamonds on Somerset Island; however, the character of the host rock and the limit-

---

4 Both the Polaris mine and the Nanisivik mine are scheduled to close in the fall of 2002, the latter because of ore exhaustion, the former because of low ore demand. – ed.

ed diamond content did not indicate economic viability. Diapros and Cominco terminated diamond exploration in the High Arctic in the late 1970s.[5]

One of the most comprehensive mineral exploration programs undertaken in the Arctic Islands was conducted by Petro-Canada from 1981 to 1985. The program involved prospecting and geological mapping of about 200,000 km² throughout the archipelago from Victoria Island to northern Ellesmere Island. The company also completed a systematic collection of heavy-mineral samples from stream sediments to a density of about one sample per 50 km². An inventory of all coal deposits was also completed.

The program resulted in the discovery of several geochemical anomalies indicating elevated levels of lead, zinc, and cobalt, none of which proved to be economically significant. At several localities, especially on Ellesmere Island, diamond-indicator minerals, including pyrope, ilmenite, chromite, and diopside, were discovered, but no kimberlites, the probable source of the minerals, were found.

From 1984 to 1986 Panarctic Oils Limited conducted a modest search for copper-silver deposits in the Victoria Island region, where the association of native copper with the Proterozoic basalts had been known for many years. The sedimentary rocks adjoining the basalts were also studied, and numerous copper-silver occurrences were found. None, however, was of apparent economic significance.

Petroleum Exploration in the Arctic Islands

The first study dealing with the oil and gas possibilities of the Canadian Arctic Archipelago was completed by Fortier et al. (1954). This was a comprehensive account of all the exploratory and reconnaissance geological information available at that time. The study included a review of the rock and fossil collections and the observations made by the early explorers, and the results, mainly by the Geological Survey of Canada, of geological investigations carried out during the few years following the opening of the Joint Arctic Weather Stations.

The season of 1958, in which Thorsteinsson and Tozer (1959a) mapped the western Queen Elizabeth Islands by Super-Cub aircraft, has already been noted. A map and descriptive notes (Thorsteinsson and Tozer 1959b) were published the year following this extraordinary mapping project, and these undoubtedly contributed to the decision by Dome Petroleum and partners to drill the first well of the Arctic

---

5 See chapter 34 for the dramatic diamond developments north of Yellowknife. – ed.

Islands: Dome et al. Winter Harbour No. 1 on southern Melville Island. The well was spudded on September 1961 in Devonian rocks of an anticlinal structure in the Parry Islands Fold Belt. Gas was recovered from below the permafrost, but in non-commercial quantities. Two additional wells, both dry, were drilled in the fold belt in 1963 and 1964.

In 1962 Triassic tar sands were discovered on northwestern Melville Island. The discovery was made independently by Alan Spector, of the Earth Physics Branch (Department of Mines and Technical Surveys, now Energy, Mines, and Resources), and by a party of J.C. Sproule and Associates Ltd. The deposits were evaluated by Trettin and Hills (1966), who noted that approximately 16 million cubic metres of oil may be contained therein.

Douglas et al. (1963) elucidated the regional geology of northern Canada with special emphasis on the petroleum prospects of the region. The stratigraphic and tectonic features of the archipelago were shown to be similar to those of proven petroleum provinces in other parts of the world.

Results of Operation Franklin in 1955 were published by Fortier et al. (1963) in a compilation that became a standard reference for oil and gas exploration in the Arctic Islands. The interpretation of Tozer and Thorsteinsson (1964) of the geology of the western Queen Elizabeth Islands was even more important for hydrocarbon exploration. These authors provided a perceptive analysis of the petroleum prospects on Melville Island and correctly identified Sabine Peninsula on Melville Island as having the best prospects for petroleum discovery in the region. Many of the targets suggested by the authors were subsequently drilled and some resulted in gas discoveries.

Commercial companies began geological exploration in the Arctic Islands in 1959. In that year, teams of geologists of Dominion Explorers Ltd. and Round Valley Ltd., under the direction of consultants A.H. McNair and W.W. Gallup respectively, were active.

Field parties of J.C. Sproule and Associates commenced working in the archipelago in 1960 and were active during the 1960s and early 1970s. Sproule's greatest contribution to oil and gas development in the islands was his role in the formation of Panarctic Oils Ltd. Discussion on development of the company proceeded from 1964 until December 1967, when nineteen companies, several individual investors, and the Government of Canada formed the Panarctic consortium and pledged millions of dollars to exploration in the Arctic Islands.

In 1969 exploration moved from the Parry Islands Fold Belt to the southern edge of Sverdrup Basin, a rift basin that contains at least 15 km of sedimentary strata ranging in age from Carboniferous to Ter-

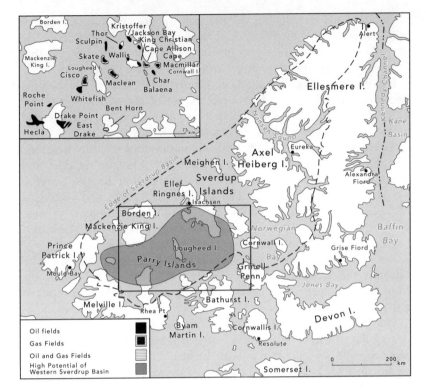

Figure 31.3
Main oil and gas discoveries in the Canadian Arctic Archipelago (courtesy N.J.
McMillan, 1897).

tiary (Figure 31.3). Most of the oil and gas discovered in the Arctic
Islands lies in the Sverdrup Basin. One of the three wells drilled in
1969, the Drake N-67, is the discovery well for the Drake Gas field, the
largest gas field in the Arctic Islands (Figure 31.3).

Between 1970 and 1973, forty-seven wildcat and delineation wells
were drilled in the islands. A gas field was found in 1970 on King
Christian Island, near the centre of the Sverdrup Basin. Twenty-one
wells were drilled in 1972, from Banks Island to Brock Island in the
west and from Russell Island to northern Ellesmere Island in the east.
Four discoveries resulted: oil and gas in Romulus C-42 on Fosheim
Peninsula; gas in Kristoffer Bay B-06 and Hecla F-62; and oil in Thor
P-38. Similar results were attained in 1973 when two gas discoveries,
Wallis K-62 and Thor H-28, resulted from twenty wildcat wells.

An innovative engineering scheme was unveiled in the Arctic Islands
in 1974 when the Hecla N-52 well was drilled from artificially thick-
ened ice more than 130 metres above the sea floor. After two months

of flooding and freezing, an ice platform was formed that was 122 m in diameter and 4.5 metres thick in the centre. A well drilled from this ice "island," a delineation well for the Hecla field, recovered gas. Two other wells were drilled as delineation wells for the Drake field.

Important progress was made on Cameron Island in 1974 when the Bent Horn oil field was discovered. A test of Devonian reefoid carbonate rocks flowed oil and water to the surface from an interval below 3,200 metres in the Bent Horn N-72 well.

An extremely active period of oil exploration took place in the Arctic Islands between 1975 and 1985, when seventy-seven wildcat and delineation wells were drilled. Gas was discovered in the East Drake I-55, Jackson G-16A, Roche Point J-43, Whitefish H-63, and Sculpin K-08 wells, and non-commercial oil was found in the Balaena D-58 well. Both oil and gas were found in the Char G-07, Skate B-80, Maclean I-72, Cisco B-66, MacMillan 2K-15, and Allison C-47 wells.

Instability in the world petroleum market began to influence exploration in the Arctic Islands in 1984. By 1986, drilling had ceased due to plummeting petroleum prices. The main base of field operations of Panarctic Oils Ltd. at Rea Point, Melville Island, was shut down.

One hundred and seventy-six wells were drilled in the Arctic Islands during the course of twenty-five years of petroleum exploration from 1961 to 1986. Jones (1981) provided a succinct review of the economic significance of discoveries up to 1980. Panarctic Oils, a participant in at least 80 per cent of the wells, showed extraordinary success and uncovered important oil and gas resources for future exploitation. Moreover, important new data on subsurface geology were incorporated into the scientific literature to assist in future exploration for oil and gas. Waylett (1979) and Rayer (1981), both of Panarctic Oils Ltd., synthesized the numerous subsurface and surface data to provide up-to-date summaries of the oil and gas potential of the area. Complementing these reports are new summaries of Arctic Islands geology and potential for oil and gas resources by Nassichuk (1983) and Procter et al. (1984).

OIL AND GAS RESOURCES

Petroleum exploration in the Arctic Islands has resulted in the discovery of eighteen hydrocarbon accumulations accounting for at least 361 billion cubic metres of marketable proved and probable gas reserves and at least 76 million cubic metres of recoverable oil reserves. Estimates by Procter et al (1984) place the total discovered and undiscovered oil resources in the Arctic Islands at 686 million cubic metres, and the total discovered and undiscovered gas resources are estimated to be 2,257 billion cubic metres.

Recoverable oil reserves of approximately 49 million cubic metres in the Cisco field make it the largest in the Arctic Islands. Bent Horn, Balaena, Skate, Maclean, Cape Allison, and Cape MacMillan add approximately 28 million cubic metres of recoverable oil to the reserves.

The Drake field is the largest gas accumulation, with at least 99 billion cubic metres of proved and probable reserves. The Hecla field has 86-100 billion cubic metres of proved and probable gas reserves and Whitefish contains at least 57 billion cubic metres of proved and probable gas reserves (Procter et al. 1984). Gas also occurs in the Cisco, Jackson Bay, King Christian, Kristoffer, Maclean, Thor, Cape Mac-Millan, Char, Roche Point, Romulus, Sculpin, Skate, and Wallis fields (Figure 31.3).

In 1985 the ice-breaking tanker MV *Arctic* transported a single shipment of approximately 16,000 cubic metres of oil from the Bent Horn field on Cameron Island to Montreal. A year later another shipment of the same amount of Bent Horn oil was delivered to Montreal. About 800 cubic metres of oil in the latter shipment was off-loaded at Resolute to allow the Northern Canada Power Commission to test the suitability of Bent Horn crude oil as a replacement for diesel fuel.

COAL RESOURCES

Lignitic to bituminous coals occur in the Canadian Arctic in strata ranging in age from Tertiary to Devonian. Coal has been exploited for short periods by small communities, for example by Pond Inlet up until 1959, but none has been mined on a large scale. Given the abundance of coal elsewhere in Canada, it is extremely unlikely that the coal in the High Arctic will be required for other than local use. Nevertheless, three major companies, Petro-Canada, Gulf Canada Resources, and Utah Mines, were granted exploration licences in 1981. Petro-Canada conducted an extensive evaluation of the thickest and most widely distributed coals in the archipelago, those in the Tertiary Eureka Sound Formation, and concluded that the deposits were not of immediate economic significance.

Conclusions

During the past forty years exploration for and exploitation of mineral and hydrocarbon resources in northern Canada have increased steadily. During that interval, about forty mines have produced gold, silver, lead, zinc, tungsten, uranium, asbestos, nickel, copper, and coal in the Yukon and Northwest Territories. At the present time eight mines are producing gold, silver, zinc, and lead in the area. Included are two lead-zinc-silver producers in the Arctic Island, Polaris, the world's most northerly mine on Little Cornwallis Island, and the Nanisivik mine on Baffin Island.

With the exception of exploration and discovery in the Norman Wells area in the early 1920s, the search for oil and gas in northern Canada was rather modest up until the late 1950s, when exploration activity began to accelerate in the Mackenzie Delta. A dramatic increase in Arctic exploration was inspired by discovery of the Prudhoe Bay oil field in northern Alaska in 1968, and since then nearly four hundred wells have been drilled and abundant oil and gas resources have been identified in the Mackenzie Delta-Beaufort Sea areas and in the Arctic Islands.[6]

Even though significant conventional oil and gas resources remain to be discovered in western Canada, where most of Canada's reserves occur, it is clear that those resources are finite and must, in the fore-seeable future, be augmented by supplies from Canada's frontier regions. Accordingly, the substantial oil and gas discoveries that have been made in northern Canada during the past forty years, and especially during the past twenty years, have shown that northern Canada will continue to be important as a frontier for oil and gas exploration.

References

Bailey, R.A. 1983. The arctic pilot project. Cold Regions Science and Technology 7: 259–72.

Blackadar, R.G. 1956. Geological reconnaissance of Admiralty Inlet, Baffin Island, Arctic Archipelago, Northwest Territories. Geological Survey of Canada, Paper 55–6.

– and Christie, R.L. 1963. Geological reconnaissance, Boothia Peninsula, and Somerset, King William, and Prince of Wales Islands, District of Franklin. Geological Survey of Canada, Paper 63–19. Report and maps 36–1963 and 37–1963.

Christie, R.L., and Kerr, J.W. 1981. Geological exploration of the Canadian Arctic Islands. In Zaslow, Morris, ed. A century of Canada's Arctic Islands, 1880-1980. Royal Society of Canada 23rd Symposium. 187–202.

Douglas, R.J.W., Norris, D.K., Thorsteinsson, R., and Tozer, E.T. 1963. Geology and petroleum potentialities of northern Canada. Sixth World Petroleum Congress, Proceedings, Frankfurt-Main, Sec. 1, Paper 7: 1–52.

Findlay, D.C., Thorpe, R.I., and Sangster, D.F. 1981. Assessment of the non-hydrocarbon mineral potential of the Arctic Islands. In Zaslow, Morris, ed. A century of Canada's Arctic Islands, 1880–1980. Royal Society of Canada 23rd Symposium, 203–20.

---

[6] The surging demand for oil and natural gas in the "Lower 48 states" of the US in 2000 has purred renewed interest in drilling in the NWT and in new pipeline construction prospects. – ed.

Fortier, Y.O., McNair, A.H., and Thorsteinsson, R. 1954. Geology and petroleum possibilities in Canadian Arctic Islands. American Association of Petroleum Geologists Bulletin 38: 2075–109.

Fortier, Y.O., Blackadar, R.G., Glenister, B.F., Greiner, R.H., McLaren, D.J., McMillan, N.J., Norris, A.W., Roots, E.F., Souther, J.G., Thorsteinsson, R., and Tozer, E.T. 1963. Geology of the north-central part of the Arctic Archipelago Northwest Territories, Operation Franklin. Geological Survey of Canada, Memoir 320.

Franklin, L.J. 1983. Arctic transportation problems and solutions. Cold Regions Science and Technology 7: 227–30.

Haglund, D.K. 1983. Maritime transport in support for arctic resource development. Cold Regions Science and Technology 7: 231–49.

Jacobsen, L.R., and Murphy, J.J. 1983. Submarine transportation of hydrocarbons from the Arctic. Cold Regions Science and Technology 7: 273–83.

Jones, G.H. 1981. Economic development – oil and gas. In Zaslow, Morris, ed. A century of Canada's Arctic Islands, 1880–1980. Royal Society of Canada 23rd Symposium. 221–30.

Kaustinen, O.M. 1983. A polar gas pipeline for the Canadian Arctic. Cold Regions Science and Technology 7: 217–26.

Nassichuk, W.W. 1983. Petroleum potential in Arctic North America and Greenland. Cold Regions Science and Technology 7: 51–88.

Olson, R.A. 1984. Genesis of Paleokarst and strata-bound zinc-lead sulfide deposits in a Proterozoic dolostone, Northern Baffin Island, Canada. Economic Geology 79: 1056–103.

Procter, R.M., Taylor, G.C., and Wade, J.A. 1984. Oil and natural gas resources of Canada. Geological Survey of Canada, Economic Geology Series, no. 1.

Rayer, F.G. 1981. Exploration prospects and future petroleum potential of the Canadian Arctic Islands. Journal of Petroleum Geology 3(4): 367–412.

Thorsteinsson, R., and Tozer, E.T. 1959a. Geological investigations in the Parry Islands, 1958. Polar Record 9(62): 459–61.

– 1959b, Western Queen Elizabeth Islands, District of Franklin, Northwest Territories. Geological Survey of Canada, Paper 59–1.

Tozer, E.T., and Thorsteinsson, R. 1964. Western Queen Elizabeth Islands, Arctic Archipelago. Geological Survey of Canada, Memoir 332.

Trettin, H.P., and Hills, L.V. 1966. Lower Triassic Tar Sands of Northwestern Melville Island, Arctic Archipelago. Geological Survey of Canada, Memoir 332.

Waylett, D.C. 1979. Natural gas in the Arctic Islands: discovered reserves and future potential. Journal of Petroleum Geology 1(3): 21–34.

Wilckens, H., and Freitas, A. 1983. Thyssen-Waas icebreaker concept, model tests, and full scale trials. Cold Regions Science and Technology 7: 285–91.

Wright, G.M. 1967. Geology of the southeastern Barren Grounds, parts of the Districts of Mackenzie and Keewatin (Operation Keewatin, Baker, Thelon). Geological Survey of Canada, Memoir 350.

Young, F.G., and Lyatsky, V.B. 1986. Beaufort-Mackenzie oil exploration and evolution of geological concepts. Oil and Gas Journal, August 25: 65–72.

32 The Montferré Mining Region Labrador-Ungava[1]

GRAHAM HUMPHRYS

The Montferré region of Labrador-Ungava covers some two thousand square miles [5,200 km²] of the southern end of the Labrador Trough, overlapping the Quebec-Labrador boundary roughly two hundred miles [320 km] north of Sept Îles. It is therefore, only a relatively small part of the interior of this vast northeast peninsula of North America. Yet it is of particular interest to the geographer, not only because production plans for the region anticipate capture of up to 15 per cent of the North American iron-ore market, but because it exhibits features directly attributable to its remote location beyond the limits of continuous settlement. Reserves of ore are measured in thousands of millions of tons, assuring the basis for a long mining life; since the developments are very recent, however, and production is not scheduled to begin until 1961, it provides an opportunity of examining the processes at work in the development of pioneer mining areas.

The region, although considered remote today, was well known to the *coureur de bois* of the seventeenth century, who, working mainly from the town of Tadoussac at the mouth of the Saguenay, traversed the whole of the southern watershed, trading and trapping furs. The

[1] The writer wishes to express his gratitude and indebtedness to the following for the cooperation and assistance in providing help and information: DrJ.A. Retty, Consulting Geologist, Montreal; Officers of the Canadian Javelin Company; Iron Ore Company of Canada Limited; Labrador Mining and Exploration Company; Normanville Company Limited; Wabush Iron Company; Quebec Geological Survey, Montreal.

SOURCE: *The Scottish Geographical Magazine* 66, no. 1 (April 1960): 38–45. Reprinted by permission of the author and publisher.

first scientific report on the local minerals was made by the federal geologist A.P. Low, who carried out a number of reconnaissance surveys across the peninsula in the 1890s. During the following forty years, several expeditions sent to report on Low's findings discovered considerable deposits of iron formation containing 35 per cent iron, but the low grade of the ore together with the remote location of the area dismissed hopes of development.

The modern mining period really began in 1936, when the Labrador Mining and Exploration Company began prospecting on a 20,000-square-mile [52,000-sq-km] concession in the centre of the peninsula, granted to them by the Newfoundland government. Their property extended from the Quebec boundary near Knob Lake, as far south as the eastern end of the Montferré region near Wabush Lake. The company actually started operations in the latter area, but with the decision in 1945 to develop the high-grade iron ore near Knob Lake, the Montferré region was once more neglected. Little more of interest occurred until 1952 when much of the activity which has continued through to the present time began.

Between 1936 and 1952 changes significant for the Montferré region had been taking place in the North American iron-ore industry. Two of these were, and are, of particular importance, the improvements in the technology of beneficiation,[2] and the depletion of the easily accessible, high-grade sources of iron ore for the North American steel industry. They are indissolubly linked; with the decrease in high-grade ore reserves, vast sums of money have been spent on perfecting cheap, efficient means of beneficiation. As mining of the Great Lakes ore became more difficult it rose in price, closing the gap between the cost of deep open-cast, unprocessed ore and shallow open-cast, beneficiated ore. Today the gap still exists, but it is more important for the companies without high-grade reserves than for others with them. The former are the ones seeking the easily mined low-grade ores to fulfill their future needs. The improvement of the techniques used to raise the grade of the ore is having repercussions hitherto unconsidered. The iron and steel industry is highly capitalized, needing the highest efficiency of operation to put the investment to good use. When beneficiation is used, the material that reaches the furnace is already a partially manufactured product, with very important advantages over natural ore. Not only is the iron content much higher (generally 65 per cent as opposed to 50 per cent in North America), but its constitution can be kept within narrow limits, and its physical structure is more suitable for furnace practice. With such a uniform high quality of raw mater-

---

[2] Beneficiation is the upgrading of ore by separating the mineral-bearing material form as much as possible of the surrounding rock mined with it, the latter generally termed the gangue.

ial, smelting can be much more easily controlled, resulting in a superior end-product. Unprocessed iron-formation is handicapped by its inferior and variable quality, and increasing costs with deeper mining. Inheriting all the advantages of mining technique and benefiting from the lack of previous exploitation, low-grade ores are finding the cost of processing a decreasing handicap. This has been reflected by the increase in the amount of beneficiated material shipped annually from the United States mines, rising from 23 per cent of the total in 1947 to over 40 per cent of the total in 1957. In the latter year, just under three tons of raw materials were used in the average United States blast furnace for each ton of iron deposits, and a resulting scramble by the iron and steel industry to secure adequate reserves in North America. The Montferré region is one of the results of this scramble.

The iron formation which has attracted so much attention to the region occurs as bands within the underlying highly contorted and metamorphosed assemblage of Upper Precambrian sediments. The iron ore, found within the formation as specular hematite with quartz or as magnetite-hematite facies, contains an average of 33 per cent iron, and is considered exploitable only where large quantities are accessible to surface mining operations. Lying as it does along the watershed of the Ungava Bay, St Lawrence, and Atlantic drainage systems, the region has a highly varied relief averaging 2,500 feet [760 m] above sea level. Near Wabush Lake, high rugged hills occur, grading off westward to an uneven rocky plateau near Mount Wright. Extensive areas of glacial drift are masked by muskeg, while in the extreme southwest deep dissection has resulted in a high relative relief of up to 1,500 feet [460 m].

The crest of the Laurentide scarp lies immediately to the south at a height of just over 3,000 feet [900 m] and has been deeply dissected by the youthful streams which cascade down glacially scoured valleys, so that the whole zone between there and the Gulf of St Lawrence offers a major obstacle to the penetration of the interior. It is in this section that construction of transportation facilities is most difficult and expensive. Prior to 1954 the only means of reaching the region were by air from Sept Îles, or by canoe up one of the many rivers. The use of planes, float-equipped in summer, sea-equipped in winter, is greatly facilitated by the many lakes, but canoeing entails so many long portages that it has been little used since 1933. The whole region was uninhabited before mining began, being crossed only occasionally by the semi-nomadic Montagnais Indians on their hunting trips from the North Shore. In 1954 the completion of the Quebec North Shore and Labrador Railway, which bounds the region in the northeast, stimulated activity around Wabush Lake by providing cheaper access. The distance involved, however, gave this route little cost-advantage over air transport in the southwestern part of the region.

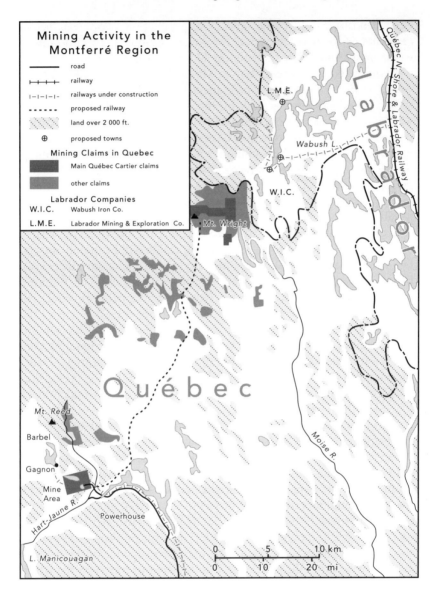

Mining Activity in the Montferré Region

———	road
+-+-+	railway
-|-|-|-	railways under construction
- - - - -	proposed railway
▨	land over 2 000 ft.
⊕	proposed towns

Mining Claims in Quebec

Main Québec Cartier claims

other claims

Labrador Companies
W.I.C.   Wabush Iron Co.
L.M.E.   Labrador Mining & Exploration Co.

The present mining activity was initiated in 1952, two years before the QNS and L Railway was completed. In that year the Quebec Cartier company (a direct subsidiary of the United States Steel Corporation) staked claims on the Quebec side of the boundary between and around Mount Wright and Mount Reed (see map). Subsequent geological surveys showed the claims to cover deposits of over 2,000 million tons of ore average 32 per cent iron, and in 1957 it was decided that the Mount

Reed deposits would be brought in to production. Work was begun immediately to survey the route for a new 197-mile [320-km] railway to the area from Shelter Bay on the St Lawrence, and in 1958 a road was completed from Shelter Bay to Mount Reed greatly facilitating the work. At the present time a permanent town-site, with a plan similar to Schefferville's, is being built on the shores of Lac Barbel. The railway is under construction, and a new dock capable of accommodating ships of up to 100,000 tons will be created at Port Cartier near Shelter Bay. Approximately $200 million will be invested in these facilities before mining can begin. To recover this large sum within a reasonable period, from such a low-cost product as iron ore, operations will necessarily be on a large scale. Initial production starting in 1961 is set at 8 million tons of concentrate a year, derived from 20 million tons of natural ore. Beneficiation will entail crushing the ore to a consistency of coarse sand and removing the gangue, which will be mainly quartz, by the Humphreys spiral system based on the different specific gravity of the iron-bearing material and the unwanted rock. The resulting product, containing roughly 65 per cent iron and 4 per cent water, will be transported by rail to Port Cartier, and from there by ship to the United States and Europe. Recent surveys suggest that navigation out of the Gulf of St Lawrence will be possible all the year round, making this the only major iron-ore region in North America shipping during the winter months. Plans for future development of the Quebec Cartier properties include the extension of the railway north to Mount Wright to exploit the northern ores.

The other two major mining developments at present nearing the production stage are both on the Labrador side of the boundary, between it and the Quebec North Shore and Labrador Railway. The Labrador Mining and Exploration Company has renewed exploration of the southwest portion of the original concession granted to them by the Newfoundland government in 1938. The area of interest is bounded by the provincial boundary on the west, and Wabush Lake on the east. By 1957 surveys had revealed local deposits of over 1,000 million tons of ore grading 35 per cent iron, and in 1958 some of the original staff from Schefferville were moved to the area to supervise the work. In 1959 the company, in conjunction with the Iron Ore Company of Canada, will invest over $1 million in exploration, and trailers will be used to provide better accommodation for the senior staff. Although no plans have been announced, it is understood that production will begin by 1962. Before this will be possible a rail link to connect the area with the Quebec North Shore and Labrador Railway some forty miles [65 km] away will have to be constructed.[3] Agreements have

3 Rail link is now completed. – ed.

already been reached with the Wabush Railway Company for the use of that part of the rail line which it is building to Wabush Lake for the Wabush Iron Company, thus saving the cost of duplicating the line. Preliminary estimates for the whole project call for the investment of some $150 million to start production at 6 million tons of concentrate a year. The ore will be exported through the Iron Ore Company of Canada's existing port facilities at Sept Îles.

The third development already well advanced is that of the Wabush Iron Company, which holds leases on the east side of Wabush Lake. Work began there in 1952, and in the following four years $8 million were invested to outline reserves estimated at over 1,000 million tons of ore averaging 38 per cent iron. The Wabush Iron Company, owned jointly by Pickands Mather and Company with United States associates and the Steel Company of Canada, leased deposits in 1957. Work on the rail line to link the area by rail to the Quebec North Shore and Labrador Railway in the east begun earlier, was continued, becoming a joint undertaking later with the participation of the Iron Ore Company of Canada, who will use the line for the movement of ore from its deposits further west (see above). Beneficiated ore will be moved by rail to Sept Îles, to be exported from there to markets in the United States and Europe. Production is expected to start at 8 million tons of concentrate a year beginning in 1962.

Two other major United States iron and steel companies, the Jones and Laughlin Steel Corporation and the Cleveland Cliffs Iron Company, have jointly financed a survey of a lease near Mount Wright, where over 1,000 million tons of ore capable of yielding 335 million tons of concentrate have been discovered. No plans for development have yet been announced. The W.S. Moore Company, which has also had experience in the United States iron trade, has optioned a lease covering small deposits of ore near Mount Wright. There is seasonal activity on most of the other claims held in the Quebec section of the region, but it is thought that production from these will take place only after the major companies begin exporting. Many of these claims seem to be held in the hope that the larger neighbours will become interested enough to develop them, with a resulting profit to the holders: they are in other words, purely speculative. Some of the claims, however, probably cover sufficient ore to justify exploitation, but only if the heavy capital costs of long distance railway lines, and other necessary facilities are borne by others. Once these are provided, the threshold for exploitation in terms of ore tonnage will naturally be lowered quite considerably, since it will then be possible to buy the services such as electricity, transport, and beneficiation from others, without the large initial investment. Once the region enters production therefore, it is to be expected that the number of operating companies will increase.

The main problem faced by all the companies is that of transport, the solution of which will have considerable bearing on the development of the region as a whole. For initial exploration work, air transport proved adequate, and is still used by those companies in the prospecting stage. For the three major companies, however, other means are needed to handle the increased freight traffic. Quebec Cartier have already built a road north from Shelter Bay to Mount Reed, to supply material to the railway during construction, and to the mine site before the latter is completed. The Wabush Iron Company has also built a road from the Quebec North Shore and Labrador Railway to Wabush Lake, but because of the greater distance involved, Labrador Mining and Exploration Company, even in the spring of 1959 were still carrying material from the track to their base camp by air. In considering the export of ore the problem is somewhat different. The cost of railway construction across the north-south oriented and dissected landscape of the western part of the region, and the distance, prohibit the use of a feeder line onto the Quebec North Shore and Labrador Railway. Instead the Mount Reed ore will be carried on an entirely new railroad being built north from Shelter Bay. As a private line there will be no obligation for it to carry other companies' material, though it is anticipated that agreements to this end will later be drawn up. The railway will, however, still have to cross the Laurentide scarp, and the engineering costs for this section will take a large share of the $200-million total expected outlay for the mining project. Eventually the line will be extended to tap the Mount Wright holdings of Quebec Cartier, providing other companies in that area with a stimulus to develop, but also introducing a tendency to integrate them into a western sub-region. For the properties on the Labrador side of the boundary, the short distance of forty miles [65 km] and the lesser relief make the construction of feeder lines to the Quebec North Shore and Labrador Railway a much more practicable proposition. Work has already begun on a line by the Wabush Iron Company, and the Iron Ore Company of Canada has negotiated an agreement for an extension to its property. The major drawback of this feeder line is the capacity of the existing line. To allow passage of the extra material further double-tracking will be necessary, particularly in the scarp section to the south, precisely where engineering problems and construction costs are greatest. Freight charges will presumably have to be increased to meet them, but since the Quebec North Shore and Labrador Railway is a wholly owned subsidiary of the Iron Ore Company of Canada, such increases would probably be felt most by the other users of the line. The total cost of the new railways into the region will be in the order of $200 million: investments which in part account for the large scale of the operations. When the region eventually has several producers,

there is little doubt that an integrated road-network will be constructed, linking up the sub-regions to produce a more coherent whole than
is at present observed. One further aspect of transport is important.
The beneficiated material will probably contain under 4 per cent water,
allowing it to be transported in winter without freezing into a solid
mass. With year-round navigation out of the Gulf of St Lawrence, there
will be no seasonal shutdown as at Knob Lake, where the 14 per cent
water-content of the ore allows it to freeze in transit to Sept Îles, prohibiting winter haulage.

Within the Montferré region at least three, and probably five, new
settlements are planned. At Lac Barbel, Quebec Cartier have already
begun building a town named Gagnon, the Wabush Iron Company
have reserved part of their property as a town-site, and the Iron Ore
Company of Canada have several potential town-sites under consideration.[4] The distance from Mount Reed to Mount Wright suggests that
Quebec Cartier will erect a subsidiary settlement to serve the latter
area, and other companies will probably house their own workers near
each mine. Labour for operations will be recruited by Quebec Cartier
in the Montreal lowlands, and it is interesting to note that married men
with families of at least two children will be preferred. The company
hopes in this way to overcome the problem of high labour turnover so
often experienced in northern mining towns. There is little doubt that
the inhabitants of the settlements in the region, as at Schefferville, will
be predominantly French speaking, further expanding the limits of
French Canada into Northern Quebec.

The actual mining landscapes will reflect to a certain extent the
remote location and type of ore mined. To repay the huge capital
investments within a reasonable period, operations will have to be on
a large scale; a fact further accentuated by the exploitation of such a
low-cost product as iron ore. The latter in particular, means that open-
cast mining, preferred anyway, will be absolutely essential. The beneficiation carried out at the mines will result in large quantities of waste
material which, unless there is careful planning, could despoil quite
large areas. Concern over this matter tends to be lessened by the dom-
ination of the developments by companies with long experience in iron
mining in the United States. Their presence can be explained by the

4 New towns at Wabush and at Labrador City were established only 5 km apart by these two
   companies on the Labrador side of the provincial boundary, while Quebec Cartier established its
   town of Fermont slightly to the south on the Quebec side. Despite the enormous costs of mining
   developments in the general area, depressed markets, declining ore reserves, and high costs have
   resulted in closures. Gagnon has disappeared. Most dramatic was the 1982 closure of
   Schefferville by Brian Mulroney, then president of the Iron Ore Company of Canada and later
   prime minister of Canada. From a modern town of 4,500 inhabitants in 1981, it has dwindled
   to about 100 inhabitants, surviving mainly as a base for hunting and fishing outfitters. – ed.

high cost of initiating exploitation, and by the nature of the North American iron and steel industry. To raise the large sums of money needed, investors have to be assured of the success of the project, normally achieved by the drawing up of long-term marketing agreements for the ore. Since there is virtually no free market for iron ore in North America, the only way to get such agreements is by the participation of major iron and steel companies, as has happened in the Montferré region. Such integration is fairly common in contemporary mining developments, and all the successful mines in Labrador-Ungava are linked with United States industry in this way.

The Montferré region seems to be moving through the second phase of pioneering associated with northern mining activity. The first phase was the seasonal activity of the true prospecting-exploration, when the area was inhabited only during the summer months. Now there is year-round occupation, though still in such a form that little evidence would remain if it suddenly ceased. Once the towns are established, however, the pioneer stage will be at an end: facilities will be provided equal to or superior to those found in settled areas further south, and it will become an outlier of civilization with an urban population in an otherwise uninhabited landscape. Though small the population will be making a major contribution to the North American economy, carving yet another new region out of the Canadian North.

33 Gold Hurry!: Hemlo's Golden Giant

JOHN WROE

The gold rush at Hemlo is over. In its place is a rush to develop the housing and infrastructure to accommodate the tidal wave of economic activity only a gold deposit worth $10 billion can generate.

In a sense, there never was a gold rush in the tradition of California or the Klondike, despite the massive public attention focused on an obscure chunk of Northern Ontario moose pasture. The real rush occurred before the summer of 1982, a year before the general public became aware of a new and huge source of income for the mining industry. And the rushing actually took place in corporate boardrooms and on the floor of the Vancouver Stock Exchange as investors and development companies hurried to get a piece of the Golden Giant, as the deposit was named.

That activity continues as the mines gear up for full production, with legal wrangling over the ownership of key claims on the deposit. The wrangling though, has done little to lessen the enthusiasm for the Hemlo site or the eagerness of communities nearby to get in on the action.

The development of Hemlo was not at all like the Klondike Gold Rush or the discovery of silver in Cobalt or the boom years of gold mines in Kirkland Lake and Timmins. In those days a man with a strong back and a bit of nerve could find transportation northward and, if not strike it rich, at least make a decent dollar doing tough physical work.

SOURCE: *Canadian Geographic* 105, no. 6 (Dec. 1985/Jan. 1986): 274-84. Reprinted by permission of the author.

But not at Hemlo. In an age of specialized technological skills a strong back is no longer a job ticket, and in these times of environmental sensitivity, there is no longer room for the small-time gold miner to find a piece of backwoods and start digging. Nonetheless, Hemlo is an awe-inspiring operation with its high visibility right beside the Trans-Canada Highway and an economic impact that will change the faces of two small single-industry towns.

Hemlo is actually three mines, all in sight of Highway 17 about 300 km northwest of the steel town of Sault Ste Marie. The most visible and most popular with tourists is owned by Noranda Inc., one of Canada's resource giants. Noranda's corporate identity is made clear through bright blue and silver buildings, contrasting with the yellow of the larger Lac Minerals operation and the dark brown of the Teck Corporation-International Corona joint venture, the smallest mine on the site. The three lie within a kilometre of each other, dividing up what is likely the richest gold deposit in Northern Ontario and, indeed, Canada.

Noranda and Teck-Corona each poured their first Hemlo gold ignot last spring and Lac expects to have its first gold shipped by the end of 1985. Three years ago there was little visible in terms of mining activity in the area, merely a lot of prospectors' trucks and a few diamond-drill rigs parked along the highway. The three companies transformed the territory very quickly into an industrial complex with roads, office buildings and, of course, the classic headframes poking up above the spruce trees north of Lake Superior. There may not be a gold rush as such, but Lac, Teck-Corona, and Noranda have shown a great deal of hurrying to get their mines into production.

The amount of gold explains why. Noranda, the most advanced operation, is now processing in the range of 1,000 tonnes of ore a day. Its facilities allow for an increase to 3,000 tonnes a day by late 1988.

Each tonne of ore contains about a quarter of a troy ounce of gold (5.8 grams), so when Noranda's mill is operating to capacity, a day's work will produce about 750 troy ounces of gold. (The internationally accepted measure for gems and precious metals, a troy ounce (23.3 grams) is equal to about $1^{1}/_{3}$ ounces avoirdupois.) A day's production will be about 17 kg of gold from 3,000 tonnes of rock.

To put that in perspective, consider converting the weights to something we all see regularly – cars. In a day's production, Noranda can process the equivalent of 2,000 mid-size cares, and from each car extract the equivalent of a cigarette lighter in gold.

During the early stages of production, it was costing Noranda about $275 to produce each troy ounce of gold. Gold at that time sold for about $300 US. By the time full production is reached economies of scale should reduce costs to about $150 per troy ounce. That means a profit of about $195,000 a day. Proven reserves indicate a mine life of twenty-five years, although Noranda expects to prove up more reserves and continue to operate for forty years. The $292 million budgeted to develop the mine can be recouped at that rate in the first five years, leaving the company twenty to thirty-five years to use the Hemlo gold revenue to expand its corporate wealth.

Just across Cedar Creek at Lac Minerals the situation is roughly the same in terms of cost, except that Lac expects to mill 6,000 tonnes per day at peak production, twice what Noranda will process. Lac also sees forty years of life for the mine. Teck has a smaller corner of the deposit and will be milling 1,000 tonnes of ore a day for the next twenty years. Add it all up and the three mines will mill a total of 10,000 tonnes of ore a day to produce about three-quarters of a million dollars of gold a day for at least the next twenty years, all from a patch of ground no larger than a good-sized farm. Small wonder the mining activity is intensive.

Officials of the companies often express exasperation at the talk of a gold rush, but there's no denying that a lot of "gold hurrying" is going on in the area. With something like $10 billion worth of gold under their feet, the hurry is understandable.

That much gold output and the profits it produces will have a ripple effect throughout the mining industry, an industry that has a history of putting its profits back into exploration. It will also profoundly affect the economies of two nearby communities, and indeed has already done so.

If the Hemlo gold deposit had been discovered fifty years ago there would now be a ramshackle town on the site, the result of makeshift accommodation erected in a hurry for miners streaming in for jobs. Kirkland Lake developed that way, with streets laid out without planning and houses built as close to the mine sites as possible. The result, sixty years later, is a haphazard town full of rough-and-tumble character but lacking proper sewage and water services and left with a legacy of winding streets and perhaps the highest concentration of bars in Canada.

The Ontario government wanted nothing like that in Hemlo and quickly banned development along the highway between White River and Marathon, the towns on the Trans-Canada closest to the mine sites. Marathon, a pulp town perched on the shore of Lake Superior 30 km from Hemlo, became the favoured site for most of the new residential development. Lac and Teck-Corona expect most of their employees to live there. Noranda chose Manitouwadge, a smaller community 50 km north of the mine site but already considered a Noranda town by virtue of its copper-lead-zinc Geco Mine five kilometres to the north.

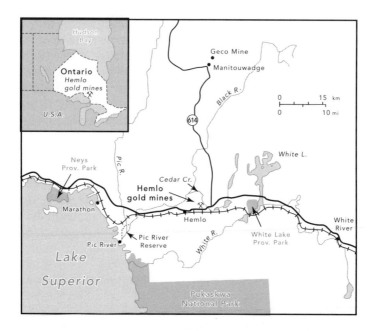

But it is Marathon where the boom is going on, and that town of 3,000 is where those looking for gold-rush jobs must turn. Mine workers who knock on the mine doors are given applications and sent on their way, but in Marathon many jobs are being created by the housing boom.

Driving into Marathon, the first thing you notice is the smell from the pulp mill which has been the mainstay of the town's economy since it was founded during World War II. Next, it's the No Vacancy signs on the motels, then a spanking new subdivision linked to the rest of the town, appropriately, by Hemlo Drive. The first phase of the subdivision has produced 880 residential units, most either complete or nearly so. Those working in the mines will have the choice of apartments, mobile homes, standard suburban houses, or more elaborate homes on large lots. And despite the gold rush, lots sell for as little as $13,000, thanks to the town's handling of the development on a non-profit basis.

Until Hemlo came along, Marathon's future looked bleak. The pulp mill was owned by American Can, and the company was ready to shut it down as a money-loser, putting an end to four hundred jobs. But in 1983 another US company, James River Paper, bought the mill for roughly the value of the log inventory and announced plans to spend $95 million on modernization and expansion.

Hemlo and James River vastly changed Gerry Layne's job. He is development manager for Marathon, and until 1983 he was striving to get some sort of secondary industry into the town. Now Layne no longer has to sell Marathon to potential investors; he is kept busy just accommodating the companies eager to settle there. The provincial government made the job easier by ruling out any sort of industrial park in Hemlo itself. All such industrial development must be channelled into the three closest towns – Marathon, Manitouwadge, and White River. Mining-supply companies must now locate close to where the workers live rather than close to the mine sites.

The result, said Layne, should turn Marathon into the major centre on the 700 km of Lake Superior shoreline between The Soo and Thunder Bay. With gold production continuing at announced levels, Layne expects the town to reach a population of 8,000 by 1990. If the major mines expand and other gold mines develop around Hemlo, the population could hit 12,000 by then. Layne discounts the question of what happens in forty years. He anticipates mining to be a permanent feature of the economy as more deposits are discovered in the area.

The boom is expected to produce a new high school, a shopping mall, a new motor hotel, and a busy construction industry. As a fringe benefit, James River is ending the rafting of pulp logs in the bay by switching to more truck and rail transport, and the town has the potential for developing a marina.

By 1990, according to Layne, Marathon could make the jump from a small one-industry town based on a pulp mill to a small city with its wealth derived from forestry, mining, and tourism. Whether tourists will spend their dollars in a town smelling like Marathon and with a skyline dominated by a belching smokestack remains to be seen.

Layne's optimistic view is not shared by all Marathon residents. Toby Maloney, editor of the Marathon *Mercury*, questions the population projection and has a bit of nostalgia for the days when Marathon had little growth, reasonably steady employment, and few newcomers to town.

The economic pressure from the influx of wealthy mining companies has produced some interesting contrasts. The Travellers Inn is unable to offer regular rooms without reservations; only accommodation in the "bunkhouse," a mobile trailer set up at the rear with small rooms and a communal washroom down the hall. It may be suitable for a travelling writer or a miner on contract, but not a family of tourists. On the other hand, the dining-room menu, encouraged by corporate expense accounts, can offer surprisingly sophisticated fare.

The town has yet to feel the influx of mining people. The bulk of the workers are now housed on the mining sites in more of those trailer units. They don't work for mining companies directly, but for a variety of contracting firms which specialize in such things as shaft-sinking and mill construction.

Despite amenities like private rooms, cafeterias, recreational facilities, and dish antennas tuned to the Playboy Channel, the trailer complexes are the closest thing the area has to a traditional mining camp. The workers usually have schedules that allow them lengthy breaks to go to their distant homes, with the time on site spent putting in as many hours as possible. The building contractors should be gone within two years, and by then all mine employees will live in Marathon or Manitouwadge.

It surprises many people to learn that total mine employment at Hemlo will likely be only in the range of 1,200. Less rich deposits discovered fifty years ago, for example, resulted in Kirkland Lake, Val d'Or, and Timmins. But the nature of the Hemlo deposit and modern technology have cut mine staffing dramatically. The Noranda mine shows it. Only two workers, ensconced in computerized control rooms, are needed to oversee the milling operation.

In a strange concession to mankind, though, the penultimate step in gold production almost totally lacks automation. In a process called electrowinning, the near-pure gold is transferred out of a liquid mixture onto metal plates and must be scraped off by hand. It is done in a totally enclosed compound within the mill and under the eyes of security staff. From there the gold is melted and poured into 1,000-troy-ounce bars for easy shipment. The gold "pours", too important to be entrusted to a machine, are also closely monitored by human eyes.

Modern-day practice also shows in the environmental controls the mines are putting in place. Disruption of the landscape is necessary at Hemlo: trees must be felled, roads bulldozed, and rock blasted away. However, the companies and the government are trying to make sure there are no environmental disasters at Hemlo.

Kirkland Lake is an example of how *not* to treat the environment. For instance, there is no lake. It was filled up by mine tailings, the crushed, muddy rock left over after the miniscule amount of gold is taken out. Today, the Lake Shore Mine stands a mile from the nearest lake shore, and the only time residents of Water Lane would hear the lapping of waves is if a water main burst.

There will be none of that at Hemlo. The mines are using earth dams to create retention areas for their tailings. Gold milling requires a lot of water, and the water will be recycled were possible so the demand on Cedar Creek will be kept to a minimum. Seepage from the tailings will be monitored. Lac Minerals speaks proudly of its record of rehabilitating tailing areas at other mines once mining is complete. Any water expected to find its way back into the environment is being treated to meet quality standards. "You can drink it," says one Noranda worker, "but I wouldn't recommend it. It tastes awful."

Although the water is drawn from Cedar Creek, the effluent is pumped over a height of land to another watershed. Small and environmentally sensitive, Cedar Creek is part of the Pic River watershed, and downstream is the politically sensitive Pic River Indian Reserve. The mining companies want to avoid a replay of the Grassy Narrows disaster.

Instead of finding its way into the Pic, the treated effluent goes into the White River system. This larger watercourse is better able to handle the impact of the treated waste water. The White also flows into Lake Superior via Pukaskwa National Park. The government people are satisfied that water quality will not be a problem. There are many retention dams, and the quality of water will be monitored at dozens of points. A saving grace for the environment is the relatively simple and consistent nature of Hemlo ore. Although cyanide is used in the extraction process, the mines claim a near perfect recovery rate for the chemical, saving themselves expense and preventing environmental damage.

The human environment, in the form of a steady flow of cars past the mines' gates, is also of concern to the companies. Noranda, with a name instantly recognized by most Canadians, gets the brunt of the traffic – people looking for tours, samples, or jobs. Security is tighter there than at Lac, its across-the-creek rival. Lac gets fewer tourists, but its mine and mill attract a lot of professional interest from other miners. The differences show in architecture, with the Noranda building looking like a showpiece for high tech: visible frame members and conduits, lots of open space, and few private offices. In comparison, the Lac office is much more modest, a spare yet utilitarian workplace that says the job is to get the gold, not impress the tourists.

Despite the differences, there is a common spirit at the mines, the quest for gold and the thrill of starting something new overriding the rivalry. And while the "gold rush" may have moved now to courtrooms where corporate lawyers argue over just who staked what claims when, the "gold hurry" continues unabated in the northern bush.

34 Diamonds under Ice

JAMIESON FINDLAY

We stand on a hillside of caribou lichen and know that winter is coming. The chill September wind of the Barren Ground blows over the tundra of autumn yellows and crab-apple reds. It buffets the nearby geologists' drill shack and the fresh core of glistening rock on the ground, fanning it dull and dry under the zinc-grey sky. By our feet are sample holes, dug by geologists who are wondering if they can get in another week of work before the snow flies. This is the visible landscape here at Munn Lake in the Northwest Territories, but the geologists – being geologists – are mainly interested in the bed of secrets beneath the tundra. They are hunting diamonds for the Toronto-based company SouthernEra Resources Ltd., and their search has focused on the clay in the sample holes.

It is rare clay, greyish with a faint tinge of blue-green, and geologist Howard Bird bends down to pick some up. Bird is the senior geologist here, a South African by birth, hospitable and wry. Earlier, he listened while the other scientists told me exactly how SouthernEra was searching for diamonds in the Munn Lake area. "Now that you know all that," he said dryly, "we have to kill you." But instead they brought me out here to their exploration drill, two kilometres north of their base camp, to look at clay.

Bird accompanies us back to the helicopter with clay in hand, and a few minutes later we are setting down at the camp. We land beside a low, rounded shelf of bedrock that has been smoothed and plucked by

SOURCE: *Canadian Geographic* 18, no. 1 (Jan/Feb 1998): 50–8. Reprinted by permission of the author.

a glacier, as an ice-cream cone is licked and bitten by a child. It is a *roche moutonnée*, says glacial geologist Lisa Sankeralli. Such patterns in rocks are arrows to the glacial geologist: by them she can determine the directions in which the ancient ice sheets moved. This is essential to the diamond hunters. They must follow the glacial paths if they want to trace the source of this clay.

Bird carries the clay over to the office tent, where he picks through it with tweezers and puts several bits into a petri dish. He places the dish under a microscope and squints into the eyepiece – a culminating moment in the day's search.

To find diamonds, geologists must first locate the body of kimberlite rock in which they lie. Millions of years ago the kimberlite bodies were fountains of thrusting molten rock. They streamed up from deep within the earth's mantle, picking up diamonds as they went from the surrounding rock, and then solidified into large rock plumes called pipes. The bluish-green clay is weathered kimberlite – a bit of the treasure chest, so to speak. The chest has been splintered and scattered across the land by glaciers. To find the treasure, SouthernEra must trace the splinters back to their source.

Under the microscope, the tiny grains are glittering stones. They have the clear-cut, faintly hallucinatory stillness that things have under a microscope. The mauve grains are pyropes, says Bird. The orange one is an eclogitic garnet. These are indicator minerals – splinters from the treasure chest. The chemical composition of indicator minerals can tell a geologist whether the kimberlite has diamonds or not. One of the main purposes of the SouthernEra camp is to collect and screen samples of glacial till, so that the indicators can be easily isolated and further analyzed at their Yellowknife lab.

There is something else here under the microscope, something smaller than the other granules. Unmagnified it is the size of a grain of sand. Magnified it is a small crystalline oval, filled with the transparent light of water.

Sometimes this happens, geologist Uwe Naeher says mildly. Sometimes you find a tiny diamond among the indicator minerals.

For this kind of gem, geologists have scoured the central Northwest Territories from the air and water and land, have trekked the paths of the vanished ice sheets, have stood on frozen lakes at −50°C, their breath freezing under their balaclavas, while they read the drill core laid out before them. They are all part of Canada's newest exploration boom.

About 100 kilometres north of the SouthernEra camp is the place where Canada's diamond rush began. It is on the shore of Lac de Gras, about 300 kilometres northeast of Yellowknife. The country there looks almost exactly the same as Munn Lake: berries abound, caribou meander, and the tundra gives under your feet like thick carpet. But Lac de Gras is now famous in the mining world. It was here in 1990 that a Canadian geologist named Chuck Fipke found "diamondiferous kimberlite." Fipke made his discovery using the standard methods of diamond hunting, such as tracking the indicator minerals. But he had to take into account the northern landscape to do it. In particular, he had to solve the colossal riddles created by glaciation.

For glaciation has made a mess of things up here – a grand, fascinating mess. This turmoil began between one and two million years ago, at the beginning of the Pleistocene Era. Huge ice sheets mumbled across the Northwest Territories, leaving a vast, confusing bed of echoes in the landforms. Underlying rock – especially soft rock like kimberlite – was eroded away. The glaciers sliced off the tops of the kimberlite pipes, dragging the indicator minerals with them. Melting, they left the indicators behind in the glacial till. The last glacier melted 10,000 years ago.

Fipke did a lot of puzzling in the early days. By all accounts he was a man built to tackle open-air puzzles – with a B.Sc. in geology, the stamina of a wolf, and the eye of a rock gypsy who had once prospected all over the world. Since 1981 he had been looking for diamonds in the NWT. Early on in the game he acquired a bush-wise partner, Stewart Blusson. Together they would travel "up ice" – against the direction of the glacial flow, to find the source of the indicator minerals.

Fipke and Blusson sifted through the glacial echoes and slowly traced the indicators to the general area of Lac de Gras. Beyond there, the minerals trailed off suddenly – suggesting that the two geologists had reached the limit of the pipe field. Now they had to zero in on it. By day they chipped rock, sifted through till, studied lakes. By night they pored over maps and prayed for good weather and readable ground – the geologists' prayer.

Judging by later events, they may also have asked the Deity to keep the diamonds away from those who already had some. De Beers Consolidated Mines Ltd., the mother of all diamond-mining companies, was also in the area. The company's agents had been looking for diamonds in Canada since the early 1960s but had found no economic blue ground (as kimberlite is called in South Africa). Neither had BP Canada Inc. or Falconbridge Ltd. or any of the other companies that had been hunting for diamonds in Canada.

All that changed one April day in 1990 when Fipke found what he had been tracking, on the bank of tiny Point Lake near Lac de Gras. He discovered a rare kimberlitic mineral called chrome diopside, which is generally found only very close to a pipe. Shortly afterwards, he and Blusson made a deal with the Australian company Broken Hill Propri-

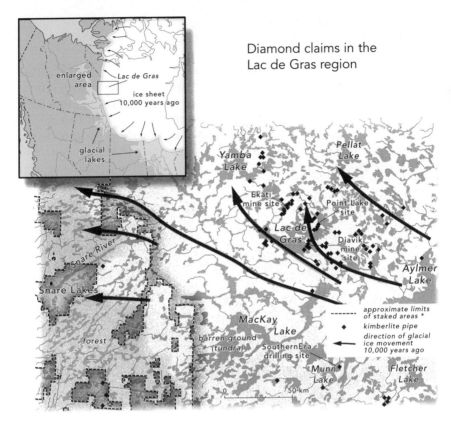

Diamond claims in the
Lac de Gras region

*Status of claims is in constant change

SOURCE: Steven Fick, *Canadian Geographic*

etary Ltd. (BHP), which brought in aerial geophysical equipment to map and "contour" the still-hidden pipes. An exploration drill was brought in and set up on the bank of Point Lake. On November 12, 1991, BHP announced that the core sample contained eighty-one small diamonds, some of gem quality.

The announcement initiated the biggest staking rush in Canadian history. Today about sixty companies are exploring for diamonds in the area, including Monopros Ltd. – the Canadian exploration subsidiary of De Beers. Many more kimberlite bodies have been found in the wake of Fipke's discovery. Four very rich pipes were discovered right under Lac de Gras itself. These are jointly owned by Diavik Diamond Mines Inc., the Canadian subsidiary of the mining giant Rio Tinto plc, and the Canadian company Aber Resources Ltd. Two of the pipes have already been subject to bulk sampling – preliminary mining done to determine the value of a project. This project, known as the Diavik project, is just across the field from the Fipke find.

Meanwhile, the area of Fipke's discovery is awash in a tide of intense labour. Hercules cargo planes land and take off on the airstrip several times a day, carrying fuel and construction supplies. Dormitories have been erected, power lines strung, a sewage plant installed. This will be the site of Canada's first diamond mine, set to begin production in the fall of 1998. Fipke's years of slogging have paid off: his company, Dia Met Minerals Ltd., owns 29 per cent of the project, while he and Stewart Blusson each have a personal stake of 10 per cent. BHP Diamonds is majority owner of the project, recently christened the Ekati mine.

The mine site is actually 27 kilometres northwest of Fipke's discovery, since, ironically enough, the Point Lake kimberlite pipe proved unprofitable to mine. But five pipes here are viable – flamboyantly so. Three of them are estimated to be among the world's richest diamond deposits. The pipes are covered by shallow lakes, which are being drained. (Because kimberlite erodes more easily than the surrounding bedrock, the pipes are often topped by depressions that have filled with water.) All pipes will be mined using open-pit methods, and the richest ones will be further plumbed by underground mining,. The revenue for the mine, which is expected to have at least a 25-year lifetime, is estimated at about $450 million a year.

And so exploration becomes extraction at this site, bringing with it an infusion of jobs and revenue for the North. But of course it is not all clear sailing. As tracts of tundra are given over to mining, the diamond enterprise comes face to face with other values, other needs.

On the wall of the new Legislative Assembly Building of the Northwest Territories hangs a painting by Native artist Antoine Mountain entitled "Mooseskin Boat." In it the boat of the title, each oar gripped by rowers, is riding fast waters through the wilderness. On both sides of the river are nebulous cliffs which, after a moment's viewing, resolve themselves into the shapes of animals. It is a simple painting, done in simple fire-and-ice colours, but the idea is suggestive. A fox barks from inside an outcrop; a grizzly walks a quartz vein. Rocks and animals share the same space; the land is alive.

Some here reject the idea that mining the rocks will endanger the animals and the rest of the living environment. The tundra is both vast and resilient, they argue; it can accommodate a few mines.

This view is frequently and forcefully expressed at The Diner restaurant in Yellowknife. The Diner is well known as the gathering place for

SLICING THE PIE

First Nations eye the profits and the risks

Canada's first diamond mine, at Lac de Gras, is being built on traditional hunting and fishing land of the Dogrib, Dene (Yellowknife and Lutselk'e), and Metis. From the lake, water flows north into the homeland of Inuit "downstream users" who depend on the rivers remaining clean and unsullied.

Before diamonds were discovered in the area, most of these groups were involved in negotiating land claims with the federal government, under Dogrib Treaty 11 and Akaitcho Treaty 8. The discovery of diamonds has complicated matters. The government says Native groups cannot include the area of the Broken Hill Proprietary Ltd. (BHP) mine as part of their land-claim settlements because the land is already in "an advanced state of exploration" – a definition set out in previous settlements. However, BHP has said that it wants Native groups to benefit from the mine, and the company is negotiating Impact and Benefit Agreements (IBAs) with the four Aboriginal groups concerned.

Since IBAS are private agreements, the parties have no obligation to make the details public. However, it is a "foregone conclusion" that they contain provisions for the hiring and training of Natives, says Ted Blondin, land claims negotiator for the Dogrib Treaty 11 Council. Two Aboriginal groups – the Dogrib and the Akaitcho – have signed IBAs with the company. The Metis and the Inuit have so far been unable to work out agreements with the company.

The closest community to the BHP diamond mine is the Dogrib community of Snare Lakes, 180 kilometres to the west. The people there depend on fish and wildlife – especially the caribou – for survival. According to Blondin, the first priority for the people is preserving the environment that supports their way of life. Since the Dogrib have been involved in granting water licences and land-use permits to BHP, they have had a say in the terms under which the work at the mine is taking place. They are also participating in studies of caribou in the area, to determine what effects the mine development is having on the herds.

old-style local prospectors. If you dream of blue ground, of the treasure silos under lakes, of pulling up drill core that is so seamed and spiced with diamonds that the grizzled old driller becomes a babbling child again – if you dream of this, come here at about 10 AM for a cup of coffee. You will be in good company. All around you will be people of a practical bent – stakers, bushrats, men in camouflaged hunting caps – from whom you can learn. When I went there I sat beside a man who had strung power lines at the Ekati mine. He was not a prospector himself, but he impressed upon me what it was like to work on the land up there.

"It's a great place for a mine," he said, "You've seen it – there's nothing there. And you've seen it at its best. In the winter it's a wasteland. A few hours of crappy daylight, and then you got night. You can't have a beard there. I had a beard but I shaved it off. It freezes because of your breath. I seen guys take their hoods off and there they were with their beards stuck right to their hoods!"

This is one view of the tundra – as a desert, as the perfect place for a diamond mine. Mining engineers would add another point in BPB's favour: diamond mines are relatively benign. They do not use harmful chemicals in the recovery process, the way gold mines use cyanide. The Ekati mine will extract diamonds by pulverizing the mined rock and then using an X-ray process to identify the crystals.

But still the mine is a large operation, and any project that size cannot help but have an impact on the environment. And, since more than one hundred other pipes have been discovered on BPB's claim block, it is safe to assume that mining activity will soon cover an even greater area. What will this mean for the vegetation and wildlife of the region?

This is the big concern of Kevin O'Reilly of the Canadian Arctic Resources Committee. His organization was one of several environmental groups that participated in the Environmental Assessment Panel's hearings on the BHP project, which were held in early 1996. BPB had already submitted its eight-volume Environmental Impact Statement (EIS), which included studies of the region's caribou, wolves, grizzlies, fish, and vegetation. O'Reilly's group, however, found the studies inadequate.

"In virtually all the areas they examined – botany, wildlife, socio-economic effects, you name it – the information was deficient," he says. "Because the EIS wasn't well done, we couldn't understand what the effects of the mine were going to be – so how could we even talk about mitigation measures."

For this reason he disagrees with the verdict of the federally appointed Environmental Assessment Panel, which concluded that "the environmental effects of the project are largely predictable and mitigable." But one good thing has come out of the hearings, says O'Reilly. An independent environmental monitoring agency has been set up to serve as a watchdog. It will review the designs of environmental studies – both BHP's and the government's – and comment on the results.

Such close scrutiny is going to become increasingly important in the next few years, when the diamond frontier expands. "There are a whole bunch of potential mines out there," says O'Reilly. "BHP is just the thin edge of the wedge. The question is, how can we assess all the environmental effects of these together?"

Construction continues at the Ekati mine. Just southeast of there, a small crew manages the Diavik site while engineers contemplate the best way to mine the pipes under Lac de Gras. The current proposal is to build a berm around the bay, drain the water, and then begin open-pit and underground mining. South of Lac de Gras, SouthernEra has moved into its winter drilling program. Snow is coming, white against the grey of Munn Lake.[1]

At dusk, if you fly across the general region of Lac de Gras, you can see the lights of exploration drills twinkling against the tundra. Even from the air it is clear that mining diamonds in this land carries with it a host of challenges – technical and environmental. But some people don't come here for those challenges. They don't even come here to get rich (although that may be a secondary consideration). They come here to track rocks. This is a challenge somewhat like tracking wolves to their dens, or caribou to their calving grounds, except that the trail leads over land that is long gone. To follow the mineral auguries, geologists must walk the ancient meltwaters, decipher the glacial glyphs. In doing so, they may find themselves being swallowed up by the chase. At night the sky will tent immensely over them, veined and foliate with stars, but this will only remind them of the things they have not yet found – the hardest things in the world, buried deep under this lichen-covered land.

---

[1] The Ekati Diamond Mine, Canada's first diamond mine, opened in October 1998. Diavik Diamond Mine is scheduled to begin production in spring, 2003, and a third diamond mine is expected to begin production at Snap Lake, south of Lac de Gras, by late 2005. – ed.

35 The James Bay Power Project

PETER GORRIE

Hydro-Québec sums up its corporate attitude about the massive James Bay hydroelectric project on the cover of one of its glossy pamphlets: *La Grande Rivière: A Development In Accord With Its Environment*. The booklet's proclamation is part of a series of messages aimed at convincing the public that more than $40 billion worth of powerhouses, dikes, transmission lines, roads, towns, and airports can be inserted harmoniously into an unspoiled northern wilderness.

For the past nineteen years it has been a relatively trouble-free selling job as the provincial Crown corporation – with enthusiastic backing from Premier Robert Bourassa and most Quebeckers – announced plans to harness the power of twenty rivers flowing into James and Hudson bays, then built the first phase of the project. For much of the 1980s there was also no debate, as economic recession cut energy demands and further phases were put on hold.

Demand for electrical power is again strong and the giant utility has relaunched an ambitious fifteen-year plan to complete the development. As it does, it is also facing renewed questions about the environmental and economic costs and the possibility that, for the first time, the huge project will be examined at public hearings.

Hydro-Québec insists the development is essential and will not cause unacceptable damage. One of its reports concluded that in the first phase "remedial measures ... have generally achieved their objectives,"

SOURCE: *Canadian Geographic* 110, no. 1 (Feb./Mar. 1990): 21–31. Reprinted by permission of the author.

and other studies offer assurances that remaining phases are "environmentally acceptable."

Premier Bourassa made his views on the project patently clear in a 1985 book, *Power From The North*: "Quebec is a vast hydro-electric plant in the bud, and every day, millions of potential kilowatt hours flow downhill and out to sea. What a waste!"

He and other supporters extol the jobs and income the project and its subsequent exports of electricity will bring Quebec. The utility has awarded billions of dollars worth of engineering and supply contracts to Quebec firms, enabling them to develop high-technology products and become international competitors. And, they say, every kilowatt of power from James Bay will cut the amount that power plants fuelled by coal, oil, or nuclear energy would have to generate at greater risk to the environment.

Critics counter that the project will create few long-term jobs while taking a devastating toll on the environment. They also worry that the 9,000 local Cree and Inuit will lose their source of food, livelihood, and identity.

The provincial government and Hydro-Québec have not ignored such concerns. Most of Quebec's environmental laws were introduced after the James Bay project was announced, in an apparent attempt to minimize damage by the development. The utility set up an environment division with officers at every construction site. Its subsidiary managing the project, the Société d'énergie de la Baie James, has a committee to advise it on environmental protection. The power company is spending hundreds of millions of dollars on impact studies and remedial measures – from creating new fish spawning grounds to landscaping tourist lookouts.

Even the New York-based Audubon Society, which says plans for future phases should be delayed, and perhaps scrapped, acknowledged in a recent report that the province "is willing to go to great lengths to reduce impacts during construction."

But all this is of little comfort to the 500,000-member Audubon Society and other critics. They complain that since James Bay is a key part of the provincial government's strategy for economic growth, environmental concerns have not been allowed to impede its progress. The provincial government and Hydro-Québec act on the assumption that the project must proceed, and only then consider how to cope with adverse consequences.

As a result, critics contend, the utility's research is inadequate or flawed, provincial reviews are cursory, and applications to proceed with various stages have been approved before impact studies were complete. The Cree and the provincial government each has two people to review and assess studies by four hundred Hydro-Québec staff and a small army of consultants.

Native people battled the project from the outset. In 1975 – after winning an injunction in Quebec Superior Court, then losing on appeal – the Grand Council of the Cree agreed to let the project proceed in return for $225 million, some control over about 75,000 square kilometres of land and an environmental review process.

But the development, described by Bourassa as "the project of the century," faces new political and legal challenges.

The Federal Court, in a case involving the Rafferty-Alameda dams in Saskatchewan, ruled last year that the federal government has a duty to review projects affecting its jurisdiction. The James Bay project falls into that category since it would affect Native people as well as migratory bird breeding habitat protected by a Canada-US treaty.

As well, the Cree have launched another legal challenge to try to force an extensive federal review and hearings, and the Inuit of northern Quebec recently asked the Federal Environmental Assessment Review Office for a public review.

In the northeastern United States, where Hydro-Québec hopes to earn billions of dollars from long-term energy sales, environmentalists are demanding their governments insist on thorough impact studies before deciding whether to approve the imports.

In response, Hydro-Québec officials have acknowledged that hearings might be worthwhile, and some in the utility are urging a two- or three-year delay.

The scope of the James Bay development is breathtaking. It would harness the energy of almost every drop of water in the rivers flowing through 350,000 square kilometres of northwestern Quebec – more than one-fifth of Canada's largest province.

The water would be collected in vast reservoirs behind powerhouses on the main rivers. While some would be released year-round to spin turbines and generate electricity, the system is geared more to winter when demand for power is at its peak. Then reservoir levels would drop as much as 20 metres as water is released and generating stations are pushed to capacity.

Cascading rivers would be dammed and diverted to create the reservoirs, flooding a combined area bigger than the surface of Lake Ontario. Some rivers would be reduced to a trickle; others simply submerged.

Hydro-Québec's latest development strategy calls for a three-stage completion of the project. If it goes ahead according to plan, by early next century it will generate up to 28,000 megawatts of power.

The project includes:

La Grande, Phase One: completed in 1985 after twelve years and at a cost of about $16 billion, it includes three reservoirs and power-houses on the La Grande River – LG2, LG3, and LG4 – with a combined production of 10,282 megawatts. In this phase, five smaller rivers were diverted into the La Grande to increase its power. Its average flow into James Bay has doubled and is four times the previous rate in winter. LG2 is now being expanded, with the addition of a 1,998-megawatt powerhouse called LG2A. This "add-on" will produce more power than the combined output of Quebec's single nuclear-powered generating station and its twenty-five plants fueled by coal or oil.

La Grande, Phase Two: its centrepiece is a powerhouse, LG1, near the mouth of the river, and five more – Brisay, Eastmain 1 and 2, and Laforge 1 and 2 – on rivers diverted in Phase One. They are scheduled to be in operation by 1996. Work on the Brisay dam and hydroelectric station was expected to start this spring.

Great Whale: north of the La Grande is the basin of the wild Great Whale River, or Grande rivière de la Baleine, which flows into Hudson Bay. This phase includes three power stations with a total capacity of 2,890 megawatts, and diversion of two other rivers. Final plans are being reviewed, but three or four reservoirs would be created on the Great Whale River by 2001.

The NBR Project: the initials represent three large rivers, the Nottaway, Broadback, and Rupert, which flow into the southern end of James Bay. The Nottaway and Rupert rivers are to be diverted into the Broadback where up to eight powerhouses would generate 8,700 megawatts. Hydro-Québec's target for completion of the first powerhouse is from 1998 to 2004, depending on demand. In addition, twelve sets of transmission lines – with a combined length of more than 5,500 kilometres and nearly 12,000 towers – will carry the power to markets in southern Quebec where it would be routed to customers either in Canada or the United States.

The scene of all this activity is a wilderness of lakes, rivers, spindly spruce and willow, lichens, and peat bogs along the east side of James Bay and the southeast coast of Hudson Bay.

Hydro-Québec reports that the region is home to thirty-nine animal species, including moose, caribou, beaver, muskrat, and lynx. The cold lakes and fast-flowing rivers teem with fish. The coastline is rich habitat for fish and birds, as well as whales and seals. Those resources are a crucial source of food and income for the Cree and Inuit. As well, the coastal waters are an internationally renowned resting and breeding ground for millions of migratory birds.

As Phase One has made clear, dams, dikes, powerhouses, and roads bring dramatic change. Damage to the natural environment is concentrated along the edges of water bodies, the richest habitats for plants and wildlife. Some rivers have been reduced to creeks. For example, downstream from its diversion into the La Grande, the Eastmain River's flow has been cut by 90 per cent.

In these shrunken waterways, riverbeds dry up leaving stagnant pools. Exposed clay and sand are eroded by rain and melting snow, and sediment chokes the mouths of tributaries. Spawning grounds are often destroyed and species such as brook trout, which live in clear, oxygen-rich rapids, can no longer survive.

Some of these rivers are subject to periodic flooding as excess water is released from reservoirs upstream. The result is heavy erosion and the destruction of new plants struggling to establish themselves in the exposed, barren riverbeds.

To date, the main remedy has been construction of weirs, or small dams, that turn sections of these shrivelled rivers into shallow lakes, with an entirely new habitat. Where weirs were considered too costly, exposed riverbeds have been planted to try to reduce erosion.

The opposite occurs in rivers that carry diverted water in a new direction. Their flow is greatly increased. For example, the Boutin now carries 15 cubic metres of water per second; when the Great Whale project is completed the little river will have swollen to 154 cubic metres per second as it carries the water from several lakes to the reservoir behind one of the main powerhouses.

Increased flows cause erosion. The resulting sediment load is deposited in places where the river slows – including reservoirs, whose capacity is gradually reduced by silting – and at the mouth, where a delta may form. Vegetation along the shore may be destroyed, eliminating habitat for ptarmigan, Canada geese, and some species of ducks. The damage is increased on fast-flowing rivers subjected to the fluctuating demands of power stations.

The new reservoirs flood rivers and submerge vast areas of forest. Shorelines become a tangled, inaccessible mess as trees and shrubs die and rot. Decaying vegetation eats up dissolved oxygen in the water and adds to the supply of nutrients, creating algae blooms. In most casts, shoreline vegetation and habitats cannot be re-established because of changing water levels.

Major changes also occur in estuaries, whether river flows have been reduced or increased. Water temperature patterns, the length and extent of ice cover in winter, and the mixing of fresh and salt water – all are altered.

Normally, rivers run highest during the spring melt; levels are lowest in winter. The James Bay development will reverse this natural pattern. Flows will be greatest in winter – up to ten times the normal volume – and the spring runoff will be diminished.

One result will be a change in water salinity at various sites during the year in James Bay and Hudson Bay. That could, in turn, wreak havoc on fish and mammals that require specific types of food or water conditions to prepare for migration, reproduce, or survive the long, intensely cold winters. In addition, nutrients that now flow into the bays will settle instead in the reservoirs.

Scientists do not have enough information to predict the consequences. But in such a complex and fragile environment to which plants and wildlife have adapted successfully but precariously over the millennia, the impact could be catastrophic.

The Audubon Society's report on James Bay cites examples of the potential damage to the bay's ecosystem. Coastal marshes and tidal flats are rich feeding grounds for many species of migratory birds, which must eat voraciously for a short time to store energy for flights to wintering areas in the southern United States and Central America.

A main source of food is a small clam that burrows in vast numbers in the mud of saltwater marshes and tidal flats. Millions of birds would have no alternative food if these feeding spots were destroyed by ice scouring or changes in salinity and temperature. Many species "would be severely threatened, possibly even to extinction," the society says.

Belugas winter in ice-free waters around islands in James Bay. The open areas appear to result from the spacing of the islands and the action of wind and tides in the channels among them. If ice patterns are affected by altered river flows, the whales could be at risk.

Hydro-Québec says it has found only small adverse changes where the La Grande River runs into James Bay. But the Audubon Society and other critics argue those results are not reassuring because too little time has passed to assess the impact. And, they say, while individual elements of the project might not have much effect, the total development could have devastating consequences. "If the damage from an individual project is marginal, the project can be approved, even though the cumulative impact of many such projects will mean the loss of the ecosystem," the Audubon Society warned in its report.

The utility is collecting volumes of information that is not of much use but suggests the appearance of action, says Alan Penn, the geographer appointed by the Cree to Hydro-Québec's environmental review committee. "it can describe things in a broad sense, but not the processes critical at certain times of the year" that determine whether species survive.

"The kind of data collection going on is not designed to focus on problems, but to provide general reassurance," Penn says. "It's what happens when you invite the developer to develop his own system of environmental monitoring."

But there is no denying one immediate and serious outcome – the release of mercury, which damages the human nervous system and can, with prolonged exposure, cause death.

Mercury is commonly found in rocks throughout the north in an insoluble form that does not affect the air and water. However, bacteria associated with decomposition of organic matter transform it into methyl mercury, which vaporizes, enters the atmosphere, then falls back into the water. From there it enters the food chain, reaching highest concentrations in fish species that prey on other fish. Local people consume large quantities of such fish – pickerel, pike, and lake trout – which are their most reliable source of high-quality protein.

New reservoirs induce a burst of decomposition that accelerates the release of mercury. On the La Grande, levels of mercury in fish downstream from the dams climbed to six times their normal levels within months of the project's completion. A 1984 survey of Cree living in the village of Chisasibi at the river's mouth found that 64 per cent of the villagers had unsafe levels of mercury in their bodies.

In time, as drowned vegetation is completely decomposed, the release of mercury should return to normal. How long that will take is not known. In studies completed up to 1981 – when Hydro-Québec put the James Bay project on hold – the mercury problem was not even mentioned. When it was finally recognized, the utility estimated that high levels would last up to six years. But a March 1988 study carried out by the utility on the Laforge 1 power station states mercury levels would remain high for ten to twenty years. It could be a generation, or longer, before fish are safe to eat again, Penn says.

In 1987 Hydro-Québec appointed a committee, with two Cree representatives, and gave it a ten-year budget of nearly $18.5 million to study the mercury hazard. To date, it has produced no practical solutions.

Decomposition can be reduced by clearing areas before they are flooded. But that is extremely expensive and poses the difficulty of disposing of the vast quantities of trees and brush. As a result, the power utility is clearing only selected areas – those that are close to power stations and other access points, and those around inlets of streams where fish spawn.

"A great deal of research needs to be done," the mercury committee concluded in its most recent report. In the meantime, it suggested weakly, Native people should stop eating contaminated fish and "anything that can be done to foster continuation of traditional pursuits would be much appreciated."

Decomposition has another by-product also causing concern – the release of methane, one of the greenhouse gases blamed for global warming. The amount of methane in the atmosphere is rising by one per cent annually. It is produced naturally by decomposition in peat bogs, wetlands, and lakes. Human activity has also made a big contribution. Large quantities of methane are generated by livestock, rice paddies, and the burning of trees and brush as forests are cleared.

Nigel Roulet, a scientist at York University in Toronto who has studied methane production in northeastern Quebec, says precise forecasts are not yet possible. But the James Bay reservoirs could be a significant new source of man-made methane.

By itself, the project will not change the earth's climate, but every contribution adds to the greenhouse effect, Roulet says.

While some proponents point out that global warming will be much worse if the power to be generated at James Bay is produced instead by coal- or oil-burning generators, the argument ignores the potential of conservation to cut energy demand. This is the view of Brian Craik, who has been involved with the project since 1972 and currently represents the Cree in discussions with the federal government.

Ultimately, the question must be asked: Are projects of this size, which basically reshape the geography of a vast area desirable?

James Bay is one of the last major undeveloped hydroelectric sites in North America. As planned, it will account for nearly 25 per cent of the continent's hydroelectric power. It will alter a huge land area in some ways that are known and others that even experts can only guess at.

So far, it has all been done without public hearings, and very little questioning.

"I'm really very upset about this," says Hélène Lajambe, an economist with the Centre for Energy Policy Analysis at the University of Quebec in Montreal. "James Bay doesn't make sense for Quebec." The province is already a wasteful consumer of electricity and demand is being fuelled artificially – through ad campaigns and price breaks – to justify the project, she says.

Last year, Quebec approved construction of three aluminum smelters which it attracted, in part, by offering the huge amounts of power they need at a cost tied to the international price of aluminum. That is an unstable yardstick, and if aluminum prices drop, Hydro could lose money on the deal, Lajambe says. In addition, she argues, power projects and aluminum smelters are expensive and environmentally damaging ways to create relatively few jobs.

The utility's cut-rate price plan for industry began to unravel late last year, however, as the low water levels of its northern reservoirs drastically reduced the generating capacity of the James Bay complex and its other hydro facilities. To head off what it termed "serious supply problems," Hydro-Québec launched another campaign, this time to convince its industrial customers to switch back from electricity to oil. That promotion quickly fell afoul of federal Environment Minister Lucien Bouchard who warned that the program could jeopardize Canada's acid rain negotiations with the United States. By encouraging the increased use of oil by industry, the utility, he said, could prevent Quebec from meeting its commitment to cut acid rain-causing emissions to 600,000 tonnes annually this year.

Even if Hydro-Québec sells James Bay power to the United States, Quebec will lose in the long run. The province will have put billions of dollars into developments that stimulate manufacturing and high-technology jobs elsewhere, Lajambe says. Quebec is investing its limited capital and best minds in projects "that chain us to an economy that depends even more on the production of resources. James Bay slows down the development process."

The access road to the NBR project will also open up from 12 million to 18 million cubic metres of marketable lumber, most of which would be exported, Brian Craik says. "The environment would be subsidizing not only the sale of hydro but also lumber to the United States."

Potential customers in the US appear to be getting cold feet about power deals. Maine has postponed signing a contract for a small long-term purchase, and the municipal council in Burlington, Vt., concerned about the environmental impact of James Bay, recently recommended that the local utility not buy power from the project.

At a conference in Montreal last summer, American energy economists argued the northeastern states could save money if they rejected James Bay power and, instead, paid for conservation programs in Quebec and then bought the electricity those measures would free up. But Hydro-Québec officials say the project will proceed, even without a US market for the power. And they remain convinced that it is environmentally sound and in the public interest.

Nevertheless, some officials are urging a delay, not because of concerns over the project's environmental consequences but because it requires a public relations campaign. And that would likely involve hearings, says Gaetan Guertin, the utility's manager of siting and impact studies. "The major conclusion we have is that maybe the public is not well prepared to react positively to these projects."

As the debate simmers, negotiations over hearings drag on and Hydro-Québec awaits approvals while engineering work is continuing. The utility is convinced James Bay power is needed and that, even with a conservation effort, demand will grow by three or four per cent annually. But none of its studies have asked: What next? James Bay is Quebec's last hydroelectric megaproject. Once it is done, the province will, like neighbouring Ontario, have no major rivers left to tame. Will it then also opt for nuclear power, creating the very problems it claims to be avoiding by developing James Bay?

If public hearings are held, they will probably focus on direct impacts. Are caribou threatened? When fish are contaminated, what will local people eat? But critics suggest the debate should centre on a much bigger question: Can humans limit their appetite for power so that such megaprojects do not need to be considered? Hydro-Québec and most other utilities assume the answer is no. Environmentalists insist it must become yes if the earth is to remain habitable. The James Bay project, they say, will not only increase the damage caused by the search for new power sources, it will also help delay the push for conservation that is likely to come only when we run out of alternatives.[1]

---

[1] Much has changed since the James Bay and Northern Quebec Agreement was signed in 1975 and the Northeastern Quebec Agreement (covering the northernmost part of Quebec, now called Nunavik) was signed in 1978. Significant environmental and social changes (both detrimental and beneficial) have followed construction of the first phase of the project; the Great Whale phase was frozen in the early 1990s. Both the Crees and the Inuit, through their active self-government bodies, have taken strong positions with the Quebec government on a variety of topics – e.g., opposing the latter's proposal of political separation from Canada. In February 2002 however the Cree signed a \$3.4 billion agreement with Quebec to accept Hydro-Québec installations on the Eastmain and Rupert rivers. – ed.

36 The Economic Impact of Northern National Parks (Reserves) and Historic Sites

DICK STANLEY AND LUC PERRON

Introduction

In 1989 Parks Canada published a study of the economic impact of northern national parks and historic sites (Thompson 1989), presenting the situation as of the 1987-88 fiscal year (1 April 1987 to 31 March 1988). This study has recently been updated to the 1992–93 fiscal year (Parks Canada, 1994) and will soon be published in its entirety. A summary of the results is presented here.

National parks and historic sites are key attractions of the tourism industry in the Yukon and the Northwest Territories (NWT) [and Nunavut]. In addition, money spent by park visitors and by Parks Canada to operate and develop these facilities has come to represent a small but stable component of the territorial economies.

Park Attendance and Visitor Spending

Large numbers of people visit the parks and historic sites in the Yukon and the Northwest Territories [and Nunavut] and the money they spend provides important employment and income benefits for the tourism sector in both the territorial and the regional economies. Nearly 192,000 visits were made to northern national parks and historic sites during 1992, and this number has been increasing steadily at about two per cent annually since 1987. Visitor spending attributable

SOURCE: *Northern Perspectives* 22, nos. 2–3 (summer/fall 1994): 3–5. Reprinted by permission of the publisher.

Visitors

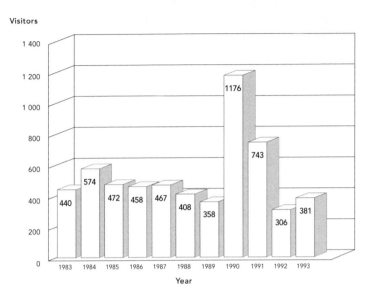

Number of Visitors to Auyuittuq National Park Reserve

SOURCE: Statistics compiled and provided by Johnny McPherson, consultant, Iqaluit, and Bob Longworth, Arctic Ridge Consulting, Iqaluit

to national parks and historic sites amounted to $15.2 million in 1992. This expenditure generated an estimated 252 person years of direct employment and $2.9 million in direct labour income for people employed in the tourism industry. Indirect effects of visitor spending (that is, employment and purchases in other parts of the economy caused by the economic activity of firms and individuals receiving the first round of visitor spending) generated an additional thirty-nine person years of employment and $0.6 million of labour income in other sectors of the territorial economies.

Overall, spending by visitors to national parks and historic sites contributed $5.58 million to the gross domestic product (GDP) of the Yukon and the NWT [and Nunavut].

Park Development and Operations

Spending by Parks Canada to operate and develop the national parks and historic sites in the Yukon and the NWT [and Nunavut] also creates jobs and income for the residents of northern Canada. In 1992 total wages and salaries paid to Parks Canada employees (including firefighters) amounted to $7.1 million. Purchases of goods and services needed to develop, operate, and maintain the northern national parks and historic sites cost an additional $7.74 million. Many of these items were purchased or produced outside the three territories, so only a portion of these expenditures remained within the territorial economies. This portion was estimated to be about 61 per cent. Salaries and operating expenditures are estimated to have created 167 person years of direct employment and an additional 57 of indirect employment.

Total Impacts

In total, Parks Canada's programs in the North created 515 person years of employment in 1992–93 and generated $12.86 million in labour income for northern residents. Spending by park visitors and by Parks Canada also increased the GDP of the territories by $15 million. This represents 0.5 per cent of total GDP from all sectors of the territorial economies (a weighted average of 1.05 per cent of the Yukon GDP and 0.25 per cent of the NWT [and Nunavut] GDP).

The impact of northern national parks and historic sites is small compared to the total size of the economies of the territories. It compares favourably, however, to that of some renewable resource sectors of the economy. The 1986 Thompson study cited data showing that the contribution of the northern national parks and historic sites to the Yukon territorial GDP was larger than the combined contribution of the forestry, sport fishing, and trapping industries.

It is also worth nothing that the $13.8 million spend by visitors to national parks and historic sites in the Yukon in 1992 actually represented 24 per cent of the total spending by Yukon tourists. Thus, Parks Canada contributes significantly to Yukon tourism.

In the NWT [and Nunavut], the economic importance of the national parks and historic sites was smaller than that of most sectors of the territorial economy. (It should be noted, however, that national parks and historic sites is not, in itself, a whole sector but merely a government program in a sector.) In 1990, for example, the value of production of the mining industry was $528 million, while the value of fishing and trapping was $4 million. The national parks and historic sites in the NWT [and Nunavut] had a total impact of $5.4 million in 1992, which compares favourably with fishing and trapping.

Conclusion

Although the economic impacts of the northern national parks and historic sites may be relatively small, they have remained stable over time, unlike some other sectors of the economy. Parks Canada's expenditures for northern national parks and historic sites have remained at about

TABLE I
SUMMARY OF ATTENDANCE, EXPENDITURES, AND ECONOMIC IMPACT BY
PARK, 1992

Park	No. of Visits	Parks Canada Expend. ($000)	Visitor Expend. ($000)	Total Expend	Economic Impacts		
					Jobs (P-V)	Labour Income ($000)	GDP ($000)
Yukon							
Kluane NPR	84,650	2,410	5,200	7,610	148.6	3,040	3,960
Klondike NHS	47,121	2,900	7,530	10,430	190.0	3,720	4,800
Yukon NHS	50,162	1,390	1,080	2,470	40.3	1,220	1,440
Ivvavik NP*	378						
Total Yukon	182,311	6,700	13,810	20,510	378.9	7,980	10,200
Northwest Territories [and Nunavut]							
Auyuittuq NPR	298	880	210	1,090	13.5	470	530
Ellesmere NPR	514	410	255	665	9.6	320	350
Nahanni NPR	1,403	920	160	1,080	16.5	590	660
Wood Buffalo NP	7,116	4,660	560	5,220	73.1	2,720	2,980
Ivvavik NP*	N/A	300	160	460	7.5	200	240
Western Arctic District	N/A	540	N/A	540	5.7	260	300
Eastern Arctic District District	N/A	460	N/A	460	4.2	170	180
Yellowknife Office**	N/A	180	N/A	180	5.6	150	170
Total NWT [and Nunavut]	9,331	8,350	1,345	9,695	135.7	4,880	5,410
Northern Total	191,642	15,050	15,155	30,205	514.6	12,860	15,610

NHS – National Historic Site
NP – National Park
NPR – National Park Reserve

*   Because Ivvavik National Park in the Yukon lies close to the Yukon/NWT border, it appears twice in this table. Visitors enter in the Yukon, but their expenditures – as well as the expenditures to develop and operate the park – are made in the NWT.

** Because current data are not available for the Yellowknife office, these expenditure data are from the 1987 study. The office accounts for only about 1 per cent of the total Parks Canada expenditure; thus the lack of current data does not cause serious underestimates of totals.

$15 million annually since 1987, helping to maintain some stability in the economies of the Yukon and the Northwest Territories [and Nunavut]. Table 1 shows details of the expenditures, visits, and economic impacts for each of the northern national parks and historic sites, including administrative offices.

TABLE 2
VISITOR EXPENDITURES AT NATIONAL PARKS (RESERVES) IN THE NWT
FISCAL YEAR 1987–88

Description	Amount	Per cent
Retail trade	$126,930	19.6
Amusement and recreation services	52,160	8.0
Accommodation and food services	180,910	27.9
Other personal and miscellaneous services	20,740	3.2
Transportation	267,570	41.3
Total	648,310	100.0

TABLE 3
VISITOR EXPENDITURES, ELLESMERE ISLAND NATIONAL PARK RESERVE,
1987

Description	Amount	Per cent
Retail trade	$25,713	17.0
Amusement and recreation services	10,285	6.8
Accommodation	27,074	17.9
Other personal and miscellaneous services	4,235	2.8
Transportation	83,944	55.5
Total	151,251	100.0

TABLE 4
IMPACT OF EXPENDITURES BY PARK VISITORS ON THE NWT ECONOMY,
FISCAL YEAR 1987-88

Category	Direct Impacts	Indirect Impacts	Total Impacts
Labour income (millions)	$0.15	$0.07	$0.22
NWT GDP (millions)	0.19	0.07	0.26
Employment (person-years)	7.4	2.6	10.0

SOURCE: Statistics compiled and provided by Johnny McPherson, consultant, Iqaluit, and Bob Longworth, Arctic Ridge Consulting, Iqaluit

References

Thompson Economic Consulting Services. 1989. *Visitor Profile and Economic Impact Statement of Northern National Parks (Reserves) and Historic Sites.* Socio-Economic Branch, Parks Canada, Ottawa.

Parks Canada. 1994. *Visitor Profile and Economic Impact Statement of Northern National Parks (Reserves) and Historic Sites 1992.* Parks Canada Investments Directorate, Parks Canada, Ottawa (to be published in 1994).

37 Indigenous Tourism Development in the Arctic

CLAUDIA NOTZKE

In the western Arctic, all stakeholders in the tourism industry appear to recognize, either implicitly or explicitly, that Aboriginal people are very important, if not the most important partners in the industry. Northern tourism confronts all stakeholders with enormous challenges. Some of the most important challenges facing the Inuvialuit, the Gwich'in [Kutchin], and other northern Natives relate to Aboriginal people's land-based way of life, to questions of how this way of life can be protected from tourism, and how this industry can be shaped to fit into this way of life. Both of these challenges have been successfully tackled by the Inuvialuit. Their approach represents an interesting example of how the provisions of their claim settlement agreement are employed to provide an operational environment for tourism, which bears Inuvialuit priorities of renewable resource harvesting in mind. Local outfitters capitalize on elements of seasonality and flexibility in both the industry and their mixed economy to combine the two. These strategies and tactics offer promising options for application and investigation in other parts of the North American Arctic. Nevertheless, a more widespread recognition on the part of community leadership and the public – that tourism (if properly controlled and realistically assessed) can really benefit communities – and a more sophisticated understanding of how these benefits can occur, are slow in coming, even in Tuktoyaktuk. The overall impact of tourism on host communities appears to be limited and generally benign, but more research is

SOURCE: Reprinted from *Annals of Tourism Research* 26, no.1 (1991): 56–76 (extract), with permission from Elsevier Science.

needed to gauge community perception and reaction. An impact survey of Tuktoyaktuk would be a useful start. It would also help to determine the feasibility of trying to improve the linkages of the industry to local supplies and services to further reduce the leakage, and it would assist in assessing tourism information and education needs at the community level.

Most people currently involved in tourism in the western Arctic look to the future with confidence, but also with some uncertainty. Northern Aboriginal communities and ecotourism are both in a state of rapid evolution, and their interface is a complicated one. Furthermore, both are operating in a highly fragile and unpredictable ecosystem. The future of the industry in this region, as visitors now encounter it, is inexorably bound to the evolution of northern local economies. In this evolution tourism has the potential for acting as an agent of change as well as an agent of preservation. The "authenticity" and "real life character" of the current tourism experience sometimes also makes it very difficult to manage. It is well-nigh impossible to predict, where the next generation is headed. As the Business Manager of Arctic Nature Tours muses, more tourists and more "professionalism" will make the industry easier to manage, but what will be lost in the process? (interview notes in July 1995). For the time being, it seems important to educate tourists about their role in northern Aboriginal people's lives, and show them that their role is appreciated. They must be made to understand that, for however fleeting a moment, they are not just witnessing, but participating in a lifestyle that deserves to live on – for the people's sake and for the land's sake.

CANADIAN ARCTIC RESOURCES COMMITTEE

The future of the Porcupine caribou herd in northeastern Alaska and
northern Yukon is threatened by legislators in Washington who want
to allow drilling for oil and gas in the herd's calving grounds. This is a
very serious issue – one that has motivated CARC repeatedly since its
inception in 1971.

The Porcupine caribou herd sweeps across much of northern Yukon
and adjacent Alaska in a never-ending annual cycle. In early June the
herd concentrates along the Arctic Coast, and here pregnant cows
renew the cycle by giving birth to their young. On the Canadian side
of the border, the calving grounds lie within Ivvavik National Park and
are fully protected. On the Alaskan side, the calving grounds are within the Arctic National Wildlife Refuge (ANWR), which sounds secure
but is not.

When the US Congress passed the Alaska National Interest Lands
Act in 1980 to establish the refuge, section "1002" of the bill set aside
the coastal plain – the prime calving grounds of the Porcupine caribou
herd in Alaska – pending assessment of its oil potential by the US
Department of the Interior. In 1987 the department concluded that oil
and gas development in the area would have "major effects" on the
herd. In response, the Canadian government urged that development
be prohibited on the "1002" lands and that the area be designated
"wilderness" by Congress. Prime Minister Chrétien has reaffirmed this

SOURCE: Members Update, Canadian Arctic Resources Committee, Spring 1995. Reprinted by
permission of the publisher.

policy, and our ambassador in Washington is using diplomatic channels to make sure that all parties understand Canada's position.

There is an interesting and important legal foundation to this issue. Both Canada and the United States have signed international and bilateral agreements that have a real bearing on this issue, including the Migratory Birds Convention, Polar Bear Convention, and the Ramsar Wetlands Convention. Most pointedly, in 1987 both countries signed an Agreement on the Conservation of the Porcupine Caribou Herd. The objectives of this agreement are

To conserve the Porcupine Caribou Herd and its habitat through international cooperation and coordination so that the risk of irreversible damage or long-term adverse effects as a result of use of caribou or their habitat is minimized.

Notwithstanding these obligations, on 26 May 1995 the Senate of the United States voted to include oil-drilling revenues from the "1002" lands in its budget package! While not opening the refuge to drilling, the vote created the assumption that authorization to approve oil leasing in the refuge would follow. Further votes in the House of Representatives and various committees will take place this summer. This issue could be won or lost in the next two months.

Generally, CARC would not suggest that its members and supporters intervene in the United States. But this issue is clearly an exception. Moreover, it is not exclusively an environmental issue. Canadian Aboriginal peoples – in particular the Gwich'in [Kutchin] – rely upon the Porcupine caribou herd for food and sustenance. They need our political help.[1]

---

[1] In December 1995, then-President Clinton vetoed proposed legislation that would have allowed drilling in the Alaskan portion of the caribou grounds, but the new Republican administration in Washington has again sought approval of the project. – ed.

TRANSPORTATION AND COMMUNICATION

39 Transportation North of 60°N

INDIAN AND NORTHERN AFFAIRS CANADA

General

Transportation is a key factor in the economic and social development of the Yukon and the NWT [and Nunavut]. Unlike the nineteenth century when travel to the North by dog team or water could take weeks or months, today most areas in the north can be reached within hours by air.

Early transportation developed primarily to support mining and national defence. The Klondike gold rush in the 1890s led to increasing numbers of paddle wheelers on the Yukon River and the construction of the White Pass and Yukon Railway to link Skagway, Alaska and Whitehorse, Yukon. The Great Slave Railway was built in the 1960s to help develop the lead and zinc mine at Pine Point, NWT.

Not surprisingly, air transportation developed more rapidly than the road network. In 1929 the first mail flight reached Aklavik, and light aircraft equipped with skies and floats became the lifeline to northern Native communities and mining camps. 'Punch' Dickins and 'Wop' May were well-known bush pilots of the 1930s who pioneered northern air transportation. Today most communities have airports and year-round scheduled air services.

The discovery of radium at Great Bear Lake in the 1930s led to the formation of the Northern Transportation Company Limited (NTCL),

SOURCE: *Canada's North, The Reference Manual* (Ottawa: Indian and Northern Affairs Canada, 1990), 10.1–10.5.3. (extract). Reproduced with the permission of the Minister of Public Works and Government Services Canada, 2001.

which has become the major marine carrier on the Mackenzie River and in the western Arctic.

It was not until the early 1940s that the first northern highway construction began. The Alaska Highway was spurred by North American defence needs during World War II. It also provided a way to extract valuable minerals from northern British Columbia and the Yukon.

In the late 1940s the then Mackenzie Highway was completed, providing the first road connection to the NWT. Extending 752 km north from Grande Prairie, Alberta, it carried freight to Hay River on Great Slave Lake, the southern end of the Mackenzie River transportation system.

In 1956 a roads policy finally emerged. Cabinet approved an annual budget of $10 million for new road construction in the Yukon and the NWT. This policy was revised in 1965 with the Territorial Roads Policy for the Future.

The roads policy was revised again in 1971 and in 1983. The annual roads budget now stands at $17.3 million. The current policy aims to maintain the established road network and complement air and marine transportation networks.

The Dempster Highway made a major improvement to the road network. The Dempster runs 726 km north from the Klondike Highway near Dawson City, Yukon to Inuvik, NWT. This highway's completion in 1978 opened up an all-weather road link between southern Canada and the Mackenzie Delta. However, for several weeks during spring break-up and fall freeze-up the highway cannot be used over the Peel River near Fort McPherson and over the Mackenzie River near Arctic Red River [Tsiigehtchie].

Bill C-13, a new National Transportation Act designed to reduce regulation in the transportation industry, became law in January 1988. Northern air and marine services are most affected by this legislation. Carriers must file their domestic service patterns and must abide by their schedules. Communities and licensed air carriers can intervene in another carrier's application if they can show that a proposed level of service would hurt a community. In marine transportation, community resupply services continue to be licensed by the National Transportation Agency on the Mackenzie River and in the western Arctic and are now licensed for an indefinite period instead of having to apply for annual licences. Charter or unscheduled service that supports northern resource exploration and development are not regulated.

Air Transportation

Because of the vast distances which must be covered, transportation by air is the choice for most travellers in Canada's North even if alternatives are available. The shortness of the ice-free season even in those parts of the Arctic Ocean where open water is experienced, makes air transportation essential. To meet this need, commercial air carriers offer scheduled service to most communities and charter service virtually anywhere.

The level of northern air service greatly increased with the 1974 Northern Air Facilities Policy. The Policy allowed for airport upgrading in all communities with populations of at least one hundred. This led to year-round air service to most communities and enabled air carriers to use larger aircraft.

Roads and Road Transportation

GENERAL

Northern roads (Figure 39.1) are often long and have little traffic. Most roads are gravel surfaces although parts of the Klondike Highway, Haines Road, Hay River Highway, and the Fort Smith Highway are now paved or have a bituminous surface treatment. More than 1,850 km of Yukon highways, including the Alaska Highway, have paved or bituminous-treated surfaces.

Because of the long distances between communities, weather forecasts and road conditions are regularly broadcast on radio. Drivers are warned of extreme weather and to carry survival equipment on long journeys.

Six ferry crossings are considered to be an integral part of the road system. They are: near Fort McPherson and at Arctic Red River [Tsiigehtchie] on the Dempster Highway; on the Canol Road at Ross River; on the Yellowknife Highway at Fort Providence; on the Mackenzie Highway at the Liard River; and on the Top-of-the-World highway at Dawson City.

Ferries usually operate from mid-May to early November in the south while at the more northerly crossings the season is from June to mid-October. Ice bridges can be started in the middle of December and kept in operation for two to three months.

In addition to the major road network, winter roads provide surface transportation to some communities. Figure 39.1 shows some established winter roads.

HIGHWAYS IN YUKON

The principal road entrance to Yukon is the Alaska Highway. Built during World War II, it is now an all-weather road which enters Yukon near Watson Lake and passes through Teslin to Whitehorse. From Whitehorse it runs roughly west to Haines Junction and then northwest, skirting Kluane National Park, to the Alaska border at Beaver

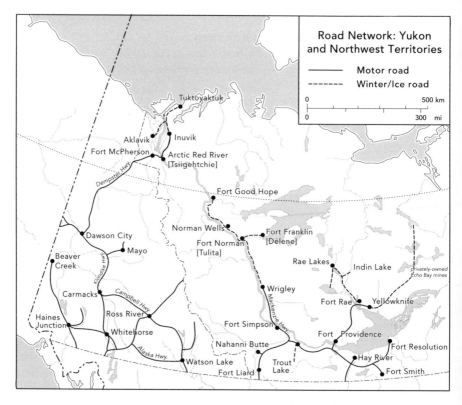

Figure 39.1   SOURCE: DIAND, Infrastructure Division, 1989

Creek. The highway is either paved or has bituminous surface treatment applied.

Another road from the south which connects to the Alaska Highway near Watson Lake is the continuation of BC Highway 37. This route runs north from its junction with the Yellowhead Highway between Smithers and Prince Rupert, BC.

The Robert Campbell Highway runs northwest from Watson Lake to a location near Ross River where it meets the Canol Road and then to its junction with the Klondike Highway at Carmacks. It also connects with the Nahanni Range Road, Highway 10, which runs northeast to Tungsten.

Three other roads cross the southern border of Yukon. Highway 7 runs south from Jakes Corner, just east of Carcross, to Atlin, BC. The Haines Road runs from Haines Junction south to Haines, Alaska. The Klondike Highway connects Skagway with Carcross, joins the Alaska Highway just south of Whitehorse, and separates from it again north of Whitehorse. It then continues north to Dawson City and joins

Highway 11 at Stewart Crossing, which provides access to Mayo, Elsa, and Keno.

The Dempster Highway branches off the Klondike Highway approximately 32 km southeast of Dawson City. The Dempster runs northeast, crosses into the NWT to Fort Mcpherson and Arctic Red River [Tsiigehtchie] and then joins the Mackenzie Highway south of Inuvik. The Dempster is Canada's first road to cross the Arctic Circle.

The Canol Road starts southeast of Whitehorse on the Alaska Highway at Johnson's Crossing and runs via Ross River to the MacMillan Pass at the Yukon and NWT border.

Boundary Road, or the Top-of-the-World Highway, runs west from Dawson City and meets the Alaska Highway at Tetlin Junction. The road is open only in the summer.

HIGHWAYS IN THE NWT

The Mackenzie Highway originates at Peace River, Alberta and is the principal highway route into the Northwest Territories. After crossing the border, it runs to Enterprise (just south of Hay River), and then runs northwest following the Mackenzie River through Fort Simpson. The Highway then runs north to Wrigley. A new bridge at the Willowlake River and a ferry at Camsell are required to provide year round access from Fort Simpson to Wrigley. These will be completed by the Government of the Northwest Territories.

Approximately 80 km northwest of Enterprise, the Mackenzie highway intersects with Highway 3 (the Yellowknife Highway). A ferry at Fort Providence provides access to Rae and Yellowknife. East of Yellowknife is Highway 4 which is designated as the Ingraham Trail.

The Liard Highway provides an all-weather connection between Fort Nelson, BC, and the Mackenzie Highway. The two highways connect approximately 45 km southeast of Fort Simpson.

The Dempster Highway originates in Yukon east of Dawson City, joins Fort McPherson and Arctic Red River [Tsiigehtchie] to the Mackenzie Highway just south of Inuvik, NWT.

Highway 5 runs east and south from Hay River to Fort Smith on the Alberta-NWT border. Highway 6 leaves Highway 5 east of Hay River and provides access to Pine Point and Fort Resolution.

In addition to the all-weather roads, winter roads provide transportation routes for many areas. These roads often provide vital transportation links necessary to provide communities with supplies needed to sustain themselves over the winter months. Without these seasonal roads, the basic supplies needed would have to be transported by aircraft. The expense is substantial. Examples of winter road links are Inuvik to Tuktoyaktuk, Wrigley to Fort Good Hope, and from Fort Norman [Tulita] to Fort Franklin [Délene].

Water Transportation

GENERAL

Water transportation was the principal means of movement through the North from the earliest exploration until the 1920s and 1930s, when air travel became possible.

During the Klondike gold rush (1896) steamers travelled the Yukon River carrying miners and their supplies from Whitehorse to the gold fields near Dawson City. These steamers continued to operate on the river until fairly recent times. At present there is no commercial water transportation on this river.

In the NWT the backbone of water transportation was and is the Mackenzie River. From Alexander Mackenzie's journey of 1789 until about 1826 the canoe was supreme. The canoe gave way to the larger shallow-draft York boats. These in turn were replaced by steamers in the late 1800s. Around 1930, tugs and barges, which now carry thousands of tonnes of freight each year, replaced the steamboats.

THE MACKENZIE WATERSHED ROUTES

Within the Mackenzie watershed there are five sectors: the Mackenzie River from Hay River to Tuktoyaktuk, including the Peel River; the western Arctic coast (Beaufort Sea area); the Athabasca River and Lake Athabasca system; Great Slave Lake; and the Liard River and Fort Nelson River system.

Navigation problems on the Mackenzie River include a short shipping season, ice conditions, low water levels (especially in the fall), four sets of rapids and decreasing daylight in the fall. Because of the rapids barge tows must stop and each barge must be carefully moved through a channel to protect the cargo. Tugs and barges are always prone to damage and this risk has increased as tugs and barges have become larger and heavier. Table 1 details the short shipping season in Great Slave Lake and the Mackenzie River by listing the average break-up and freeze-up dates at different locations.

Six shipping companies operate in the watershed. Four are licensed by the federal government to provide resupply services to communities along the Mackenzie River and in the Western Arctic. Cooper Barging Services operates in the Liard-Nelson River system and resupplies communities between Fort Simpson and Wrigley. Coastal Marine Limited resupplies communities between Inuvik and Tuktoyaktuk. Beluga Transportation Limited resupplies Inuvik, Tuktoyaktuk, and Aklavik. Arctic Transport Ltd. supports oil and gas development in the Beaufort Sea. Len Cardinal Transport services exploration and development activities between Inuvik and Tuktoyaktuk.

TABLE I
AVERAGE BREAK-UP AND FREEZE-UP DATES, GREAT SLAVE LAKE AND
MACKENZIE RIVER

Location	Break-up*	Freeze-up**	Days***
Hay River	May 7	Oct.30	176
Yellowknife	June 1	Oct. 29	150
Fort Resolution	June 8	Nov. 14	159
Reliance	June 28	Nov. 18	143
Fort Providence	May 18	Dec. 6	202
Fort Simpson	May 16	Nov. 15	183
Wrigley	May 22	Nov. 21	183
Fort Norman [Tulita]	May 28	Nov. 14	170
Norman Wells	May 26	Nov. 12	170
Fort Good Hope	May 30	Nov. 9	163
Aklavik	May 31	Oct. 12	134
Inuvik	June 3	Oct. 18	137

*   Break-up is the date the water is completely free of ice
**   Freeze-up is the date the water is completely frozen
***   Number of days between break-up and freeze-up. However, there will be fewer navigable days.

SOURCE: Environment Canada, Atmospheric Environment Service, Canadian Climate Centre, Downsview Ontario

The average sailing time between Hay River and Tuktoyaktuk (1,806 km) is:

	Round Trip	Delays due to darkness
Spring	14 days, 6 hours	–
Fall	17 days	4 days, 9 hours

Note: Dates are based on break-up dates for an average 22 years and freeze-up dates for an average 42 years. However, most locations are missing records for at least some years. Reliance is missing records for eight years.

The dominant carrier is the Northern Transportation Company Limited (NTCL), which moves between 80 per cent and 90 per cent of the tonnage in the Mackenzie watershed. NTCL has provided resupply services throughout the Mackenzie watershed since 1934, along the Western Arctic coast since 1957, along the North Slope of Alaska since 1963, and in the Keewatin [Kivalliq] (from Churchill) since 1975. (See Figure 39.2). Barges, with tanks below deck, are designed to handle large volumes of bulk petroleum products. Cargo carried on deck includes general merchandise, construction materials, steel containers,

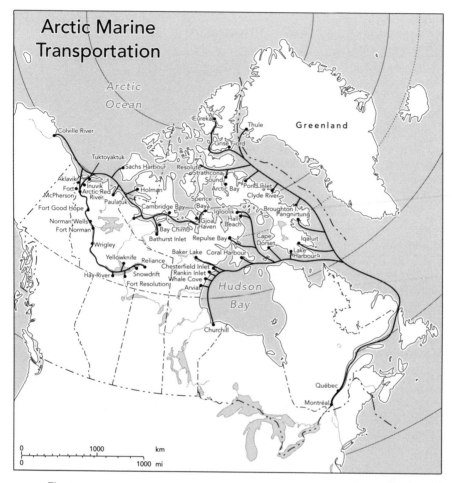

Figure 39.2

SOURCE: National Transportation Company Ltd., Canadian Coast Guard Eastern Arctic Sealift

highway trailers, drummed fuel, drilling rigs and pressure silos of bulk drilling muds, and cements. Resupply tonnages have remained fairly constant throughout the years. Keewatin [Kivalliq] tonnages have averaged about 30,000 short tons (27,216 metric tonnes) during the last three to four years. This is in addition to the community resupply tonnages on the Mackenzie River. Cargoes related to oil and gas exploration have decreased since the mid-1980s.

With the closing of the Eldorado uranium mine at Uranium City, Saskatchewan in 1982, NTCL stopped operating in the Lake Athabasca sector. NTCL has a capacity of 450,000 short tons (408,233 metric tonnes).

NTCL was purchased from the federal government in 1985 by the Inuvialuit Development Corporation and Nunasi Corporation. These organizations are wholly owned by Natives. The Government of the NWT prefers using NTCL to move its materials.

CENTRAL, EASTERN, AND HIGH ARCTIC SHIPPING

Much of the freight carried on the Mackenzie system is transferred to ocean vessels at Tuktoyaktuk for distribution to the Beaufort Sea area and to coastal points and islands as far east as Spence Bay [Taloyoak]. NTCL carries almost all of the freight into the area. (See Figure 39.2).[1]

The eastern Arctic is served by the Eastern Arctic Sealift, coordinated by the Canadian Coast Guard. Cargo, originating from many Canadian and foreign points, is assembled principally at Montreal. Consignees include federal and territorial government departments, Crown agencies, the United States Air Force, private companies, and individuals. Tonnage carried varies from year to year in volume and make-up but bulk fuel has predominated, at least in recent years.[2]

The Coast Guard provides ice routing and escort services to the ships carrying this cargo. Areas served include Iqaluit, Strathcona Sound, Resolute Bay, Rae Point, Little Cornwallis Island, Eureka, and sites in Foxe Basin and other points as far north as Grise Fiord.

NTCL service to the Keewatin [Kivalliq] consists of shipments moved by rail north to Churchill, Manitoba, marshalled in Churchill, and shipped by tug and barge to communities along the west coast of Hudson Bay.

Rail Transportation

GENERAL

Only two railways are located in Canada's North at the present time. A line from Grimshaw, Alberta, which runs through Hay River to Pine Point, NWT and a line from Skagway, Alaska to Whitehorse, Yukon.

THE WHITE PASS AND YUKON RAILWAY

A long-established narrow-gauge railroad connects Whitehorse to tidewater at Skagway, Alaska. The White Pass and Yukon Railway was built in 1898–1900 to serve the transportation needs of the prospectors and miners of the Klondike. Its route is from Skagway, Alaska, through

---

1   In 1993 NTCL secured the contract to supply and transport bulk fuel to Alaskan Arctic coastal communities as far northwest as Point Hope on the Chukchi Sea. – ed.

2   NTCL provides fuel supply service to Eastern Arctic communities. Dry cargo transport to the same is provided by Nortran Inc., a wholly owned subsidiary of NTCL. The NTCL fleet consists of fifteen river- and ocean-going tugs, two Arctic Class 2 supply tugs, and 100 dual purpose barges. – ed.

TABLE 2
TONNAGE CARRIED BY NORTHERN TRANSPORTATION COMPANY LTD.,
1979–1988

Year	Short Tons*
1979	314,000
1980	306,000
1981	348,000
1982	303,000
1983	297,000
1984	216,381**
1985	237,122**
1986	192,963**
1987	153,315**
1988	148,000** (p)

* One short ton equals 907 kg (2,000 lbs.)
** Includes only community resupply or regulated tonnages along the Mackenzie River. Not included are industrial tonnage, i.e., materials for use in oil and gas exploration and development, and Keewatin tonnages.
(p) Preliminary

SOURCE: Northern Transportation Company Ltd., *Annual Reports*
National Transportation Agency of Canada

the White Pass into British Columbia and thence north to Whitehorse, Yukon.

This once vital link has suffered economically in recent years. The railway closed in 1982 after the closure of its main customer, the Cyprus Anvil Mine, in Faro. The railway has started a summer tourist service from Skagway to the White Pass Summit and Fraser, BC.

THE GREAT SLAVE LAKE RAILWAY

The Great Slave Lake Railway was constructed by Canadian National Railways in 1965. The purpose was to develop the section of Canada from northern Alberta to the area of the NWT south of Great Slave Lake and in particular to permit the exploitation of the lead/zinc deposits there. The route runs from Grimshaw, Alberta, north to Hay River, NWT, and thence east to the mine site at Pine Point.

Although this line was designed for the movement of lead-zinc ore concentrates from Pine Point Mines, there have been considerable grain and lumber shipments as well. The increase in petroleum exploration in the Arctic generated northbound traffic on the line to Hay River, the staging area for the Mackenzie River barge system. With the 1988 closure of the Pine Point Mine, the section from Hay River to Pint Point [has closed].

TABLE 2
TONNAGE CARRIED BY EASTERN ARCTIC SEALIFT, 1979–1988

Year	Dry Cargo (tonnes)*	Bulk Fuel (tonnes)*
1979	9,682	34,500
1980	10,033	33,400
1981	11,862	27,094
1982	10,393	28,881
1983	10,018	27,363
1984	12,594	36,126
1985	16,192	31,066
1986	14,956	32,511**
1987	14,704	32,452**
1988	14,128	36,747**
1989	15,612***	33,915**

*   One tonne equals 1,000 kg (2,205 lbs.)
**   Excludes coast Guard bunker fuel tanker tonnage
*** Preliminary

SOURCE: Transport Canada, Canadian Coast Guard

References

Department of Indian Affairs and Northern Development. 1985. *Western Arctic Transportation Study* (prepared by Prolog Planning Inc., Calgary, Alberta). Ottawa: Department of Indian Affairs and Northern Development.

Government of Canada and Government of Manitoba. 1986. *Port of Churchill Resupply Operations*. Ottawa: Transport Canada.

Government of the Northwest Territories. 1989. *Keewatin Port Facility Study*. Yellowknife: Department of Public Works and Highways.

Government of the Northwest Territories. 1990. *Eastern Arctic/Baffin Region Ports Facility Study*. Yellowknife: Department of Transportation.

Minter, Roy. 1987. *The White Pass*. Toronto: McClelland and Steward.

National Transportation Agency of Canada. 1989. *Annual Review 1988*. Ottawa: Minister of Supply and Services Canada.

Northern Transportation Conference. 1982. *PROCEEDINGS: Northern Transportation Conference – The Challenge of the Eighties*. Whitehorse, Yukon.

Transport Canada. 1987. *Mackenzie River Economic Study*. North Vancouver, British Columbia: Canadian Coast Guard.

40 Communications North of 60°N

INDIAN AND NORTHERN AFFAIRS CANADA

Telephone Service

Telephone service is provided in the North by two carriers. Bell Canada serves most of the area east of 102°W longitude. West of this line and in most of the Arctic islands, NorthwesTel is responsible. (NorthwesTel is owned by BCE Inc., the management holding company which also owns Bell Canada). The administrative offices of NorthwesTel are in Whitehorse, Yukon. Bell Canada maintains an office in Iqaluit. Telephone company rates in the North are regulated by the Canadian Radio-television and Telecommunications Commission (CRTC).

BELL CANADA

In the eastern sector, there are twenty-two communities including Little Cornwallis Island. Bell provides communications completely via satellite to all of these because they are in remote locations. They have full direct dialing. The Bell Canada directory, which covers the North-west Territories, is trilingual – English, French, and Inuktitut. Mobile service is available in some areas, served by bases at Iqaluit, NWT and Alma, Quebec. A range of telecommunications services is available, including data transmission and telex.

SOURCE: *Canada's North, The Reference Manual* (Ottawa: Indian and Northern Affairs Canada, 1990), 11.1–11.3. (extract). Reproduced with the permission of the Minister of Public Works and Government Services Canada, 2001.

In the western Northwest Territories and all of Yukon, NorthwesTel provides service to forty-nine communities. Some terrestrial microwave is used for trunk routes in addition to satellite facilities. Eight small communities are served by radio telephone.

Serving NorthwesTel's 2.35 million km² operating area are 8,690 km of microwave systems. An 1,800 channel system extends from Grande Prairie, Alberta to the Alaska-Yukon border near Beaver Creek. Other major systems serve Yellowknife and extend down the Mackenzie Valley, serving communities along the way. These two systems are connected with a 960 channel system from Fort Simpson to Fort Nelson, British Columbia. There are a total of 32,000 network access lines, all of which have direct distance dialing capability.

There are over a hundred manual mobile telephone base stations, located to provide maximum coverage of the major highways and the Mackenzie River transportation corridor. NorthwesTel also provides automated mobile radio coverage in the 400 MHz band in the Yellowknife, Lower Mackenzie, Dawson City, and Inuvik areas. As well as voice transmission facilities, the network also provides telex, telegram, data transmission, manual and automated mobile service, facsimile, and computer communications. In addition, program quality channels, i.e., bundles of channels which improve the quality or dynamic range of voice transmissions, are provided for the Canadian Broadcasting Corporation (CBC).

Radio

Radio in the Yukon and the NWT means both broadcast transmissions for information and entertainment and radio-telephone, a voice communications medium. All are subject to CRTC regulations and licensing procedures. Because of the totally different nature of the transmissions, these are treated separately here.

RADIO-TELEPHONE

Although many telephone conversations are carried over part of the route by radio link, the term radio-telephone is usually reserved for simplex systems. These are characterized by the use of a single channel shared in turn by the persons communicating, so that A listens while B speaks and vice versa. Both Bell Canada and NorthwesTel provide mobile radio-telephone service from many centres throughout the North. This provides an effective extension of the switched telephone network.

In addition to mobile telephone service, stand-alone radio-telephone is widely used. Equipment ranges from hand-held, battery-operated transceivers operated under the provisions of the General Radio Service (often erroneously called Citizen's Band, the American equivalent) to sophisticated equipment to keep parts of a widely dispersed oil exploration company in touch with one another and with field offices.

Both FM (frequency modulated) and AM (amplitude modulated) signals are employed. Except for the very lowest power transceivers used in General Radio Service, all equipment must be licensed by the Department of Communications.

RADIO AND TELEVISION BROADCAST FACILITIES

Radio and television broadcast facilities in the NWT and the Yukon are provided by two classes of station. Typical radio broadcast stations do much of their own programming, but also use network programs to provide more national information. The other class includes radio and television rebroadcast or repeater transmission stations. Both types of stations are licensed by the Canadian Radio-television and Telecommunications Commission (CRTC).

[Licensed radio broadcast] stations have production facilities to originate their own programming. Some of this programming may be rebroadcast to other communities through repeater transmitters. The NWT [and Nunavut have] sixteen radio broadcasting stations; the Yukon has ten.

Also, Yellowknife, Iqaluit, and Whitehorse have television production centres. Most television programming produced in the North is first sent to Toronto and then transmitted to TELESAT's ANIK D satellite through the CBC Network Control Centre. A facility in Iqaluit transmits/uplinks a television signal to a satellite. Repeater stations throughout the North access signals from the satellite.

[Northern radio and television rebroadcast/repeater station] transmitters may also rebroadcast through a land line or microwave link. There are fifty-five rebroadcasting stations in the NWT; twenty-eight in the Yukon. Also the NWT has seventy-eight television repeater stations; and the Yukon has twenty-eight.

Cable service is relatively new in the North. In the late 1970s, tapes produced in Vancouver were sent to Yellowknife and Whitehorse where they were transmitted through cable systems. When satellite technology emerged in 1981, it became possible for a cable system to access satellite signals. As a result, closed-circuit cable systems have increased in the last five to eight years. In 1989 nine communities have cable systems.

Radio and Television Programming

Providing an acceptable range of radio and television services for Canada's North is difficult. Community populations are small, often numbering in the hundreds. Whitehorse is the largest community with fewer than 20,000 people.[1] Nevertheless, seventy-three other smaller communities over 3.9 million km^2 have access to radio and/or television services. It is not enough to broadcast international and national news to these remote places. Northerners also need accurate and current information to help them make decisions on issues such as major resource development, resolution of Aboriginal rights and Native land claims, and the constitutional development of the territories.

Most northern regions had no television in 1968. When CRTC was formed in 1968, the North was identified as a "special" concern. The CRTC recognized that satellites could play a major role bringing television to the North and Native broadcasting would reinforce the unique cultural and linguistic distinctions of Canada's Native people. In 1979 CRTC established the Committee on Extension of Services to Northern and Remote Communities, which produced the "Therrien Report" in July 1980. The report provided a public forum to examine and debate the broadcasting needs of many interest groups. The Northern Native Broadcast Access Program (NNBAP), approved by the federal Cabinet in March 1983, and administered by the Department of the Secretary of State, funds thirteen Native communication societies across Canada. In 1986 the CRTC Action Committee on Northern Native Broadcasting was formed to respond to conflicts on the issue of access to various kinds of transmission facilities used by Native communications societies. In December 1984 the Commission announced the formation of a Northern Native Broadcasting Committee, which was to identify broadcasting-related problems of the Native communication societies. This committee comprises representatives from Native communication societies, the CBC, the Department of Communications, and the Department of the Secretary of State.

Canadian Satellite Communications Inc. (CANCOM), private radio stations, community radio stations, and cable companies distribute Native communications societies' programming.

Since 1983 there have been efforts to enhance and protect Aboriginal languages and cultures through Native broadcasting. Under the Northern Native Broadcast Access Program (NNBAP), four Native

---

[1] The 2001 Census population numbers for Whitehorse were 19,058 (city) and 21,405 (area), compared with the populations of Yellowknife (16, 541) and Iqaluit (5,236). – ed.

communications societies have been funded. Support is provided for capital equipment, ongoing operational expenditures, and staff salaries. In 1988–89, $4,950,000 was given for operations and $63,515 for capital equipment. The total Native listening audience is about 40,000. Amounts provided for operations and capital have varied according to need over the years.

As of February 1989 there were about thirty Aboriginal community stations in the NWT [and Nunavut] and fourteen in the Yukon. Aboriginal stations provide news and information in languages indigenous to these communities.

RADIO PROGRAMMING

The North has three radio programming services: CBC Northern Service, Northern Native Broadcasting, Yukon (NNB,Y) and the Native Communications Society of the Western NWT.

CBC Northern Service

CBC Northern Service was established in 1958 as a branch of CBC. It is responsible for providing radio and television services to the North. The Service is funded through the parliamentary appropriations given to the CBC.

CBC Northern Service broadcasts in eight languages and in four time zones. It also provides transmitters to communities of more than five hundred people. Communities with fewer than five hundred people receive transmitters from the territorial governments.

Each week CBC produces about 220 hours of local programming that represents the voices, ideas, and concerns of about five hundred northern citizens. About 25,000 stories are presented each year, in the regular daily programs.

Radio programming from Whitehorse is directed to a mostly English-speaking audience; little Native language programming is produced. CBC Inuvik produces English, Inuvialuktun, and Gwich'in (Loucheux) [Kutchin], and Yellowknife provides English, Chipewyan, Slavey, and Dogrib. Residents of the central and eastern Arctic receive English and Inuktitut from CBC operations in Rankin Inlet and Iqaluit. Because fewer non-Natives live in these regions, Inuktitut is the main language of production. Specials are also produced on topics ranging from concerts of northern musicians and dancers to the Arctic Winter Games and territorial election results.

Many settlements have a radio society. A society can arrange with the CBC to use the local repeater to broadcast locally produced programming to its own area. Local groups can broadcast their own music, news in the local dialect, or other items of community interest.

The facility has also been used to coordinate searches for lost hunting parties.

Northern Native Broadcasting, Yukon (NNB,Y)

In the Yukon, Native language programming is different. A Native communications society, Northern Native Broadcasting, Yukon (NNB,Y) has established a network of stations at Haines Junction, Old Crow, Pelly Crossing, Ross River, Whitehorse, Burwash Landing, Carcross, Carmacks, Mayo, Teslin, Upper Liard, and Watson Lake. Regularly scheduled programming includes Native Languages and translation, community information, music, and other features that enhance Native heritage and culture. The Northern Native Broadcast Access Program (NNBAP) funds NNB,Y. Radio programming is about sixty hours per week for an audience of approximately 3,000.

Native Communications Society of the Western Northwest Territories

The Native Communication Society (NCS) of the Western Northwest Territories was incorporated in 1975. It has a listening audience of about 13,000 Dene/Metis in thirty communities from Hudson Bay to the NWT-Yukon border. Station CKNM has the second-largest audience in Yellowknife and reaches twenty-three communities by satellite. The production format includes open-line shows, news, weather, sports, music, and children's programs.

Production is increasing in the five Dene languages: Chipewyan, Dogrib, Gwich'in (Loucheux) [Kutchin], and North and South Slavey. Radio programming is about eighty-four hours per week. NNBAP and the Government of the Northwest Territories (GNWT) financially support NCS.

TELEVISION PROGRAMMING

Television programming for the North is carried out by the CBC Northern Service, the Inuit Broadcasting Corporation (IBC), the Inuvialuit Communications Society (ICS), and NNB,Y communications society.

References

Canadian Radio-television and Telecommunications Commission. 1986. "CRTC Action Committee on Northern Native Broadcasting." *Public Notice CRTC 1986–75.* Ottawa.
Canadian Radio-television and Telecommunications Commission. 1985. "Northern Native Broadcasting." *Public Notice CRTC 1985–274.* Ottawa.
Canadian Radio-television and Telecommunications Commission. 1980. *The 1980s: Decade of Diversity – Broadcasting, Satellites and Pay-TV.* Report

of the Committee on Extension of Services to Northern and Remote Communities. Cat. no. BC92–24/1980. Ottawa: Supply and Services Canada.

Department of the Secretary of State of Canada. 1989. *Guide to Native Citizens' Programs*. Cat. no. S2–195/1989. Ottawa: Supply and Services Canada.

Murin, Deborah Lee, ed. 1988. *Northern Native Broadcast Directory*. Ottawa: Runge Press Limited.

Murray, Catherine A. 1983. *Managing Diversity: Federal Provincial Collaboration and the Committee on Extension of Services to Northern and Remote Communities*. Kingston, Ontario: Institute of Intergovernmental Relations, Queen's University.

Tourigny, Patrick. 1983. *Community Television Handbook for Northern and Underserved Communities*. Broadcast Programs Analysis Branch, Canadian Radio-television and Telecommunications Commission. Cat. no. BC92–29/1983. Ottawa: Supply and Services Canada.

NORTHERN SETTLEMENTS

41 A Winter at Fort Norman

MAURICE R. CLOUGHLEY

Halfway down the Mackenzie River, where its wide, muddy waters are joined by the clear, cold stream of the Bear River draining out of Great Bear Lake, lies the attractive little Indian village of Fort Norman [Tulita]. No one seems to know how the place got its name, but Norman himself was probably some functionary of the old fur-trading North West Company. A hundred and eighty people lived there in 1957, not counting a small assortment of White outsiders like myself, posted to the place on various pretexts. For Norman had a two-room school and it was here that I took up my first-ever teaching position in Canada.

The Native people, legally at least (and these things mattered in 1957 as well as long after) were quite a mixture, ranging from Treaty Indian, through Non-Treaty, to Metis, and dwindling on through varying shades of almost (but not Quite) White. It was a delightful little place, and towards each other, the people including the Quite Whites were all mostly friendly and well-intentioned if not always well-informed or even well-acquainted. The settlement was sited above a bank sloping steeply up from the North's Old Man River, the mighty Mackenzie. Above this bank there were two terraces, one well above the other, and by long established (but unintended) precedent, these ancient formations served to define quite distinctly one's status: who you were, where you came from, where you belonged. All the Native

SOURCE: Maurice R. Cloughley, *The Spell of the Midnight Sun* (Victoria, BC: Horsdal & Schubert, 1995), 1–2 (extract). Reprinted by permission of the publisher.

people of whatever category lived in their tired old log cabins or tents on the lower terrace nearer the river, and all the Whites lived in or adjacent to their workplace on the upper level. This quite unplanned arrangement seemed to suit everyone. There wasn't a lot of room for many more on the lower terrace and the Indians like being close to the river, while the Quite Whites seemed to be happy having their own larger enclave at some distance and elevation from The Rest.

A more-or-less road meandered along the upper level connecting all the elements of this higher enclave. (Like all roads, this one had started out as just a dirt track, but there were now two tractors at Fort Norman, so that made it a road.) Two Mounties manned the stockade at the western end of this road, and not far from their detachment was the Roman Catholic church which normally had two Oblate priests in residence. Then came a Department of Transport radio station followed by the Indian Agency. The Hudson's Bay Company post was across the road overlooking the Native village and the river. Next came the federal day school where Frank Frey, the principal, lived. The game warden's office was across the road, and at the end was the nursing station.

It was to this elevated stratum that I clearly belonged. The Whites decided things and ran things, even including each other, and while in our various capacities we all dealt with the local people on a daily basis, there was always the unspoken but clear understanding that socially we confined ourselves to our own kind. Oh dear! This was hardly why I came north and secretly I was more than a little disappointed to find myself bound by these expectations, but I wasn't about to change the world.

A small house, a design known across the North as a 512 for its square footage, was being shipped in for me to Fort Norman, but as this wasn't built yet, I had to board with the Freys for a few weeks. They lived upstairs in the school building itself, which was a common design in the North in the early 1950s: two classrooms side by side separated by a folding partition that facilitated the conversion of the school into a community hall but did nothing to dampen the noise of either class in the adjacent classroom. A hallway separated these two rooms from furnace room, toilets, storage space, office, and corridors, and directly above these rooms was the three-bedroom apartment of the principal.

Eventually, the old *George Askew*, last stern-wheeler on the Mackenzie River system, nosed a load of barges gently against the river bank. Piles of building materials were offloaded on the beach, and within a short time my 512 got built at the edge of the school playground, and I moved into my first-ever home of my own. It felt great and I was very happy with it. I got a Metis man to provide me with a 45-gallon [200-litre] steel gasoline drum with the top hacked out and replaced with a plywood cover. This drum I painted inside and out, and that was my water tank. It was filled periodically in the summer season by pump and canvas fire hose from the river, but on the advice of my neighbours who warned me gloomily of problems ahead, I laid in a stock of ice blocks against the house in case of need and especially for use in spring when the river water would be filthy.

School started right away in September and I busied myself for a few weeks in my work as the leaves turned colour and fell, and, falling with them, the enchanting snow and the temperatures. All this was humdrum to everyone else but was novel and exotic to me and I was enthralled. The ground froze. The lakes froze. We flooded a rectangular skating rink in the school playground, and that froze.

I bought an outdoor thermometer at the Bay and nailed it at my cabin entrance so I could monitor the temperatures as they stepped steadily day by day down the old Fahrenheit scale until I could add the unaccustomed term "below zero" in my letters home to New Zealand. I was thrilled and wonderstruck at the amazing cold. The river began at first to carry thin, colourless, skim ice along on its surface. This soon congealed into small, rounded, ice pans, and then larger and larger pans until they floated everywhere over the wide, vapour-smoking surface of the great, silent river, leaving us cut off for six weeks from the bush planes that normally ran a scheduled service into our settlement every two weeks from Norman Wells. Finally and suddenly the running ice stopped running as it jammed up tight from one bank to the other, a jumbled mass in most places, but flat and smooth enough in some areas to allow us a winter airstrip.

42 Fermont: A Design for Subarctic Living

WILLIAM O'MAHONY

Quebec-Labrador is a vast and barren territory, hostile to life. A meagre 10 to 12 cm of poor soil rests on top of sand, permafrost, or rock, and is covered by yellow-brown caribou moss.

The wealth of this immense territory rests not on its soil but in its rich mineral deposits. In Quebec-Labrador, iron ore formations range from towering mountainous outcrops to deep deposits that sink 300 m or more underground. This is the "brown gold" that has drawn man to the northlands and cast its lure over investors and miners alike, in spite of an eight-month winter and the briefest of summers.

Attracting and keeping workers in these isolated territories has always been a problem for mining companies. One way in which they have tried to solve the problem is by building innovative communities such as the town of Fermont, established by the Quebec-Cartier Mining company when it began to develop the gigantic Mount Wright iron ore deposits in northern Quebec.

In general the construction of towns in northern Canada has been directly connected with mining and processing our mineral wealth. The examples are many: Sudbury is synonymous with nickel; Gagnon with iron ore; Val d'Or with gold. In the early days of the mining industry in Canada, these communities were planned and established by the companies to provide homes for their employees.

For one who has never lived in the North, it is extremely difficult to understand the effect of isolation on workers and their families. The

SOURCE: *Habitat* 21, no. 3 (1978): 17–20. Reprinted by permission of the publisher.

plaintive melodies of Robert Service, and Jack London's novels on the North poignantly express this sort of isolation. It is the feeling of being captive, of having little opportunity to get out of the town, to get away from people with whom one works and lives. It is the feeling of having virtually no privacy, even in one's own home. It is the absence of family; the lack of simple things, like fresh vegetables or a choice of schools.

Today, in towns like Fermont, however, there is a new approach to the problem of isolation and life in the Subarctic.

The first northern settlements were temporary shelters inhabited by the indigenous people of the North who practice a food-gathering and hunting economy. They were followed by a second generation of settlements established by the first miners and prospectors; rude and haphazard collections of tents and shacks clustered at the edge of the mining area. There was no town planning and less social organization. In fact, these collections of shelters were less towns than temporary tent camps for transients.

As mines were developed which proved large enough to support a permanent community, the settlements became more stable, and thus a third generation of settlements evolved. "New towns" were planned and built by the mining companies, based on the typical suburbs to the south. But problems emerged: driveways and roads became an economic burden because of the heavy snow removal costs. The expansive lawns were scarred and unsightly for all but the short summers. Suburbia transplanted to the North proved wasteful of both land and energy.

A fourth generation of settlements followed the suburban patter, replacing the scattered institutional and commercial buildings with a 'town centre' – a cluster of communal facilities linked by a temperature-controlled mall. But there was still room for improvements, and these came with towns such as Fermont.

In order to develop a comprehensive overall design, Quebec-Cartier Mining Company (a subsidiary of United States Steel Corporation) hired architect-planner Norbert Schoenauer, of Desnoyers and Schoenauer of Montreal, as town planning consultant. At that firm's suggestion, Ralph Erskine, a Swedish authority on arctic town planning and design, was also added as a project consultant.

The site of the new town was carefully chosen on a southern slope overlooking a lake for maximum exposure to the winter sunshine and shelter from cold northerly winds. Had it been situated in a valley, it would have been a settling place for cold air; on top of a hill it would have been exposed to winds.

Around the northeast side of the settlement, the dominant architectural feature consists of a five-storey building which gives shelter from the north winds to at least two-thirds of the town. All the community facilities have been grouped together in this multi-purpose windscreen building. They consist of an elementary and secondary school; a shopping centre with hotel and entertainment establishments; and an indoor swimming pool, ice-hockey rink, curling club, and bowling alley. The building also houses apartments, thus giving residents access to the community facilities by means of a climate-controlled walkway. In addition to this man-made windscreen, stands of black spruce have been left in some areas to provide shelter for low-density housing.

The town has been developed to accommodate 26 persons per acre [0.4 ha], a much higher density than in most northern settlements. The streets have been kept short and curved in order to cut down on wind tunnelling effects and to reduce walking and driving distances. This has meant savings on capital investment in paved roads, sidewalks, sewers, street lighting, and power distribution, and has also cut down on the cost of road maintenance, snow clearing, and policing.

The community has a variety of housing types. Approximately one-third of the housing consists of apartments in the windscreen building and its adjoining wings, another third is made up of townhouses and semi-detached houses, and the remainder are single detached houses. The latter were included at the request of citizens, some of whom desired more privacy. Except for the apartment blocks, all the units were prefabricated near Montreal and transported to the site by rail and truck. To provide more variety, housing included several models in each category. A committee of women whose husbands were to be transferred to Mount Wright worked closely with the architects to design house and apartment interiors suited to northern living.

In order to bear the heavy snow load, roofs were built with a low pitch. The dwelling units were oriented so that living room windows faced southeast, south, or southwest in order to gain full benefit of the winter sun. The windows also look out on the lake and the tree-covered mountains beyond.

All houses, including their basements, are electrically heated. Indeed, Fermont is one of Canada's first electrical towns. When the houses were being constructed, strict insulation standards ensured maximum use of heat from the sun's rays. Garage floors are concrete slabs on gravel beds 2.4 metres thick. Driveways are no more than 7.3 metres long, thus reducing the surface that must be cleared of snow. In an area where the annual snowfall ranges between 3.6 metres to 4.8 metres or more, this is no small consideration. Garages extend toward the street and beyond the house line in order to further reduce the demand for snow removal.

Unlike the plan of some company towns, housing for corporate executives has not been kept apart, but is scattered throughout the built-up area.

Since a vast natural landscape lies within a short walking distance of the town, parks, open spaces, and large front and back yards have not been incorporated into the design. In any case, the northern climate and poor soil make gardens and greenery difficult to maintain. However, each dwelling has been provided with its own small garden, patio, terrace, or balcony.

Fermont is considered by many to be superior in concept and design to the town of Swappavara in Sweden, where the principle of wind-screens has also been applied. The Swappavara windscreens consist of long strips of three-storey apartment buildings. However, the housing is more dispersed than was proposed in the original plan, and an intended town centre was never built.

In general, Fermont boasts a variety of services and recreational facilities which are not usually found in northern European towns. As a result of careful planning, and the use of energy-conserving measures, the inhabitants are able to enjoy a relatively high standard of living, with many of the amenities provided in a southern town. At the same time, they are protected as far as possible from the inhospitable climate of the North.

43 Rankin Inlet: From Mining Town to Commercial Centre

STACEY J. NEALE

Between 1957 and 1962 the community of Rankin Inlet was a thriving mining town that had grown out of southern investment concerns designed to exploit the nickel deposits found in the area. The importance of the Rankin Inlet Nickel Mine Limited, which broke ground in 1953, cannot be overlooked since it illustrates how industrial change came north and was solely responsible for the development of this settlement. The new community would come to symbolize the dramatic shift in lifestyle that confronted the Inuit during the post-war period. In time, Rankin Inlet became like any other mining town in Canada, boasting such institutions as churches, schools, and a Hudson's Bay Store that serviced the needs of the rising population of southern and Inuit workers. By 1960 the community was a beacon attracting Inuit from across the Arctic with promises of stable employment.

In keeping with most northern development projects, the company brought with it everything it could possibly need in order to run a successful mine. In all, 1,500 tons of supplies were unloaded from a meticulously packed ship and, within a short period of time, these raw materials were converted into buildings which eventually became a community. Operations did not run smoothly, however, and finances became a concern, but by August 1956, additional supplies were unloaded and the renamed mine, the North Rankin Nickel Mine Limited, proceeded with its plans. For the five years that the mine was in full operation, the economy of Rankin Inlet was booming and there

SOURCE: *Inuit Art Quarterly* 14, no. 1 (spring 1999): 21–2. Reprinted by permission of the author and publisher.

was little concern, on the part of the owners, of the consequences should the mine close. Inuit employment with the company first started with the off-loading of the supply ship in 1953 and by the time it closed in 1962, 80 per cent of the workforce was Inuit. Considered a bold undertaking at the time, North Rankin Nickel Mine Limited employed Inuit in all aspects of its operation and this experiment introduced the Inuit to the ways of the industrial world.

Despite cultural differences between the Inuit and the white owners of the mine, solutions were found that accommodated the Inuit need for time on the land and the company's need for a smoothly run operation. The Inuit proved to be good employees and, while the pay was no doubt low, it provided them with steady work.

With so many Inuit dependent upon the wage economy and easy access to southern goods in the community, swift acculturation took place. The period of prosperity was short-lived, however, and the closure of the mine had a dramatic effect on the fate of this little town which was based on southern economic standards. The government, eager for the Inuit to find gainful employment, was pleased with the mine's success, but was ill prepared for the catastrophe that was about to take place. When the mine closed in 1962, most Inuit had grown accustomed to the wage economy and were reluctant or unable to return to the land. The community was devastated by the closure and alternative forms of employment had to be found as most of Rankin Inlet's citizens were collecting social assistance.

Ultimately the government's favourite plan was firm: "This little town was 'phasing out,' and funds were not available for art or anything creative, new, or practical" (Williamson 1980: 18). A government report, prepared after the mine's closure, concluded that Rankin Inlet was overpopulated. As relocation was a primary preoccupation for the bureaucrats in Ottawa, who had little understanding of life in the North, it is not surprising that moving the residents to other mining towns, other underpopulated communities, or back to the land was offered as the best solution for the Inuit.[1]

This plan was met with strong resistance by the several hundred people who had made a home for themselves there and refused to leave. The move to settlement living had provided Inuit with a regular income and access to "improved living conditions and health, education, and

---

[1] At the time, relocation was a popular solution to the various problems plaguing northern communities. For a comprehensive overview of this controversial policy, please see Alan Rudolph Marcus, *Relocation Eden: The Image and Politics of Inuit Exile in the Canadian Arctic* (Hanover, New England, Dartmouth College: University Press of New England, 1995) and Frank Tester and Peter Kulchyski, *Tammarniit (Mistakes): Inuit Relocation in the Eastern Arctic 1939–63* (Vancouver: UBC Press, 1994).

retail facilities [which] all combined to motivate the majority of the Rankin Inlet Eskimo population toward a way of life which would perpetuate the enjoyment of such advantages" (Williamson 1974: 127. From this perspective, it is not surprising that returning to the uncertainty of nomadic living was not appealing. Also, as Robert Williamson, a longtime resident of Rankin Inlet and supporter of Inuit art, pointed out at the time, many Inuit "had relinquished their dog teams, disposed of their skin clothing and indeed some of the weapons they needed for the hunt" (Williamson 1974: 127)

The option of establishing small industry in Rankin Inlet was seriously considered. A cannery and an arts and crafts centre were built and attempts to create a tourist industry also took place. Throughout the 1960s, these activities met with varying degrees of success. Also at this time, the Department of Indian Affairs and Northern Development was working on plans to spread the arts and crafts industry across the Arctic and "decisions on where to establish projects were generally based directly on economic need rather than on any special interest by the local Inuit to produce arts and crafts" (Goetz 1984: 43). There is no doubt that Rankin Inlet was in need of this assistance for economic reasons and to improve morale.

This era saw the development of more structured programs such as printmaking. These new projects called for "a special building and a planned program ... [which] imposed a structure foreign to the experience of the new townspeople they were designed to serve" (Goetz 1984: 44). In Rankin Inlet, however, the adaptation to working in the mine had left the Inuit with a clear impression of structured employment. Being a member of the new arts and crafts program was seen as a positive sign that you were receiving a steady income, which enabled you to provide for the needs of your family through the western wage economy.

The community of Rankin Inlet benefitted from the presence of the Inuit art industry: an arts and crafts officer was hired, sewing and carving programs were established, participants received wages in relation to their contribution to the program, and morale improved among the participants. In time, an arts and crafts centre was built to service the needs of this expanding endeavour.

Slowly, Rankin Inlet began to rebound from its earlier setback as the infrastructure that the mine created became an important element in the community's survival. Housing, schools, and other southern conveniences were in place to support the next phase of northern development. In the early 1970s, after the transfer of power from the federal government to the Northwest Territories, the administrative centre for the District of the Keewatin [Kivalliq] was moved from Churchill to the old mining town. This new prosperity, however, had an adverse effect on the ceramics project as senior artists like Donat Anawak and Robert Tatty left the program to take better-paying jobs. Rankin Inlet as a government town continues to thrive. Today, with a population of roughly 1,400,[2] it has become an important communication and transportation centre for the eastern Arctic and services the needs of all its northern citizens. Unlike the communities of Baker Lake and Cape Dorset, Rankin Inlet has never been able to define itself as an art centre even though many of its artists gained international reputations during the productive 1960s and art continues to be an important factor in a town that is dominated by western economies.

References

Goetz, Helga. 1984. "The Role of the Department of Indian and Northern Affairs in the Development of Inuit Art." Unpublished manuscript. Ottawa: Department of Indian Affairs and Northern Development.
Williamson, Robert. 1980. "Creativity in Kangirlliniq." *Rankin Inlet/Kangirlliniq*. Winnipeg: Winnipeg Art Gallery, 11–23.

---

[2] In 2001 the population was 2,177. – ed

44 Arctic Housing Update

GABRIELLA GOLIGER

Once a year in late summer a boat pulling crate-laden barges arrives in the tiny western Arctic settlement of Holman Island.

For a hectic few days, another year's supply of goods from the south is unloaded on the dock. It is a scene repeated in all the Inuit villages scattered along the Arctic's vast coastline. Besides packaged dry goods, food staples, appliances, and hardware for the Hudson's Bay stores, the barges are also loaded with crates of building supplies and components. The Arctic has few indigenous construction materials, so virtually every board and nail must be imported from the south.

Since the early 1950s when the Inuit first began to live in permanent settlements, they have relied on prefabricated housing units built in the south and shipped north for reassembly. In an earlier age they simply built small but efficient winter shelters – snow houses – and lived in skin tents or sod houses in summer. But as the Inuit abandoned their nomadic existence for a life in fixed communities, their housing needs changed drastically.

The first communities were crude shacks built of scrap materials left by the white men who had come to the North. Whole families crowded together in cramped quarters. Contagious infections and pernicious diseases such as tuberculosis were rampant.

In response to this crisis, the federal Department of Indian and Northern Affairs shipped small, one-room houses some 26 square metres the new settlements. Later, these "matchboxes" were followed by larger one- to three-bedroom units of up to 66.8 square metres. Both

SOURCE: *Habitat* 24, no. 1 (1984): 24–8. Reprinted by permission of the publisher.

the "matchboxes" and the larger bungalows were provided to the Native people at low rents under the federal government's Northern Housing Rental Program.

Although better than scrapyard shacks, the buildings from this early housing program were crude dwellings by southern standards and certainly not the final answer to Native housing needs.

In 1974 the Northwest Territories government, through its newly formed Northwest Territories Housing Corporation (NWTHC), took over responsibility for northern housing from the federal government. Its mandate was to "make available an adequate standard of housing to all residents of the Northwest Territories." It was a formidable task for the fledgling organization considering the harsh Arctic environment, the lack of building experience in the North, and the special needs of Native peoples adapting to a foreign lifestyle.

Temperatures may plummet to an icy −30°C or colder during much of a winter that lasts seven months. Gale force winds howl through settlements unprotected by trees or other natural barriers. It takes a sturdy house to withstand the constant battering of fierce Arctic storms.

But by far the biggest environmental headache for designers of northern houses is the permafrost, or permanently frozen ground, that underlies most of the Northwest Territories. Permafrost areas have two distinct layers – an upper or active layer that freezes and thaws with the seasons, and a lower layer that may be more than 100 m thick in places and remains frozen.

A slight change in the environment and the loss of insulating vegetation, for example, can start the permafrost thawing and turn once-solid ground into a muddy bog. Foundation problems arise when the heat from a house placed directly on the permafrost causes the frozen soil to melt. As a result, the house starts to sink into the mud.

The permafrost problem is usually tackled in one of two ways. First, by constructing houses on piles sunk deep into the frozen ground – an effective but expensive method. Or, by the more common method of erecting housing on insulating gravel pads. However, ice often remains embedded in the gravel, creating an unstable foundation as it gradually thaws, and wreaking havoc with joints and woodwork.

Another major obstacle for northern builders is the short construction period in the Arctic, a mere four or five months at the best of times. Moreover, there are few indigenous materials with which to build houses in the tundra, nor are there as yet enough skilled tradespeople in the North to build housing components. This means that materials must be imported over great distances from the south to be assembled before the first winter storms.

To offset these problems, northern houses are usually prefabricated in the south and shipped to the Arctic in easy-to-assemble panels.

Needless to say, the transportation costs, handling accidents, and shipping delays all add to an already inflated construction bill.

Houses built by the NWTHC since 1974 have been, in general, more in keeping with southern standards than those built under the first crisis-oriented housing programs.

Constructed under emergency pressure and under short-sighted programs, a large number of northern homes are ill-suited to the environment, and are deteriorating as a result.

The prefabricated buildings are especially susceptible to movements of their gravel pad foundations. For many a northern family this means gaping cracks between floors and walls, ill-fitting doors, and broken window frames that let in icy drafts. Many houses, moreover, are insufficiently insulated, having been built at a time when fuel was much cheaper than it is today.

Insulation will, of course, help bring down energy consumption in northern homes that currently burn up huge amounts of fuel. In some settlements, families might consume a staggering 9,092 litres of oil a year, while a comparable home in the south might use only a third of that quantity. But there is also a tendency for northerners to be less concerned with conservation since few pay fuel bills out of their own pockets. Indeed, few could afford to do so.

The Territorial government subsidizes rent-to-income public housing and, in the case of government employees, low-cost staff housing schemes. Government officials hope that with the gradual increase of homeownership in the North, and with greater community involvement in housing programs, residents will become more conservation-conscious.

With the support of the federal government through Canada Mortgage and Housing Corporation, the NWTHC is attacking this and other northern housing problems in several ways. It has initiated a major rehabilitation program to bring existing houses in the North up to standard and to make them more energy-efficient. Architects in consultation with northern communities are creating innovative building designs, specifically adapted to northern conditions. Through extensive training programs, the NWTHC is fostering the growth of a local construction industry as a means of lowering the exorbitant cost of building in the North.

Unless rehabilitated, the average northern house has a life expectancy of about fifteen years. This is due to the wear and tear caused by nature, and also to heavy use by large families crowded into their homes through much of the year. The rapid deterioration of buildings compounds the already existing housing shortage. Year after year, demand outstrips supply for, with improvements in health services and living conditions, the Native population has been growing rapidly.

NWTHC officials estimate that another 4,000 units will be needed in the Territories by the end of the 1980s to augment the existing stock of roughly 10,000 units. Current plans call for rehabilitation and insulation by 1989 of most of the 2,360 houses built under the Northern Housing Rental Program.

Much of the rehabilitation work on northern homes will be done by local Native people who are currently receiving extensive training in the building trades. "We're encouraging local people to take more and more control over their own housing," says George Forrest, managing Director of the NWTHC.

"Our goal is that within three years, housing will be turned over to the Native people through their local housing associations and through the district housing federations. The Corporation will become a resource for technical assistance, financial funding, and control teaching."

To realize this goal, the NWTHC has launched a five-year training program whereby local residents learn the construction, maintenance, and management trades. In 1980 alone the Territorial government spent $420,000 on instruction for 130 trainees. The NWTHC also plans to train local contractors in tendering procedures, program, and finance management.

In addition, community housing associations are participating in the planning, design, and siting of new homes in consultation with NWTHC officials and architects. The hope is that a home-grown construction industry will eventually lower the high cost of building in the North. Obviously it will be cheaper in the long run to have northern homes built by local tradespeople than by importing skilled labour. The growth of a local industry will also bring badly needed jobs to northern communities.

The Inuit, many of whom still rely on hunting for a livelihood, feel that community involvement in housing is long overdue. For years they have been frustrated by the fact that their homes were designed, built, and administered by outsiders who did not always understand the Native lifestyle.

"Our basic premise now is that a house must effectively represent the needs of the hunter," says George Forrest. Innovative homes now being built in the North include details such as a porch for the hunter to butcher meat or repair a snowmobile. Forrest is especially proud of a series of demonstration homes just completed in seven communities in the Keewatin district of the Northwest Territories [now Kivalliq in Nunavut]. Most striking about these demonstration houses are their energy-conserving features. Each building is a highly insulated, airtight cube with walls 30 cm thick. Almost all windows face south to bring in as much sunlight as possible and a large porch across the front of the building provides a buffer against strong winds. The porch also acts as a passive solar collector to capture the sun's heat on bright days.

The demonstration homes are heated by centrally located space heaters controlled by non-electric thermostats. In the event of a power failure, not uncommon during fierce Arctic storms, the heating system can still function. Their triple-glazed windows, three outer doors, and airtight construction help to seal the Keewatin [Kivalliq] homes against cold drafts and the escape of precious heat from indoors. The airtight seal, however, necessitates the use of a special ventilation system to bring in fresh air and to prevent a buildup of moisture in the house. An air-to-air heat exchanger allows moist hot air from the house to trickle outdoors between sheets of plastic, and permits cool air to enter. The architects estimate that sound design measures will reduce fuel consumption by 90 per cent over conventional housing units.

Michael Pine, a CMHC policy analyst who has kept a close watch on the project, puts it more graphically: "I wouldn't be surprised if those houses couldn't be heated to a liveable level with just body heat and candles – I don't mean comfortable – but you could live in them."

The Keewatin [Kivalliq] homes have also been designed to accommodate the Native lifestyle. For example, the large cold porch can be used for preparing meat and skins, and the kitchen merges with the living rooms, enabling the housewife to keep an eye on her children or chat with visitors while she is cooking.

NWTHC officials are enthusiastic about the experimental houses and expect them to be in great demand in other settlements. In fact, a new generation of demonstration units has been planned for the Baffin region.

Michael Pine cautions, however, that there might still be some kinks to iron out in the new units.

"This is still a demonstration house. There might still be some problems with it. The airtight seal on the house, for example, may cause bad condensation problems in spite of the air exchanges. We're going to find that out in the coming year."

As a 1978 statement on housing by the Inuit Tapirisat of Canada (the national Inuit organization) points out, there is no shortage of work to do. "Two-thirds of the houses in Inuit settlements in the Northwest Territories are in need of major repair or are in condemned condition; over half the households live in overcrowded conditions, and basic community sewage and garbage disposal are drastically in need of improvement. To compound the problem, the supply of new housing cannot keep pace with new family formations."

Since 1978 progress has been made. The Housing Corporation and the people who occupy these homes are now taking part in hardheaded, clear-thinking, long-sighted planning so that all northern families can live in comfort.

POLITICAL CHANGE

45 Integration of Territory into the Administrative Map of Quebec

JOHN CIACCIA

These people are inhabitants of the territory of Quebec. It is normal and natural for Quebec to assume its responsibilities for them, as it does for the rest of the population. And that is what the Quebec Government will be in a position to do as a result of this Agreement. It will be the guarantor of the rights, the legal status, and the well-being of the Native peoples of its northern territory. Until now, the Native peoples have lived, legally speaking, in a kind of limbo. The limits of federal responsibility were never quite clear, nor was it quite clear that Quebec had any effective jurisdiction. The land these people inhabited was in Quebec, after 1912, and yet Quebec's title was not properly defined. This agreement will remove an grounds for further doubt or misunderstanding. Jurisdiction will be established in a precise and definitive manner. Until now, Quebec's presence in the North has not been complete. Today we are completing and reaffirming this presence.

The Government of Quebec has taken the opportunity presented by these negotiations to reorganize the territory, and to set up the institutions and structures that will give substance to the role that it intends to fulfill. The Native communities will have local administrations substantially in the manner of local communities throughout Quebec, and regional administrators will exercise municipal functions in areas beyond the old established communities. In districts inhabited by both Native and non-Native populations, Cree representatives and representatives of the Municipality of James Bay will form a joint adminis-

SOURCE: *The James Bay and Northern Quebec Agreement* (Quebec City: Editeur Official du Québec, 1976): xv–xvi (extract). Reproduced by permission of Publications du Québec, 2001.

tration to be known as the Zone Council. On this subject, let me draw your attention to the fact that with this Agreement twenty-one new municipalities are to be created. All will be subject to the Ministry of Municipal Affairs. Thirteen of the municipalities will answer to the Ministry through an administration to be called the Kativik Regional Government, a novel instrument suited to the conditions of the region. In all, 250,000 square miles [647,500 km²] of newly organized territory will come under the jurisdiction of the Ministry of Municipal Affairs when this Agreement is implemented.

Why do we want to do all this? Simply because there are people living in the North, who need public services, who are counting on good administration of their affairs, and who have a right to participate in that administration. The principles of sound and rational administration prompt us to act in this manner. The well-being and the interests of the people require that we do it.

The inhabitants of Quebec's North, like everybody else, have to have schools. They have to be able to depend on health services. They have to have the security of justice and a system of law enforcement. This Agreement responds to these needs, and provides the structures through which they can be met. There will be local school boards, health and social services boards, police units, fire brigades, municipal courts, public utilities, roads, and sanitation services. All of these agencies will answer to the appropriate ministry of the Quebec Government. The proper jurisdiction of all ministries, such as, for example, the Ministry of Education, will remain intact. The services will all be provided through structures put in place by the Government of Quebec.

This means that where facilities such as schools and hospitals already exist under federal jurisdiction in Native communities, they will be transferred to the jurisdiction of Quebec. In the case of certain federal programs, already operating, the Quebec Government will assume the responsibility for them,

These are all steps that would have to be taken, these are all services that would have to be provided and developed anyway, regardless of whether or not there was a James Bay project. What the Government of Quebec is doing here is taking the opportunity to extend its administration, its laws, its services, its governmental structures throughout the entirety of Quebec; in short, to affirm the integrity of our territory.

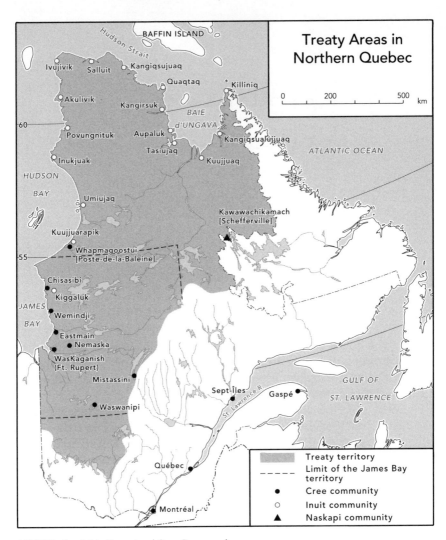

Treaty Areas in Northern Quebec

0   200   500 km

BAFFIN ISLAND

Hudson Strait

Ivujivik   Salluit   Kangiqsujuaq
Quaqtaq   Killiniq
Akulivik   Kangirsuk
BAIE
d'UNGAVA
60
Povungnituk   Aupaluk   Kangiqsualujjuaq
Tasiujaq   ATLANTIC OCEAN
Inukjuak   Kuujjuaq
HUDSON
BAY
Umiujaq
Kawawachikamach
[Schefferville]
Kuujjuarapik
Whapmagoostui
55   [Poste-de-la-Baleine]
Chisasibi
Kiggaluk
JAMES   Wemindji
BAY
Eastmain
Nemaska
WasKaganish
(Ft. Rupert)
Mistassini
Sept-Îles   GULF OF
Gaspé   ST. LAWRENCE
Waswanipi
St. Lawrence R.

Québec
Treaty territory
Limit of the James Bay territory
● Cree community
○ Inuit community
▲ Naskapi community
Montréal

SOURCE: after Sylvie Vincent and Garry Bowers, eds.

46 Aboriginal Land Claims

PARKS CANADA

A Brief Overview

In Canada, the common law concept of Aboriginal rights and title has been recognized by the courts. The existing Aboriginal rights of Aboriginal peoples have also been recognized and affirmed under section 35 (1) of the *Constitution Act, 1982*.

The evolution and development of the federal government's land claims policy has been linked to court decisions. The first claims policy statement in 1973 was initiated by a decision of the Supreme Court of Canada (the 1973 *Calder* decision) which acknowledged the existence of Aboriginal title in Canadian law. In order to address uncertainties created by the decision, the federal government announced its intention to negotiate claim settlements. As the policy developed, claims were divided into two types:

1 Comprehensive claims – based on the concept of continuing Aboriginal rights and title that have not been dealt with by treaty or other legal means; and

2 Specific claims – arising from alleged non-fulfillment of Indian treaties and other lawful obligations, or the improper administration of lands and other assets under the *Indian Act* or formal agreements.

SOURCE: New Parks North, Newsletter 9 (March 2000): 2-4. Reprinted with the permission of the publisher.

In recent years, an unnamed third category of claims has developed to deal with Aboriginal grievances that fall within the spirit of the comprehensive and specific claims policies, but do not meet strict acceptance criteria.

Comprehensive Claims

The primary purpose of comprehensive claims settlements is to conclude agreements with Aboriginal groups that will resolve the legal ambiguities associated with the common law concept of Aboriginal rights. The objective is to negotiate modern treaties which provide clear, certain, and long-lasting definition of rights to lands and resources. Negotiated comprehensive claim settlements provide for certainty for governments and third parties in exchange for a clearly defined package of rights and benefits for the Aboriginal beneficiaries codified in constitutionally-protected settlement agreements.

Comprehensive claim agreements define a wide range of rights and benefits to be exercised and enjoyed by claimant groups. These may include full ownership of certain lands, guaranteed wildlife harvesting rights, participation in land and resource management throughout the settlement area, financial compensation, resource revenue-sharing, and economic development measures.

If a national park is established in a settlement area through the claim process, the claimant group continues to exercise its traditional harvesting activities within this protected area. As well, a management board may be established, with representation from the Aboriginal community and government, to advise the Minister on the management of the national park. Finally, the land claim agreement sets out what economic opportunities associated with the national park will be enjoyed by the claimant group. These may include employment provisions and contracting opportunities.

Significant amendments to the federal comprehensive claims policy were announced in December 1986, following an extensive period of consultation with Aboriginal groups. Key changes to the policy included the development of alternatives to blanket extinguishment of Aboriginal rights, as well as provision for the inclusion in settlement agreements of offshore wildlife harvesting rights, resource revenue-sharing, and Aboriginal participation in environmental decision-making. The 1986 policy also provides for the establishment of interim measures to protect Aboriginal interests during negotiations, and the negotiation of implementation plans to accompany final agreements.

The 1997 Supreme Court of Canada decision in *Delgamuukw* has initiated calls from within Aboriginal communities to once again review the comprehensive claims policy. The *Delgamuukw* decision is the first comprehensive treatment by the Supreme Court of Canada of Aboriginal title.

Self-government negotiations may take place parallel to, or at the same table as, the comprehensive claims negotiations. The federal government is prepared to consider constitutional protection of certain aspects of self-government where the parties to the agreement concur. Self-government must be negotiated in keeping with the 1995 *Framework for the Implementation of the Inherent Right and the Negotiations of Self-Government* policy.

Specific Claims and Treaty Land Entitlement

Specific claims relate to the fulfillment of treaties and to the federal government's administration of Indian reserve lands, band funds, and other assets. The government's primary objective with respect to specific claims is to discharge its lawful obligation to First Nations.

Treaty Land Entitlement (TLE) is a large category of claims that relate primarily to a group of treaties that were signed with First Nations, mainly in the prairie provinces. Not all these First Nations received the full amount of land promised. Claims from First Nations for outstanding entitlements are categorized as TLE claims and are handled separately from other specific claims.

Parks Canada is currently involved in TLE discussions that concern Wood Buffalo National Park. The Salt River First Nation indicated its wish to select land within Wood Buffalo National Park as part of its TLE negotiations. The Minister of Canadian Heritage agreed to consider this request in 1997. Salt River First Nation is in the process of being split into two bands – Salt River First Nation and Smith's Landing First Nation. The two bands are negotiating separately with Canada for their respective TLEs. Smith's Landing First Nation signed a Memorandum-of-Intent (MOI) with Canada and the Government of Alberta in December 1999. As part of this MOI, three reserve locations in Wood Buffalo have been set aside, totalling approximately 10 km². Guidelines for the use and management of the reserve lands within the park, and a land and resource management framework with Park officials, are set out in the MOI. Negotiations with Salt River for Indian Reserve land inside Wood Buffalo are also well advanced.[1]

---

[1] In June 2002 the Salt River First Nation signed a final agreement with the governments. – ed.

Other Claims

The federal government is reaching or negotiating settlement of a number of other Aboriginal grievances, which have sometimes been referred to as claims of a third kind. These grievances fall within the spirit of the comprehensive and specific claims policies, but do not meet strict acceptance criteria.

One such proposal now under negotiation involves the Metis of the South Slave Region of the NWT. When the Dene and Metis Comprehensive Land Claim Agreement was rejected by the Aboriginal communities in 1990, the federal government decided to enter into regional claims with Aboriginal groups in the NWT. However, in the South Slave District, Dene people have opted to seek fulfillment of their Treaty 8 entitlement. This left eligible Metis in this area without a vehicle to press for their concerns. A Framework Agreement was signed in August 1996 that outlines a two-stage negotiation process – land and resources and, after the signing of an Agreement-in-Principle, negotiation of self-government issues. A pause in negotiations has been called by the parties in this process. When negotiations resume, they may impact on Wood Buffalo National Park.

47 Tree Line and Politics in Canada's Northwest Territories

WILLIAM C. WONDERS

Introduction

In these days of increasing specialization in geography as in many other disciplines, the older emphasis upon interaction between the physical and the cultural facts of place is often downplayed. Ironically, a growing public concern about issues such as environmental conservation and Aboriginal rights, has pointed up to relevancy of knowledge about that interaction, yet it often is non-geographers who have moved into this field traditionally occupied by geographers. Canada's Northwest Territories (NWT) presently provide an example of such interaction on a vast scale.

The Northwest Territories' 3,426,320 square kilometres is over ten times the size of Great Britain (Baffin Island alone is twice the latter's area), and makes up 34 per cent of the total area of Canada. In the Territories' enormous extent there are only 52,238 inhabitants (1986 Census)[1] – less than 0.1 per cent of the population of Great Britain and only 0.2 per cent of the total Canadian population. Yet among the Canadian provinces and territories the situation in the Northwest Territories is unique in that Aboriginal peoples constitute the majority (58 per cent) of the total population, making them a dominant political force in territorial policies and elections.

[1] In 2001 the NWT's population was 37,360, and that of Nunavut was 26,745. – ed.

SOURCE: *Scottish Geographical Magazine* 106, no. 1 (April 1990): 54–60. Reprinted by permission of the publisher.

Recent and current political developments in the Northwest Territories point up the importance of one element of the natural landscape – tree line – in the cultural, economic, and political life of the inhabitants. While this is particularly true for the Aboriginal peoples, it also has great potential future importance for the local white population and even for Canada generally as regards future resource development policies. Hitherto tree line has been of interest chiefly to a small number of specialists, but recently it has come into prominence as a matter of much greater consequence. This paper focuses on its political significance for the Aboriginal peoples of the Northwest Territories and for the Territories' future.

Boreal Forest Zonation and Tree Line

'Tree line' has been long identified as one of the most significant features of the high latitudes by botanists, climatologists, geographers, and anthropologists. It is often designated on a variety of maps of the northern hemisphere and of specific northern lands. Closer study, however, has shown that the 'line' is not as precise as many generalizations suggest.

As much as fifty years ago the Finnish geographers, Tanner (1938, 1944) and Hustich (1939) stressed the transitional nature of the change from boreal forest to tundra in Labrador. About the same time Halliday extended his concepts of forest sections, based on his work in the Canadian Prairie Provinces, into the larger national forest regions, resulting in an authoritative forest classification for all Canada. In this an extensive 'Northern Transition Section' was designated between an Arctic tundra area and a 'Mackenzie Lowlands Section' of the Boreal Forest in the Northwest Territories. In the transition section he noted that "areas of open swamps and tundra associations are intermixed with the stunted forest cover, and the trees become confined to narrow fringes on the river-banks, and finally disappear" (Halliday 1937: 24).

In subsequent years much additional research was carried out on the boreal forest and 'tree line' in the Northwest Territories as well as elsewhere in the North. In 1959 Rowe revised Halliday's work, though maintaining the essential original framework, with a further slight revisions in 1972 (Rowe 1972). In Rowe's revised regional delineation (Figure 47.1), the Mackenzie Lowlands forest was greatly contracted westwards, restricted essentially to the actual physiographic lowlands of the Mackenzie River system. He further subdivided it approximately at Great Bear River into southern and northern sections ('Upper Mackenzie' and 'Lower Mackenzie' respectively). Most of the forest east of the lowlands and most of Halliday's 'Northern Transition Section', were included in a broad 'Northwestern Transition' section

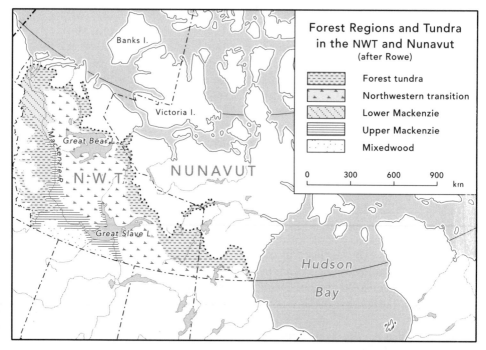

Figure 47.1

by Rowe – "a zone of open Subarctic woodland which extends from Hudson Bay almost to the delta of the Mackenzie River ... areas of bog, muskeg, and barren rock are intermixed with open stands of dwarfed trees, although on local patches of sheltered, deep, frost-free soil the density and height growth of forest patches can be surprisingly good. Characteristic of the park-like coniferous stands on upland sites is a ground cover of light-coloured, foliose lichens" (Rowe 1972: 55). Between this transition section and the tundra, Rowe introduced a new 'Forest tundra' section of the tundra 'barrens' and patches of stunted forest.

The three northwest-southeast trending vegetation zones delineated in the NWT south of the tundra proper generally conform with zonal divisions recognized by other scientists nationally and internationally, though the terminology may vary (Atkinson 1981; Timoney 1983). The 'forest-tundra', 'open woodland', and 'closed forest' zones of Hare and Ritchie (1972) correspond with Rowe's 'forest tundra', 'northwestern transition', and 'upper Mackenzie' sections. The forest tundra zone is subdivided by some into a forest tundra sector proper and a 'shrub tundra' sector between it and the true tundra.

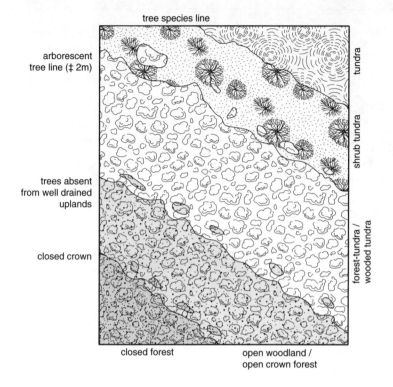

Figure 47.2
Forest-tundra transition (after Timoney)

While there may be general acceptance of the basic zonal patterns of boreal tree growth, there still is considerable variation in precise definition of 'tree line' and of 'forest tundra', and for that matter in defining a 'tree'. After reviewing the latter variations Timoney accepts Hustich's concept of a tree at its Subarctic limit as being >2 m in height, providing the trunk projects above the snow cover (Hustich 1953). This marks the northward limit of the forest tundra – the 'arctic tree line' of Hare and Ritchie, the 'tree-form line' of Atkinson, and the 'arborescent tree line' of Timoney (Figure 47.2). It should be noted however that both upright and shrub *krummholz* forms of conifers may extend into the shrub tundra. The southern boundary of the forest tundra is termed 'tree line' by some and the 'northern forest line' restricted to the northern limits of the closed crown coniferous forest. Without delving further into the complexities of definition, 'tree line' clearly is not the precise reality that the widespread usage of the term might suggest.

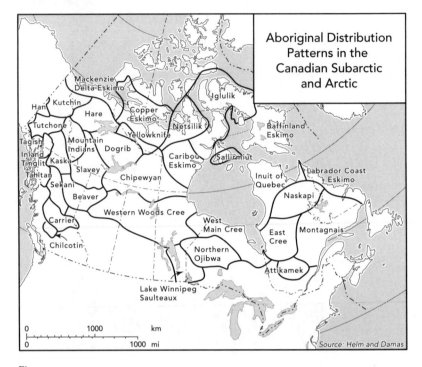

Figure 47.3

One further complication which should be noted is that the geographic location of tree line is not a fixed, permanent thing. Major controls of its location are climatic and ecological, with their relative importance still under investigation (Elliot-Fish 1983). There is ample evidence however that during the warmer summer temperatures (5°C warmer than today) of the Main Climatic Optimum of the Hypsithermal Interval, woodland could survive several hundred kilometres north of present tree line and extend into the Thelon Basin of the Northwest Territories, for example (Atkinson 1981; Nichols 1976).

Aboriginal Distribution Patterns

There still are significant discrepancies between government and other authorities on the actual numbers of Native peoples in Canada, partially reflecting variations in definition (Wonders 1987). The 1986 census recorded 30,530 Natives in the total population of 52,238 in the Northwest Territories (Statistics Canada 1987). Two major Aboriginal peoples are involved: the Inuit (or Eskimo) 18,360, and the Dene (Indian) 9,380, with an additional 3,825 Metis or mixed blood people, mainly Dene-white. (Some individuals identify with more than one Native origin).[2] Traditionally the Inuit have been characterized as tundra dwellers and the Dene as forest dwellers, by the people themselves as well as by anthropologists and for the most part this is a satisfactory distinction (Helm 1981; Damas 1984). 'Tree line' therefore has separated the two Aboriginal peoples and today still continues to do so (Figure 47.3) At the same time it has been clearly established that there have been frequent occasions on which members of one group have ranged well into the other group's area, for example Mackenzie Delta Eskimo [Inuit] coming as far south as the Mackenzie River Ramparts to obtain flint for their weapons and the Copper Eskimo [Inuit] to Great Bear Lake for wood for their komatiks, while Chipewyan Dene ranged out into the tundra in hunting the barren-land caribou. In some areas such as that east of the Mackenzie Delta the same area was used by Dene in winter and by Inuit in summer.

Historically the relations between the two peoples have ranged from hostile to tenuous, with mutual avoidance the norm. The lucrative fur trade of the Mackenzie Delta earlier this century drew both into close proximity and the region still remains the only major zone of contact between Dene and Inuit. Elsewhere the general movement of Native peoples off the land and into settlements has markedly reduced the current areas of overlapping land use by the two peoples (Wonders 1984; Figure 4). Today therefore there is an even more marked association of Inuit with the tundra/Arctic region and of Dene with the forest/Subarctic region (Figure 47.4) Overt hostility between the two is a thing of the past, but there still remains a lingering reserve between them.

Treaties and Politics

As part of Canada's sparsely settled wilderness frontier the matter of boundaries of most sorts has been of little real consequence until recently. This changed fifteen years ago when the Supreme Court of Canada acknowledged the existence of Natives' Aboriginal title to their traditional lands. The Natives' moral and legal rights have been strengthened further through subsequent inclusion of 'Aboriginal rights' in the new Canadian Constitution of 1982, though precise definition of these remains to be established. Since the court decision the Government of Canada has been attempting to settle comprehensive land claims with the Native peoples. (By July 1997 ten comprehensive

---

[2] In 1996, 62 per cent of the NWT's population were Aboriginal (38 per cent Inuit, 17 per cent Indian, and 6 per cent Metis), while 84 per cent of the population of Nunavut were Aboriginal (83 per cent Inuit). – ed.

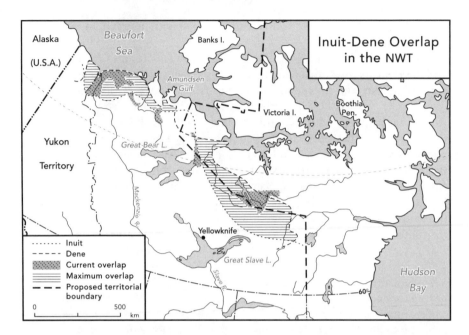

Figure 47.4

agreements had been finalized in the northern territories, including those with the Inuit, Inuvialuit, and several Dene and Yukon First Nations.) Some of the complications in reaching that agreement and other similar agreements have been pointed out previously, including the requirement that Natives reach agreement amongst themselves in cases of overlapping land use (Wonders 1985). Since 1984 the focus has shifted eastwards in the NWT as agreement is sought between the Dene and Metis on one hand and the main body of the Inuit on the other. Most of the same complications persist, but farther reaching political implications are involved.

All Native peoples of the NWT have emphasized the importance of land and control of it to their cultural survival. This has been stressed particularly by the Inuit. They felt that though the 1967 proclamation of Yellowknife as territorial capital and subsequent transfer of most government offices had met the criticism of 'remote control' for most Territorial residents (who live in the Subarctic), it still was far removed from many of the Arctic residents. Moreover, the Inuit believe that they have very different attitudes and needs even from those living in the Subarctic where the much larger numbers of whites has resulted in simply importing southern white values and superimposing them on the resident Aboriginal peoples. The Arctic, however, remains almost entirely Inuit in its ethnic composition.

As long ago as February 1976 the Inuit political organization, the Inuit Tapirisat of Canada, called for the separation of the area north and east of tree line as a new territory and future province, 'Nunavut' ('Our Land') to be governed by the Inuit. In October of the same year the Indian counterpart, the Indian Brotherhood of the NWT (now the Dene Nation) also called for formation of an Indian province 'Denendeh' ('Our land'), centred on the Mackenzie Valley and of two other new provinces for Inuit and whites respectively. The Prime Minister of Canada rejected the proposals in a policy statement in August 1977, on the grounds that "Legislative authority and government jurisdiction are not allocated in Canada on grounds that differentiate between people on the basis of race ..." (*The Edmonton Journal*, August 4, 1977).

Territorial Division

The Inuit refused to let the issue die, and continued to press the matter politically. In November 1980 the Legislative Assembly of the NWT agreed in principle to the division. A Territorial plebiscite on the issue in April 1982 resulted in 82 per cent support in the Eastern Arctic, but the much lower support in the Subarctic reduced the overall territorial approval to 56 per cent.[3] In November 1982 Ottawa approved the split in principle but saw no provincial status "for the foreseeable future" because of the small populations and weak economies. Amongst prior conditions imposed before any actual division were that all land claims in the NWT must be settled, that the location of the boundary between the two territories had to be agreed upon, and territorial powers had to be suggested.

Ottawa established two regional constitutional groups including representatives of all peoples of the NWT, to work together as an alliance on the division process – the Nunavut Constitutional Forum (NCF) dominantly Inuit in the Arctic, and the Western Constitutional Forum (WCF) mixed Dene/Metis/White in the Subarctic. Any ultimate division agreement had to be approved by a majority of Territorial residents. At

---

3 As generally used in Canada and the NWT, 'Subarctic' means the area south of tree line and 'Arctic' the area to the north. Until recently 'Eastern' and 'Western' Arctic were usually divided on the basis of sea accessibility. Since the formation of the Committee for Original Peoples' Entitlement (COPE) in 1976 amongst the Inuvialuit and their 1984 agreement with the Government of Canada, the latter division is generally made at the agreement boundary line northwest of Coppermine [Kugluktuk] (see Figure 47.5). The area west of Hudson Bay is sometimes now referred to as the 'Central Arctic'.

the opening of the Territorial Legislature in early February 1985 the federal government announced that it hoped division of the NWT would be effected by 1987. Such has not proved to be the case, despite the expenditure of some $4.5 million to date by the Government of Canada towards the process.

The establishment of the boundary between the proposed new territories has proved the most difficult problem as it has in the matter of settling comprehensive land claims. Initially the NCF insisted upon tree line as the only acceptable boundary, in order to include all Inuit within the same territory. This was unacceptable to the WCF because the Western Arctic area is linked strongly with the Mackenzie Valley economically and in its transportation ties. Keller (1986) has stressed the cost of air travel between the widely spaced small communities which characterize the settlement pattern of the Northwest Territories, and the importance of this factor for any possible political division of the area. The potential long-term financial resource of the Beaufort Sea oil and gas revenue is a major attraction for both new territories. It is the Inuit of this Western Arctic (the Inuvialuit) who were the first Aboriginal people to settle their comprehensive land claims.

Bitter wrangling often characterized the meetings between the two constitutional groups and their supposed "alliance" was questionable at times. So strong was the Inuit push for division that the NCF finally agreed in 1985 to abandon their goal of cultural unity and to the inclusion of the Inuvialuit within the proposed western territory despite the objections of the latter group. Four Central Arctic communities were to vote on which territory they would join. Within weeks, however, the tentative agreement collapsed when eastern Assembly members withdrew their support, and it was a year before the two forums resumed negotiations in earnest with new key players.

In January 1987 the two constitutional forums signed an agreement which had been reached in May 1986 for the proposed division of the Northwest Territories (Anon. 1987). The proposed boundary was the eastern edge of the Inuvialuit Settlement Region – an arbitrary line which separates the Inuit of Holman on Victoria Island from their immediate relatives at Coppermine [Kugluktuk] and Cambridge Bay – and the line between the Dene/Metis and Inuit comprehensive claims regions. The latter line was ratified under their claims overlap agreement, but included over most of its extent an approximation of tree line (Figure 47.5). (The Territorial Legislative Assembly had previously voted to make the land claims boundary the political boundary between the proposed two new territories.) Target date for division was set for October 1, 1991. Prior to that the leaderships of member organizations of the forums and the Territorial Legislative Assembly had to approve the agreement, the boundary had to be ratified by a Territorial-wide plebis-

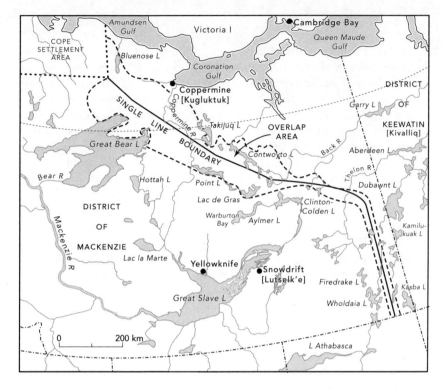

Figure 47.5
Single Boundary Line and overlap area in the Central Arctic

cite, and the residents of each new jurisdiction had to develop and rat-ify their new constitutions (Intercom 1987). It is anticipated that the latter will introduce new concepts reflecting the attitudes of Native peoples which may differ significantly from those of the pro-vinces, where European traditions prevail.

Having apparently reached consensus, the agreement foundered once more despite intensive last-minute efforts to salvage it. March 31, 1987 was the deadline imposed by the Territorial Legislative Assembly as one of several conditions for the plebiscite, and without agreement the plebiscite was cancelled. The immediate cause for the 1987 defeat was that the Dene/Metis communities rejected the boundary agreement of their negotiators. They proposed instead to extend the line to the north and east, thereby adding some 156,000 km² to the Dene/Metis claim area. The focus of their dissatisfaction were the hunting areas beyond tree line, in the vicinity of the upper Thelon River northeast of Great Slave Lake, and around Contwoyto Lake east of Great Bear Lake.

Conclusion

Despite the latest set-back talks are continuing between the two groups about political division of the NWT. It is difficult to know whether such division will occur even if tentative agreement is eventually reached. Although the main group of Inuit have been strong advocates of division, the mixed people of the more populous western part of the NWT (including the 22,000 whites) have been much less supportive. Moreover, the disillusioned Inuvialuit of the Western Arctic and many Inuit of the Central Arctic are resentful of a boundary which would reduce them to an ethnic minority in the western territory and divide some communities with close family ties. Start-up costs of effecting division have been estimated at between $332 million and $400 million and the total additional annual operating costs at between $97 million and $114 million (*The Edmonton Journal*, February 1987). With the Government of Canada providing about 70 per cent of the NWT income, and only some northerners pushing for division, it may be that the federal government would ask northerners to cover these additional expenses if division occurs.

Significantly, in the autumn 1987 elections for the 24-member Territorial Legislative Assembly the issue of territorial division received scant attention from any of the candidates, even in the Eastern Arctic. The voters were more concerned about employment and economic development. An Eastern Arctic member was elected government leader in the Assembly and it remains to be seen whether this evidence of greater influence by the people of this region (overwhelmingly Inuit) will deflect pressures for division or reinforce them.[4]

Recent decades have demonstrated very forcefully that tree line, with its many complexities, is no longer merely of interest to the scientific community. It has always been of major concern in the traditional lives of the Aboriginal peoples of northern Canada. In the light of its long-term political and economic implications it has now become even more vital for Native northerners and for all Canadians.

References

Anon. 1987. *Boundary and Constitutional Agreement for the Implementation of Division of the Northwest Territories between the Western Constitutional Forum and the Nunavut Constitutional Forum, January 15, 1987, Iqaluit, Nunavut*. Ottawa: Canada Arctic Resources Committee.

4 In fact, division of the former NWT did occur, and Nunavut became a reality in 1999 (see chapter 48). The remaining portion of the NWT decided to retain its former name. – ed.

Atkinson, K. 1981. "Vegetation zonation in the Canadian Subarctic." Area, 13, 13–17.

Damas, D. ed. 1984. *Handbook of North American Indians, Vol. 5: Arctic*. Washington: Smithsonian Institute.

Elliot-Fish, D. 1983. "The stability of the North Canadian tree-limit." *Annals, Association of American Geographers* 73, 560–76.

Halliday, W.E.D. 1937. "A forest classification for Canada." *Forest Service Bulletin* 29. Ottawa: Canada Dept. of Mines and Resources.

Hare, F.K., and Ritchie, J.C. 1972. "The boreal bioclimates," *Geographical Review* 62, 333–65.

Helm, J., ed. 1981. *Handbook of North American Indians, Vol. 6: Subarctic*. Washington: Smithsonian Institute.

Hustich, J. 1939. "Notes on the coniferous forest and tree limit on the east coast of Newfoundland-Labrador, including a comparison between the coniferous forest limit on Labrador and in northern Europe." *Acta Geographica* (Helsingfors), 7(1).

Hustich, J. 1953. "The boreal limits of conifers." *Arctic* 6, 149–62.

Intercom. March 1987. Ottawa: Indian and Northern Affairs, Canada.

Keller, C.P. 1986 "Accessibility and areal organizational units: geographical considerations for dividing Canada's Northwest Territories." *The Canadian Geographer* 30, 71–9.

Nichols, H. 1976. "Historical aspects of the northern Canadian treeline." *Arctic* 29, 38–47.

Rowe, J.S. 1972. *Forest regions of Canada*. Ottawa: Canadian Forestry Service, Dept. of the Environment.

Statistics Canada. 1987. *1986 Census of Canada*. Ottawa: Supply and Services, Canada.

Tanner, V. 1938. "Naturförhallenden pa Labrador. Iakttagelser under den finländska expeditionens färder och forsikningar ar 1937." *Arsbok XVII* B(1). Helsingfors: Societas Scientiarum Fennica.

Tanner, V. 1944. "Outlines of the geography, life and customs of Newfoundland-Labrador." *Acta Geographica* (Helsinki-Helsingfors), 8(1).

Timoney, K. 1983. Definition of treeline. Research Proposal (unpubl.). Edmonton: Department of Botany, University of Alberta.

Wonders, W.C. 1984. *Overlapping land use and occupancy of Dene, Métis, Inuvialuit and Inuit in the Northwest Territories*. Ottawa: Indian and Northern Affairs, Canada.

Wonders, W.C. 1985. "Our land, your land – overlapping native land use and occupancy in Canada's Northwest Territories." In Leidelmair, A., and Franz, K., eds. "Environment and human life in highlands and high latitude zones," *Innsbrucker Geographische Studien, Band 13*, Universitat Innsbruck, 189–97.

Wonders, W.C. 1987. "The changing role and significance of native peoples in Canada's Northwest Territories." *Polar Record* 23, 661–71.

48 Nunavut: Canada's New Arctic Territory

WILLIAM C. WONDERS

On April 1, 1999, the political map of Canada was changed dramatically when a new territory – Nunavut ("Our Land") – came into existence. It had been half a century since the last major political change, Newfoundland and Labrador joining the Canadian Confederation. Unlike that event, which added to the national area, Nunavut was created by division of the existing Northwest Territories (NWT).

The initiative for this creation lay with the political determination of the Inuit population of the Canadian Arctic to obtain a territory of their own. A new generation of younger, formally educated Aboriginal leaders was concerned that Inuit values and interests were increasingly influenced by White man's government and by the threat to the natural environment posed by large outside companies seeking to develop Arctic energy and mineral resources primarily for the benefit of distant shareholders.

The opportunity of effecting a significant change arose from a 1973 decision of the Supreme Court of Canada acknowledging the existence of Aboriginal title to the Native population's traditional lands. Since 1974 the Government of Canada, therefore, has negotiated land claims settlements with its Aboriginal peoples, and provided funds to them to cover the costs involved. Though much government responsibility had been devolved from Ottawa to Yellowknife in 1967 when the latter centre had been proclaimed capital of the Northwest Territories, many

SOURCE: Tom L. McKnight, *Regional Geography of the United States and Canada*, 3rd ed. (Upper Saddle River, N.J.: Prentice-Hall, 2001), 494–5. Reprinted by permission of the publisher.

Inuit still felt it was too removed physically and to a considerable degree philosophically from their Arctic focus.

In 1982 a NWT-wide plebiscite on division of the Northwest Territories passed, due to the overwhelming support by the Inuit and, thereafter, creation of a new territory and their own territorial government became an integral part of their claim. In June 1993 the Nunavut Land Claim Agreement Act was passed into law by the Canadian Parliament – the largest native land claim settlement in Canadian history. Major elements included giving Inuit title to more than 350,000 square kilometres (136,000 square miles) of land, of which 36,000 square kilometres (14,000 square miles) includes mineral rights, capital transfer payments of $1 billion over fourteen years, equal representation of Inuit with government on a new set of wildlife management, resource management, and environmental boars. In addition, it provided for the establishment of a new territory, Nunavut, and through this a form of self-government for the Nunavut Inuit.

The new territory has been carved out of the central and eastern Arctic parts of the former Northwest Territories. (The residents of the remaining western part prefer to retain the historic "Northwest Territories" name.) Nunavut includes almost two million square kilometres (735,000 square miles), about one fifth of Canada's total land mass (almost three times the size of Texas), with half of it on the continental mainland west of Hudson Bay and the other half in the Arctic Archipelago to the north.

The Government of Nunavut is elected by all residents of the territory, regardless of their origin, but with Inuit making up 85 per cent of the total population, it is not surprising that fifteen of the nineteen members elected to the first territorial legislature were Inuit. Though its public services is open to all, the intent is to increase the Inuit component to approximate its percentage in the territorial population by 2008. (The Government of Canada committed $40 million to help recruit and train Inuit employees for the Nunavut public service.) Inuktitut, the Inuit language, is the working language of government, although government services are also offered in English and French.

The government includes many unusual features: There are no political parties, decisions are reached by consensus, the premier is elected by the members of the legislature, traditional knowledge of Elders is incorporated into government systems, so that Inuit cultural values are an integral part of decision-making processes, etc. It is highly decentralized with ten departments in eleven different communities. The territorial capital is Iqualuit (formerly Frobisher Bay), and there are three internal administrative regions: Kitikmeot in the west centred on Cambridge Bay, Kivalliq (Keewatin) in the southwest centred on Ran-

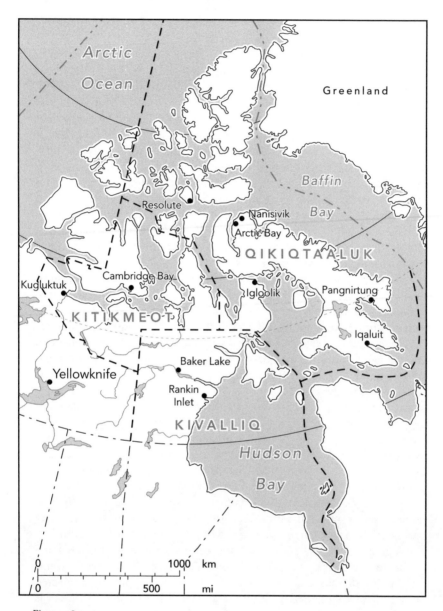

Figure 48.1

kin Inlet, and Qikiqtaaluk (Baffin) in the north and east centred on
Igloolik (Figure 48.1).

The great hopes for the future of the territory are tempered by great
difficulties. Though its Aboriginal residents have learned through many

generations how to live with the challenges of the Arctic environment, they now must balance their traditional values against the realities of the outside world to which they are increasingly tied. In this huge territory, there are only 25,000 people [26,745 in 2001] in twenty-eight small communities scattered widely over great distances. (The largest centre is Iqaluit, with a population of 4,500 [5,236 in 2001]. Though only one community (Baker Lake) is not on a seacoast, those coasts are ice blocked for most of the year. Total road mileage is only 21 kilometres (13 miles) between Arctic Bay and the mine at Nanisivik on northern Baffin Island. Air transportation is available but expensive, not only between communities, but for essential links with the large service centres of southern Canada. Cost of living is often twice that of southern Canadians.

Nunavut's resource base includes minerals, but again distances and harsh environmental conditions are limiting factors. At present, only two lead-zinc mines are operative in the High Arctic (closed in 2001) and a gold mine 400 kilometres (250 miles) northeast of Yellowknife, with gold prospects encouraging in the Rankin Inlet area. It is hoped that diamonds may be found east of the deposits in the Northwest Territories, which presently employ several dozen Inuit from Kitikmeot Region communities. Hunting, trapping, and fishing still are important in providing traditional foods in most communities though the anti-fur lobby has adversely affected commercial sales for many residents. Arctic char, shrimp, and scallop fisheries have been a recent development, while an estimated 30 per cent of Inuit now derive some of their income from arts and crafts. Under the terms of the Nunavut Land Claim Settlement, three new national parks were created, and it is expected that these will further expand the tourism industry. Nevertheless, government in one form or another is the chief employer in Nunavut.

About 56 per cent of Nunavut's population is under the age of twenty-five, and population is increasing at three times the national average. Providing employment opportunities for these growing numbers of young people is the greatest single challenge facing Nunavut. School dropout rates are high, and social problems (substance abuse, criminal activities, and suicides) are far above the national average. A third of its residents presently are dependent on social support funds. Some critics have labelled Nunavut a "welfare case."

Of Nunavut's first budget of $610 million, fully 90 per cent was provided by the federal government. Even its staunchest supporters consider it will take at least a generation before Nunavut is self-sufficient. Whether it can achieve the great goals envisaged by those who strove to see it established, only time will tell.

49 The Quest for Provincial Status in Yukon Territory

STEVEN SMYTH

Introduction

The quest for provincial status has become a Yukon tradition, dating from 1905 when the Yukon's Commissioner, William McInnes, proposed it in a speech in Dawson City (Morrison 1968: 72). Economics of the day dictated a different course of events. Soon after a wholly elected Council had been established, the federal government reduced the size and powers of the Council, and the Yukon became just another arm of the federal government. Despite Yukon Act amendments in 1960, which sought to establish elected representation in the Yukon's budget process, a 1962 court case indicated that in legal terms the Yukon still had colonial status. Mr Justice Sissons (quoted in Shortt 1967: 3) commented:

The Yukon is still a Crown Colony. The legislation and administration are controlled by the Dominion Government. There is no Legislative Assembly. The Executive Body and the Legislative Body are one and the same. The Council is to advise the Commissioner. It is not a Legislative Assembly and is not responsible to any Legislative Assembly. I know of no Government of the Yukon Territory distinct from the Commissioner or the Commissioner in Council and the home government of the colony is the Government of Canada.

SOURCE: *Polar Record* 28, (no. 1649) (1992): 33–6. Reprinted by permission of the publisher.

Demands for responsible government and provincial status continued. The Yukon Territorial Council, in Motion 40 passed 5 May 1966 and Motion 1 passed 23 January 1968, called on the federal government to begin processes, including amendments to the Yukon Act, which would have led to provincehood.

Public demands (for example Shortt 1967), paved the way for the appointment of an Executive Committee in 1970, and various program transfers and delegations of responsibilities to elected representatives through to 1979. In 1976 the Leader of the Opposition, Joe Clark, promised Yukoners the opportunity to opt for provincial status during his first term of office as Prime Minister (*Yukon News* 14 July 1976). The Yukon Legislature's Standing Committee on Constitutional Development, established in the following year, in its second report dated 5 December 1968 recommended provincehood through the adoption of a new Yukon Act. In 1979 the Yukon's demands again appeared to be under serious consideration. Clark reiterated his promise of granting provincial status during his first term of office, providing Yukoners demonstrated that they wanted it (*Whitehorse Star*, 7 May 1979: 1–2). Unfortunately his government fell before the question could be put to the people of the territory.

Barriers to Provincehood

Challenges and blocks to achieving this objective have become increasingly onerous. The constitutional amending formula was changed in 1982, over the protests of northerners. Under the Constitution Act of 1082 (S.38) the process for becoming a province passed from negotiation with the federal government and ratification by Parliament to one of obtaining concurrence from eight governments. The Meech Lake Accord threatened to make the process even more difficult by requiring the consent of 11 governments, at the same time stripping northerners of the right to nominate people for Senate and Supreme Court appointments (1987 Constitutional Accord, 3 June 1987).

In addition to the hurdle of a more onerous amending formula, the federal government could specify further conditions by policy, which would have to be met before the Yukon could become a province. For example in 1982 the Minister of Indian and Northern Affairs, John Munro, stated that amendments to the Yukon Act granting further constitutional development would be processed only after a land claims settlement had been achieved (Yukon Legislative Assembly, Sessional Paper 82-2-11, 22 November 1982: 2). Other preconditions could include the ability to be totally self-financing, at least to the level achieved by other provinces, and the achievement of an arbitrarily determined population base. The Report of the Royal Commission for Canada (Macdonald and others 1985, III: 354) noted that:

Over the past two decades, the Northern Territories have evolved from virtual colonial status to the acquisition of responsibility for a wide range of 'provincial' services. The logical end of this process is provincehood, although four barriers might delay progress towards provincial status for a decade or more. These are the Territories' small populations, their uncertain revenue base, their unresolved internal disputes, and the practical considerations of a national interest in the North.

Two conclusions are apparent from this analysis: (1) the obstacles to provincial status seem to increase with the passage of time; and (2) northerners are being required to overcome hurdles to achieve provincial status which no other provinces entering Confederation had to overcome.

Overcoming the Barriers

THE '7 AND 50' RULE

Perhaps the least fair of any requirements to attain provincial status is the '7 and 50' rule. Paragraph 42(1)(f) of the Constitution Act, 1982, specifies that new provinces can only be established pursuant to the provisions of subsection 38(1) of the Act. This subsection states:

An amendment to the Constitution of Canada may be made by proclamation issued by the Governor General under the Great Seal of Canada where so authorized by (a) resolutions of the Senate and House of Commons; and (b) resolution of the legislative assemblies of at least two-thirds of the provinces that have, in the aggregate, according to the then latest general census, at least fifty per cent of the population of all the provinces.

The table on the following page shows current estimates of Canadian national and provincial populations. In applying this formula the population figure for Quebec is excluded from the Canadian total. Thus the 50 per cent population figure (9,642,050) could be achieved by a combination of the seven remaining provinces excluding Ontario, or by any six provinces including Ontario.

When originally proposed, the application of this formula to the creation of new provinces was vehemently opposed by the Yukon's Member of Parliament, Erik Nielsen, and other northern leaders. The Meech Lake Accord, threatening the more onerous requirement of unanimous consent, was also loudly protested by northern Canadians. Eventually the First Ministers agreed to discuss the requirement in an

NATIONAL AND PROVINCIAL POPULATION FIGURES.

Canada: total	25,923,300
Quebec	6,639,200
Difference	19,284,100
50% of difference	9,642,050
Alberta	2,401,100
British Columbia	2,984,000
Manitoba	1,084,700
New Brunswick	714,000
Nova Scotia	883,900
Ontario	9,430,400
Saskatchewan	1,011,200

SOURCE: Canada Year Book 1990: 2–20.

attempt to end the impasse over the Accord. One might conclude that the reasonableness of the northern governments' position on this issue was tacitly recognized by the First Ministers. However, it is unlikely the issue will again be addressed by the First Ministers until some constitutional accommodation is reached with the Province of Quebec.

This constitutional requirement is particularly offensive when one considers that the Premier and Government Leader of the Territories are not entitled to attend First Ministers conferences, not even if the issue to be addressed is provincehood for the northern territories. The Yukon has few options for overcoming it. Territorial governments could mount a lobbying campaign to try to persuade federal and provincial governments to amend the formula now that they have publicly supported the concept of reviewing it (in the First Ministers' Meeting on the Constitution, final communique, 9 June 1990, subsection 4(1)). Alternatively the Yukon government could lobby federal and provincial governments to grant the Yukon provincial status under the existing amending formula. The territorial governments could take the issue directly to the people of Canada through a public information and education campaign, to embarrass the federal government into supporting provincial status for northern territories.

The feasibility of the second option will be enhanced should Quebec decide to separate from the rest of Canada. Provisions of the amending formula could be met if six provinces including Ontario, or seven excluding Ontario, and the federal government passed resolutions supporting provincial status for the Yukon.

UNCERTAIN REVENUE BASE

Perhaps the most difficult obstacle to provincial status is the development of a strong, sustainable economy capable of generating sufficient revenues to provide essential services to residents and reduce reliance on federal transfer payments. Robertson (1985) and Stabler (1987) have concluded that the Yukon would not be able to generate enough revenue to enable it to qualify for equalization payments under the formula used to fund provinces. Stabler's analysis suggested that Northwest Territories (NWT) could achieve this objective if sufficient revenues were generated from Beaufort oil and gas production, but did not include the possibility of the Yukon obtaining revenues from Beaufort production. The Yukon Government is negotiating with the NWT Government and the federal government over a northern oil and gas accord, which will entitle the Yukon to a portion of the royalties generated from Beaufort production. Thus there is significant potential for the Yukon to reduce its dependency on federal transfer payments.

Furthermore, the Yukon Government has recently completed its Yukon Economic Strategy, which provides a blueprint for developing and diversifying the Yukon economy and will reduce the Yukon's dependence on non-renewable resources extraction in the long term. Finally, fundamental economic and revenue transfer issues will ultimately be addressed in the agreement negotiated between Canada and the Yukon at the time the Yukon formally enters Confederation. We can look to the resource transfer payments between Canada and Alberta and Saskatchewan, and the 'Terms of Union' agreement between Canada and Newfoundland, for clues as to what resource and revenue agreements can be written into such constitutional agreements. It should be evident that the unique problems associated with the Yukon economy and the high costs of living and doing business in the north would justify special financial and revenue-sharing arrangements. These should provide lead time for achieving revenues from taxation and resource loyalties sufficient to satisfy the formula which applies to the other provinces.

SETTLEMENT OF LAND CLAIMS

The requirement that land claims be settled as a precondition for constitutional amendment and provincial status was particularly unfortunate for Yukoners, resulting in the whole issue of constitutional development becoming a 'bargaining chip' instead of a common, unifying cause for all Yukoners. Negotiations to achieve a land claims settlement and to develop responsibilities to the territorial government are both empowering processes that meet legitimate needs and demands of Yukoners. Linking the two made one an obstacle to the other, and both processes suffered as a consequence. The argument that Native and non-Native interests in the constitutional development of the territory are different or distinct can no longer be sustained. Yukon Indians now participate fully in the political process, and have sought election as candidates in each of the Yukon's political parties. Once elected, they have been appointed to the highest positions within government, including the posts of minister, House Leader, and Speaker. These elected officials have promoted devolution, constitutional development, and the settlement of land claims as fervently as their non-Native counterparts.

All Yukoners have interests in the local management, ownership, and control of land and resources, regardless of their location within the territory. Yukon's Indians will be as affected by hydroelectric projects and resource extraction activities as non-Indians. They will be as powerless as non-Natives to affect decision-making processes approving or reject such projects, so long as land and resources remain under direct federal ownership and control. Yukoners who wish for full equality within the Canadian Confederation need to work together to achieve a fair and equitable land claims settlement and to obtain the same rights as are guaranteed to the residents of the provinces.[1]

SMALL POPULATION BASE

The small population in Canada's north has been used as an argument to deny provincial status. However, people who live in the smallest provinces have the same rights as those in the largest. All are free to move from one province or territory to another, but individuals lose rights when they move from a province to a territory. Many choose not to live in the north, as is their right. But should those who choose to live there lose rights to self-government?

Any population size criterion will be an arbitrary figure with little validity. Populations ebb and flow for reasons that governments may not control: it is doubtful that a province would lose its status if its population fell to that of the Yukon. There is no provision within the federal constitution to justify granting or removing provincial status on the basis of population, and Yukoners should reject arguments that favour this requirement.

Consequences of Not Being a Province

These are numerous and substantial. Firstly, as residents of a territory, Yukoners are not entitled to representation where decisions are made that directly affect them, for example at First Ministers' conferences. The Yukon's Premier was specifically excluded from the negotiations leading up to the signing of the Meech Lake Accord, despite the fact

[1] In May 1993 the Council for Yukon Indians, representing fourteen Yukon First Nations, signed an umbrella final agreement with the Government of Canada and the Yukon Territorial Government. Since then six individual agreements have been signed. – ed.

that those negotiations traded away the rights of Yukon's residents. When the Premier is invited to a First Minister's conference, he is permitted only a brief statement for the record, with no opportunity to question or debate Yukon interests.[2]

Secondly, the Yukon Government has no ownership over land, water, and resources: the federal government manages most Yukon lands and all its water, forests, minerals, gravel, oil, and gas resources. Yukoners are advisers in some resource allocation decisions, but their advice can be rejected by federal officials and ministers who make final decisions. Federal ministers and Parliament have passed such legislation as the Canada Oil and Gas Act and the Northern Pipeline Act over the protests of northern Canadians, and approved mining projects and land use activities without the consent of Yukoners. Similarly the Yukon Government cannot prosecute offences under the Criminal Code of Canada: prosecutions must be handled by a person appointed by the federal government.

The Yukon Act is a federal statute which can be amended without the consent of the Yukon legislature. The Yukon's constitution, form of government and political boundaries, and the political rights of its residents, can all be varied without the consent of those affected by the changes. Regulations pursuing to the Yukon Act can be promulgated without consent of Parliament, and the Minister of Indian and Northern Affairs can at any time direct the Commissioner to take unilateral action to block territorial legislation, alter the form of the Yukon Government or take other administrative actions he deems necessary.

Finally, the Yukon Government has no guaranteed federal funding to match that of the provinces. It receives grants from the Department of Indian and Northern Affairs through a formula agreement, but there are no constitutional requirements to provide such grants. Consequently the Yukon Government is subject to the vagaries of the federal budget process and federal government priorities, and territorial programs may be drastically reduced.

Such practices are morally, if not legally, repugnant in a democratic society which prides itself on its record of granting equality to all its citizens.

Conclusions and Recommendations

If Yukoners clearly indicated their wish for provincial status, would the people of Canada object? Apparently not. In 1982 an opinion poll on the issue of granting provincial status to the northern territories, spon-

---

[2] The premiers of Canada's three territories now attend First Ministers' conferences as full members. – ed.

sored by the Department of Indian and Northern Affairs, indicated that 87 per cent of southern Canadians who were polled believed that the Yukon and NWT should be granted provincial status as quickly as possible; 82 per cent agreed that territories should be granted the same resource ownership rights as provinces (*Whitehorse Star* 1 December 1982: 1).

But the Yukon has more than public opinion in its favour. In 1985 the report of the Royal Commission on the Economic Union and Development Prospects for Canada (Macdonald and others 1985: 355) recommended:

...though even territorial leaders who aspire to provincehood are not demanding it immediately, the people of the North are making a legitimate request for *de facto* status. Commissioners believe that the federal government should indicate its commitment to some form of provincehood for the Territories as an ultimate goal and should grant Northerners all the benefits of Canadian citizenship.

More specifically they recommended (ibid.: 481):

On the basis of federal commitment to the ultimate goal of some form of provincehood in the northern territories, the governments involved should establish a timetable for the transfer of provincial-type responsibilities in areas such as health, labour relations, inland waters, renewable resources, and the institution of criminal proceedings. Additional measures should be taken to:
– advance the process of transferring to territorial governments responsibilities for Crown lands that do not bear directly on the national interest and that have not been ceded to the Native people through claims settlements;
– institute resource-revenue/sharing arrangements comparable to the types of agreements worked out with Nova Scotia and Newfoundland;
– confirm participation of the territorial governments in federal-provincial forums where matters of direct concern to Northern residents are being discussed. Joint-management arrangements may be valuable transitional procedures.

Federal policy has evolved to the point where legitimate aspirations of northern people are acknowledged. A recent northern policy statement (Department of Indian and Northern Affairs 1986: 6), adopted in 1987, states:

Northerners want to join the Canadian political mainstream. They want greater control over land and resources and over Programs which, in all other regions of Canada, are the responsibility of the provinces. Northerners also expect to shape their own political and economic future and to be the architects of their own constitutions in ways that reflect the unique challenges of the North. In a real sense, the North is the 'unfinished business' of Canadian nation building.

The policy also recognized that, while the federal government could no longer unilaterally grant provincial status to northern territories, it could 'support and encourage' this result (ibid.: 7):

Northerners expect their government to continue evolving towards full provincial status ... [however] the federal government can no longer unilaterally confer provincial status. But it can support and encourage the evolution of responsible government by transferring responsibility for the administration and management of the remaining provincial-type programs.

The following course of action is recommended.

1   The Yukon Assembly should again endorse a resolution favouring provincial status for the Yukon – a clear message to federal and provincial governments consistent with positions taken historically by the Legislature.
2   The Assembly should establish a permanent Standing Committee on Constitutional Development to promote the goal of provincial status and to monitor the Yukon Government's progress towards this goal.
3   The Yukon Government should make constitution development (provincial status) its primary goal and establish a secretariat dedicated to achieving it.
4   The Yukon Government should develop a detailed strategy for achieving the goal of provincial status which would include: (i) accelerating negotiations on devolution, land claims and a northern energy accord; (ii) opening negotiations with the federal government on amendment to the Yukon Act, 'Terms of Union' agreement, and the wording of a resolution to be placed before Parliament to effect the entry of the Yukon into Confederation; (iii) opening discussions with the provincial governments to obtain their support to amend Section 38 of the Constitution Act to allow new provinces to be admitted with federal consent alone; and (iv) establishing a timetable for the achievement of provincial status.
5   The 'Terms of Union' agreement for the Yukon should be based on some variation of the Newfoundland Agreement and incorporate provisions respecting the transfer of non-renewable resources to Yukon control at an appropriate time.

Acknowledgment

This paper was delivered to the Yukon Legislative Assembly's Select Committee on Constitutional Development on 25 February 1991.

References

Department of Indian and Northern Affairs. 1986. *A northern political and economic framework.* Ottawa: Department of Indian and Northern Affairs.

Macdonald, D., and others. 1985. *Report of the Royal Commission on the Economic Union and Development Prospects for Canada.* Ottawa: Minister of Supply and Services Canada.

Morrison, D.R. 1968. *The politics of the Yukon Territory, 1898–1909.* Toronto: University of Toronto Press.

Robertson, G. 1985. *Northern provinces: a mistaken goal.* Montreal: Institute for Research on Public Policy.

Shortt, K. 1967. *Blueprint for autonomy: eight steps to provincehood.* Whitehorse: Whitehorse Daily News.

Stabler, J.C. 1987. "Fiscal viability and the constitutional development of Canada's northern territories." *Polar Record* 23(146): 551–68.

NORTHERN CHALLENGES

50 Davis Inlet: A Community in Crisis

STEPHEN B. JEWCZYK

Davis Inlet, known to the Innu as Utshimassit (The Place of the Boss), is an unincorporated community of 500 residents,[1] located approximately 295 kilometres north of Goose Bay on the Labrador coast. Situated on a site that does not meet its social, cultural, economic, or health and welfare needs, Davis Inlet is a community in crisis.

Historically, the Naskapi Innu were nomadic people who spent their summers fishing along the Labrador coast and the rest of year travelling across the interior of the Labrador Peninsula in search of caribou. After the establishment of a trading post on a nearby island, the Innu spent an increasing amount of time in two temporary tent settlements near Davis Inlet: one, around a Mission Church located on the mainland; the second, near a government store on an island approximately three kilometres distant.

As the years passed, the Innu spent so much time at these seasonal, temporary settlements that they became year-round communities. Unfortunately, the sites could not sustain permanent settlements, which were heavily dependent on the use and consumption of local natural resources. The result was unhealthy living conditions for the inhabitants, most of whom continued to live in tents.

In 1967 the federal and provincial governments – in conjunction with the Roman Catholic Church – decided to move the two Innu com-

[1] In 2001 the hamlet population numbered 465, with 480 in the area. – ed.

SOURCE: *Plan Canada* (January/janvier 1994): 13. Reprinted by permission of the publisher.

munities to the Davis Inlet site on Iluikoyak Island. According to the government documents, an abundance of trees and an adequate water supply made this site an ideal choice for a permanent community. Although attempts have been made to supply the community with water and sewer services, sewers have not yet been installed and there is an inadequate supply of potable water. In addition, the population of Davis Inlet has increased beyond initial expectations, which has resulted in a dwindling supply of local natural resources, unpleasant living conditions, and ongoing health and social problems.

In recent years the Innu of Davis Inlet have come to realize that the site chosen for them twenty-six years ago does not adequately meet the needs of their growing community. The site is isolated and lacks the very resources for which is was initially selected. The Innu believe it is the governments' responsibility to provide them with the resources necessary to identify and develop a site that can accommodate an adequate level of service infrastructure and that will support their culture and traditional way of life.

The plight of the Innu did not attract the attention of the Canadian public, however, until the problems of substance abuse among Innu children gained national media coverage last year. Their current situation is reflective of the constitutional and jurisdictional confusion that has surrounded the Innu for years.

When Newfoundland joined Canada under the Terms of Union in 1949, no reference was made to the Innu. The provincial government has since treated the Innu as residents of the province, making them eligible for provincial programs and assistance. It is the view of the provincial government, however, that constitutional responsibility for the Innu should be shared with the federal government as outlined in Section 91(24) of the Canadian Constitution. Any additional or special programs and resources provided to the Innu (above and beyond what is available provincially) should come from the federal government. Because the Innu are not status or registered Indians, however, the federal government does not recognize them as a federal responsibility. In spite of this, some federal government funding has reached the Innu by way of federal/provincial agreements. Using this avenue, the two levels of government have responded to Davis Inlet's relocation crisis by cost-sharing a feasibility study on two different settlement options.

In August 1992 the federal and provincial governments jointly appointed the consulting firm of Terpstra Associates Limited to undertake an engineering study of the Davis Inlet site and Sango Bay – an alternative mainland site not far from Davis Inlet and identified by the Innu as a preferred location for settlement. Although the Sango Bay site does not offer any additional economic opportunities[2] and the capital cost to provide hydroelectricity and water and sewer infrastructure would be extremely high, Terpstra recommended this site over Davis Inlet. The group further recommended that relocation to Sango Bay be coupled with a transfer of control to the Innu, suggesting that a move toward self-government would reduce social problems and help the Innu people rebuild their self-esteem.

Since the release of the Terpstra study, the provincial government has stated that it cannot support the Sango Bay option because it does not meet a number of site selection criteria put forth by the federal and provincial governments and the Innu people. It claims that the Sango Bay site is as isolated and bereft of natural resources as Davis Inlet. Instead, the provincial government feels a number of sites should be examined before a choice is made to ensure that the selected site meets the long-term needs of the Innu community. The provincial government has also made it clear, however, that it would not offer any objection if the federal government wished to relocate the Innu to Sango Bay – with the condition that the federal government finance the move and any associated site development costs.

What will happen, no one knows. The previous federal government (led by the Conservatives) was still weighing its options prior to the October 1993 election. It had indicated, however, that it would be prepared to provide assistance for a move to Sango Bay provided the Innu register themselves under the Indian Act. The Innu have not agreed to do so and continue to resist registration under an act that they consider to be "anachronistic," "paternalistic," "colonialistic." And "outdated." Although the new federal (Liberal) government has not yet made its position known, an announcement is expected in the near future.[3]

---

2 The Innu argue that the Sango Bay (Natuashish) site is closer to the caribou hunting grounds and that historically it was a traditional gathering place of the Innu. – ed.

3 In late 1996 the Innu signed an agreement with the federal and provincial governments to relocate to their preferred new site at Sango Bay (Natuashish) at government expense, though as late as September 2000 the move had not been effected. In late 1999 the Innu Nation signed an agreement in principle with the two governments, which stresses self-government and economic development by the Innu. Finally, the first families moved into their new houses in December 2002, with final relocation scheduled for March 2003. – ed.

51 A Critical Look at Sustainable Development in the Canadian North

FRANK DUERDEN

Introduction

There are a number of reasons for interest in the possibilities for sustainable development in the Canadian North. The most obvious is the symbolic value of the notion itself – the mantra "think globally, act locally" (World Commission on Environment and Development 1987) is an easily articulated basis for empowerment and action for communities and populations that have felt the adverse environmental and social impacts of resource exploitation. The tendency to imbue North American indigenous populations with the characteristics of a society that is environmentally conscious and caring is not new (Martin 1978), and the assertion that indigenous populations are role models for the sustainable development movement is a logical extension of this view. The obvious relationship between the notion of an environmentally sensitive society and sustainable development has already generated a number of papers promoting the merits of the northern indigenous populations as practitioners of sustainability (Kassi 1987; Usher 1987; Mulvihill and Jacobs 1991).

The pragmatic basis for consideration of the possibility of sustainable development in the North arises from the conflict between expanding populations and narrow ecosystems and from pressure placed on northern communities and the environment by increasing integration into the global economy. Northern communities are faced with the challenge of supporting increasing populations, promoting community health in the

SOURCE: *Arctic* 45, 3 (September 1992); 219–25. Reprinted by permission of publisher.

broadest sense, and providing meaningful, dignified employment. The means identified to bring about such development include community involvement in decision-making (Kassi 1987), promotion of small-scale and often labour-intensive activities (Pell and Wismer 1987), promotion of ecologically sound activities (Rees 1988; Weeden 1985), and local investment of non-renewable resource revenues (Robinson and Pretes 1988).

Defined simply, sustainable development means activity that is sensitive to the environmental context and takes into account the well-being of future generations. Weeden (1989), borrowing from Firey (1966), provided a catholic definition of sustainable development for the North, arguing that a sustainable activity is one that is economically feasible, environmentally sound, and socially acceptable. Naess (1988) encapsulates the perspective of the deep ecology movement that sustainable development should avoid speciesism and consider long-term needs and place priority on the global context.

This paper reviews suggested approaches to sustainable development in the North. It proceeds by outlining the manner in which evolution of the North's economy has conditioned both the need and the possibilities for sustainable development. It then examines the role of communities in sustainable development, moving to look at the possibilities in various land-use sectors, and concludes by critically examining the various approaches in both local and global contexts.

The Economic Landscape of the North

The economic landscape of the North is the immediate spatial context of the sustainable development debate. While this landscape defines the range of possible sustainable activities, the conflicts and changes that accompanied its evolution give the notion of community empowerment that is assumed to accompany sustainable development great appeal.

Market economies and non-renewable resource activity introduced to the North over the past two hundred years were accompanied by intractably linked social, technological, and demographic changes. The introduction of an exchange economy incorporating distant industrial societies meant that products from the land could generate imported material goods along with the traditional requirements for survival, and the fervour with which fur-bearing animals and marine mammals were slaughtered created the first market-generated environmental stress in the North (Ross 1981; Ray 1984; Martin 1978).

As the demands of industrial societies grew, so, in some regions, mining superseded renewable resource harvesting in importance. Mining is a patently unsustainable activity and often detracts from the quality of other land uses. In the North its ascendancy was marked by local displacement of Native subsistence harvesting, either through direct environmental degradation or displacement due to population pressures (Cruikshank 1974; Dimitrov and Weinstein 1984; McCandless 1985). The legitimacy and dominance of mining, and more recently energy extraction, in the North was institutionalized through legislation such as the Yukon Placer Act and in recent times the Canada Oil and Gas Act, designed to encourage or facilitate extraction, essentially giving it precedence over other land uses.

The manner in which industrially oriented activity and its associated populations brought increased pressure on the land and transformed attitudes towards resources and consumption is generally well known. What is not perhaps so clearly acknowledged is the extent to which early industrial society approached self-sufficiency, with trade goods as a veneer over strong dependency on the land for food and energy. The extreme classic example of this is the Klondike Gold Rush. Although essentially urban in character and accompanied by all the trappings of late Victorian industrial technology, the early twentieth-century population of the Dawson region depended heavily on local resources. Building materials and fuel were obtained from the local region, and it is estimated that 12 million board feet of timber were utilized in 1898 just to build Dawson City (Heartwell 1988). Several farms emerged providing the basic vegetable requirements of the population (Robinson 1945), and McCandless (1985) has estimated that in the early 1900s the Dawson population obtained at least 33 per cent of its meat supply from the land. This approach towards self-sufficiency was unsustainable. The quest for fuel ranged over several hundred square kilometres, resulting in major deforestation, with attendant biophysical impacts. Up to 1,200 moose and 4,600 caribou were harvested in 1904 alone in the area surrounding Dawson City, probably with profound ecological consequences (McCandless 1985).

The significance of the late-nineteenth-century developments in the Yukon lies in the fact that the types of activities introduced and the conflicts they generated characterize contemporary debates about sustainability in the North. Agriculture and forestry persist in the North and are promoted because of their possible substitution effects in local economies (Government of Yukon 1986; Hodge 1982). Non-renewable resource extraction continues to have deleterious impacts on Native food harvesting, and the re-assertion of local and traditional control over resource management is seen as the key to sustainability (Usher 1987).

The Klondike Gold Rush was a unique, spectacular, and short-lived event, and it was not until after the Second World War that the increasing tempo of industrialization in the North provided a basis for inter-mittent debate regarding desirable futures for the region. It was with the energy crisis of the early 1970s and its attendant plethora of mega-project proposals that this debate intensified. The duality of the north-ern economy was brought into sharp focus, with the distant energy needs of the continent's urban industrial infrastructure pitted against concerns about local economies and well-being. New impetus was given to the transformation of northern economies, and the desire to remove impediments to industrial development combined with a landmark legal decision (*Calder et al. vs. Attorney General of British Columbia*) resul-ted in increased emphasis on the resolution of Native claims (Task Force to Review Comprehensive Claims Policy 1985).

Social Impact of Change

The impact of industrial incursions on Native society is well docu-mented. As the supply of consumer goods increased and new forms of technology were introduced to the North, a dichotomy developed be-tween a societal structure traditionally geared towards low-technology harvesting and dependent on locally obtained resources and (thanks to trucks, snowmobiles, and outboard motors) a new ability to harvest effectively with (relative to the past) less expenditure of time or effort. Over the past few years satellite dishes and VCRs plugging isolated communities into the outside world have introduced a plethora of urban-based lifestyle role models and associated exhortations to con-sume, further strengthening the hold of industrial society. The resultant structure of Native economic life was best characterized by Brody (1981) as a mix of harvesting, part-time employment, and transfer pay-ments.

Societal change and economic shifts have been accompanied by demo-graphic trends, which are an important backdrop to any discussion of sustainability. The population dynamics of contemporary indigenous and non-indigenous groups contrast sharply. Among indigenous popu-lations there is a high birthrate, accompanied by a tendency to migrate towards major urban centres in the North on either a permanent or temporary basis. This migration rate is not high enough to offset the impact of the high birthrate in small communities, especially in the Eastern Arctic. The population of Baffin Region increased by 20 per cent between 1981 and 1986, and some 27 per cent of the population is below the age of nine – comparative figures for Canada are 4 per cent and 14 per cent respectively (Statistics Canada 1987). Increases in indigenous populations in areas with a narrow ecosystem place pres-sures on both local sources of food supply and waste disposal systems.

Constant non-Native populations levels are largely illusory and are maintained by high population turnover rates. In reality a substantial portion of the non-Native population stays in the North for only short periods of time (Duerden, 1983). Because of its transient nature, much of the non-indigenous population may have no vested interest in maintaining the quality of the local environment.

Sustainable Development – The Community Perspective

A major emphasis by proponents of sustainable development in the North is that it should take place at the community level, with community values taking precedence over other factors in determining appropriate courses of action. This emphasis is a reaction to the type of problems identified that emanate from activities and values drawn from distant regions and is reflected in the range of countervailing considerations that various authors stress should be incorporated into sustainable development – ecologically sensitive and appropriate development (Rees 1988; Weeden 1989), emphasis on the role of indigenous knowledge (Usher 1987), promotion of sunrise industries utilizing local skills (Pell and Wismer 1987), subsidy of traditional harvesting activity (Quigley and McBride 1987), community empowerment (Kassi 1987) and restructuring of north-south interdependence (Mulvihill and Jacobs 1991).

Usher (1987) encapsulates the arguments in favour of community-based approaches towards sustainability by arguing that by returning control of resource decision-making to the community level, the presence of social constraints and collective community interest avoids the deterioration of common property resources. Hardin (1968) characterized this "tragedy of the commons" as having its roots in the lack of collective interest in maintaining the quality of resources, a tragedy that in the case of land was ameliorated through private as opposed to collective ownership. Usher's view is the antithesis of this approach, arguing that the collective interest manifest in the contemporary and traditional land management and harvesting practices of indigenous populations is an appropriate basis for the promotion of sustainable development in the North. Because of their vested interest in the quality of the renewable resource base as a source of food and a cornerstone of cultural well-being, Native populations will manage land in such a way that it would not become degraded.

Essentially Usher's (1987) argument has two components: First, distant control and distant interest exacerbate the tragedy of the commons. Second, in the past culturally based land management practices of Native groups reflective of collective interest in sustenance and well-being resulted in long-term maintenance of a healthy resource base. There is ample evidence (see, for example, Brody 1981, or the background to the Dene, Inuvialuit, or Yukon land claims) that attempts to assert indigenous land management interests have been impeded by the manner in which effective control often lies in the hands of distant and often faceless interests, such as corporations or government.

There is some controversy regarding the notion of indigenous populations as conservationists or beneficial resource managers. Berkes (1987) clearly illustrated the manner in which culturally based management practices of the Chisasibi Cree of James Bay have continued to maintain the health of the Native food fishery; on the other hand, Brightman (1987) argues that Algonquins had no conscious notion of conservation. People-land relationships, certainly in less isolated areas, have been transformed radically in the last two hundred years, with explanations ranging from Martin's (1978) argument that the mystical basis for conservation was shattered by events accompanying proto-contact to Farb's (1986) economically deterministic view that indigenous attitudes towards land were radically transformed with the realization that it had an exchange value rather than a use value. There is little dispute that transformations of relationships with land started with the fur trade but alienation from traditional ways was subsequently exacerbated (as Usher [1987] acknowledges) by wage employment, increased accessibility to consumer goods, government relocation and settlement rationalization programs and dramatic changes in communications. Wolfe and Walker (1987) provide quantitative evidence for the adverse impact of such changes on the volume of country food harvests in Alaska.

Native Harvesting as a Sustainable Activity?

Maintenance or enhancement of Native harvests is a function of the interplay among population levels, the biophysical resource base, and competition from other forms of land use. Fuller and Hubert (1981) examined the prospects for increased sustainable yields in the Northwest Territories (NWT) and concluded that, given contemporary population trends, the projected population increase to the end of the century would only be supported at present participation rates. Although Fuller and Hubert may be unduly pessimistic, it does point to the eventual inability of the country food base to support increasing populations. There would be considerable costs associated with maintaining or improving on contemporary levels of activity, and resources would have to be harvested over a wide area, incurring increased transport energy costs. Fuller and Hubert developed a scenario in which every major river in the NWT would have to be harvested in order to provide increased sustainable yields. Thus population pressures mean environmental degradation or increased energy costs, or both.

The imputed value of country food harvests has been used extensively to depict the value of Native harvesting in conventional economic terms (Brody 1981; Dimitrov and Weinstein 1984; Rushforth 1977; Stager 1974; Usher 1971, 1976). At one level such studies make impressive reading. In Sanikiluaq, for example, 57 per cent of community income was in the form of imputed value of country food (Quigley and McBride 1987). Valued at some $2 million for a community of 435, this indicates a very healthy local harvesting economy. Brody (1981) calculated that for East Moberly approximately 33 per cent of community income was in the form of replacement value harvest from the land. In the Yukon, Dimitrov and Weinstein (1984) valued food country harvests for Ross River at in excess of 50 per cent of community income; for the community of Teslin the figure was 25 per cent (Duerden 1986). However these figures can also be interpreted as indicative of the weakness of the other sectors of the community economy and the obvious limitations on Native harvesting as an income generator. Communities where the proportion of expenditure on food greatly exceeds the average Canadian family expenditure on necessities are in reality cash poor; at the same time there is an obvious upper limit on the value of country food harvests – self-sufficiency. The economically precarious nature of Native harvesting is further highlighted by the observations of a number of proponents of Native harvesting that communities depend on a variety of transfer payments to cover the capital cost associated with wildlife harvesting (Brody 1981; Duerden 1986; Fuller and McTiernan 1987; Staples and Usher 1988; Quigley and McBride 1987).

Thus it is not too difficult to portray Native harvesting as a barely viable pursuit, clearly falling outside Weeden's definition of sustainable development. The environmental basis is precarious as populations expand, and the economic viability is in doubt. The conventional economic approach, however, ignores the manner in which harvesting and culture are interwoven, virtually synonymous with its collapse. Community cohesion, community stability, development and maintenance of self-esteem with attendant diminution of social problems are seen to flow from enhancement of country food harvests and the assertion of the primacy of traditional land use. Ultimately the maintenance of high harvest levels in indigenous communities is most legitimately seen not as an economically sustainable activity from a conventional standpoint but rather a community and cultural sustaining activity promoting community well-being. This argument can also be cast in brutally positivistic terms; it is less costly to sustain such harvesting activities than it is to bear the costs, both tangible (unemployment benefits, welfare, hospitalization) and intangible (social justice, human waste) of societal breakdown.

Conventional economic arguments also ignore various non-monetary costs associated with surrogates for country food, and it is with these costs that the real essence of the sustainable development debate lies. Country food is supported by the biosphere and the local ecosystem, and the energy exchanges involved are natural; probable surrogates (for example, beef) involve the utilization of fossil fuels at various stages of the production process, and their production involves net energy deficits. Lovins et al. (1984) estimate that in the United States almost ten calories of energy are required for production, processing, distribution and preparation for each calorie of food consumed. In the North additional energy is required for importing foodstuffs over vast differences. The major energy use related to Native harvesting is for bush or river transportation, and the upper limits on its sustainability would be that point where pressure on the land brought about diminishing returns and/or energy costs per unit exceeded those for imported food. Both are obviously strongly related insomuch as harvesting over larger areas would be a probable response to local resource depletion. Thus in terms of community well-being, relative energy consumption, and historical compatibility with the environment, Native harvesting appears as an eminently sustainable activity. However, it must be recognized that population growth threatens to place strong pressure on the physical resource base and thus the long-term quality of the environment.

Non-Traditional Approaches to Sustainable Development

Because of its symbolic relationship to cultural values, enhancement of country food harvests is perhaps the most widely discussed approach to sustainable development. Other possible renewable resource endeavours include greater reliance on fuel wood, expansion of northern agriculture, and non-consumptive tourism.

Small Native communities south of the tree line are or can be close to self-sufficient in terms of wood-based energy, and Hodge (1982) made the case that there is potential for increased sustainable reliance on wood fuel at an even larger scale. It was calculated that in the Yukon an average household requires twelve cords of wood per year, and that 60 acres of land are required to harvest this on a sustainable basis. The city of Whitehorse would require 270,000 acres of trees to provide wood-based energy for its population on a sustainable basis. Chipping and increased insulation may result in a reduction of acreage required by a factor of 3.5, and there are 80,000 acres of woodland in the Whitehorse region. The impact of these approaches in promoting sustainability is unknown, although it can be conjectured that the ecological impact of large-scale firewood harvesting could be profound. Given regeneration periods in the North, it is questionable whether harvesting of northern forests can really be considered to be a renewable resource activity, while large-scale utilization of firewood may have a deleterious effect on air quality in northern communities. The Riverdale subdivision of Whitehorse recorded the most intense domestically generated air pollution in Canada (Environment Canada, pers. comm. 1986).

Agriculture has been a persistent, if precarious, activity in parts of the middle and near North since the turn of the century. Increased agricultural viability is of interest in the NWT (Resource Management Consultants 1985), while there is sufficient arable land for the Yukon to be self-sufficient in some agricultural commodities (Redpath 1979). A growing local market as populations expand, the local economic advantages of substituting for imported produce, and the alleged superiority of local produce are arguments used by lobbyists who believe that northern agriculture is sustainable.

Increased agricultural activity would be manifest in increased hothouse operations, extension of arable land and greater emphasis on cattle farming. Ecological modifications would be attendant on land clearance, and (if experience in southern Canada is replicated) attitudes towards "wild animals" would probably become more hostile as predation increased. Land-use conflicts between agriculture and wildlife harvesting and commercial and non-commercial wildlife harvesting may ensue. At the time of the Klondike Gold Rush farming in marginal areas was made profitable by the presence of a relatively large population and isolation from the major agricultural areas of North America (Robinson 1945); today in both time space and cost space the North is a lot closer to the major producing areas. Irrigation, hothouses and improvement of marginal soils are all major costs of northern agricultural operations. In the North these costs have usually been measured in conventional monetary terms (Alaskan Agricultural Action Group 1983), but a more appropriate approach may be to measure energy consumption in northern agriculture relative to energy consumption associated with using the land for country food or to energy invested in imported agricultural produce.

There are no direct data regarding energy consumption by agriculture in the NWT, but the author's manipulation of data depicting energy consumption in the agricultural sector and value of agriculture production for eco-regions of the Canadian provinces (Statistics Canada 1986) reveals that the energy required to produce one dollar's worth of output per hectare increases considerably with nordicity. The highest energy input requirement was in the Boreal Plain eco-region (northern Saskatchewan, northern Manitoba), followed by the Cordillera and the Boreal Shield (northern Ontario). The Boreal Plain required about three times the energy input of agriculture on the Pacific coast, about twice as much as southern Ontario, and about 50 per cent more than on the prairies.

Non-consumptive tourism has many of the characteristics nominally associated with sustainable development. It can be characterized as a sunrise industry most compatible with the seasonal cycle of indigenous life and has the greatest potential to utilize indigenous land-related skills. It often involves small-scale enterprises and a range of services (food, accommodation, guiding) and thus has the greatest potential for putting money into circulation in communities. Because of physical constraints, many communities north of the tree line have limited municipal services, and development that would not aggravate environmental stress would perhaps only occur if much of the tourism infrastructure were insulated from the immediate physical environment. In an extreme version of this schema tourism would perhaps constitute the basic sector of the economy, tourists would be brought from distant regions, increasing volumes of food would be imported and, given the inadequacy of community sanitation programs, waste exported. Transporting tourists from urban centres to distant locations is energy consumptive – the promotion of tourism may optimize sustainability in small communities but detract from a global objective of sustainability.

The Role of the Non-Renewable Resource Sector

Thus far this paper has dealt with questions regarding the renewable resource sector because renewal is a key element of sustainable development. Exploitation of non-renewable resources is the strongest link between the North and the global economy, and for many outsiders mining is synonymous with "northern development." Mining is an activity that is by nature unsustainable and that through negative impacts on other land uses (for example Native harvesting) may detract from the sustainability of other activities.

However, while it is obvious that mining is not compatible with sustainable development, it is also highly unlikely that it will disappear from the northern landscape. The only ways in which its adverse impacts could be ameliorated would be if it did not have priority over other uses and if the northern economy were structured in such a way that more of its economic benefits were channelled to enhance environmentally compatible economic activities. This approach has been advocated by Robinson and Pretes (1988). They suggest the use of trust funds generated through non-renewable resource revenue to enhance community-based renewable resource activities. Asch (1982) had recognized the link between non-renewable resource revenues and funding for traditional renewable resource harvesting, arguing that taxation of mining ventures should generate a capital base for a culturally appropriate renewable resource economy. However, he also expressed concerns that this approach may make the capital base for harvesting dependent on the pace of industrial activity.

If the scale is shifted from a discussion of sustainable development in the North to sustainable development and the North, moving from the local scale to the continental or global scale, the perspective changes somewhat. Natural gas extraction, for example, may have deleterious effects on the northern environment and thus on the prospects for sustainability at the local and regional levels. Compared with other forms of energy, natural gas is relatively clean (Canadian Gas Research Institute, 1989). Its utilization as a surrogate for other fossil fuel sources of energy in urban areas, along with energy conservation measures, may go some way towards improving the quality of the physical environment for large populations and thus contribute towards a global goal of sustainability.

Discussion

From the foregoing it is obvious that there are considerable difficulties in nailing down the practical dimensions of sustainable development in the North and moving beyond the important symbolic value of the notion to identify appropriate approaches. At the outset there are a number of approaches that from time to time have been associated with the notion of sustainability. These include sustainable self-sufficiency (pre-contact society), strong dependence on local resources (initially characterized by the Klondike era and now by promotion of substitution effects in the northern economy), and sustaining development (the maintenance of community-based renewable resource activities in order to promote economic/social well-being).

A major element missing from the debate on sustainable development for the North is the emergence of either a coherent view of the appropriate interests to be served by such a policy or a strategy for resource utilization that reflects long-term goals and the capacity of the physical environment. In critiquing the World Conservation Strategy (IUCNNR 1980), Daly (1980) and Tisdell (1983) pointed to the lack of any clear notion of whose interests conservation is serving or a cohesive view of time frame. These observations are equally valid for the North. Apart from Fuller and Hubert's (1981) work, there is scant discussion and no consensus regarding optimum resource consumption rates for present populations, what the desirable sustainable standard of living is, and thus what the rate of future resource use should be. The promotion of forestry, agriculture, and tourism appears to be divorced from any broad contextual considerations regarding their suitability as candidates for sustainability. Often it seems to be assumed that because an activity is nominally renewable or offers the possibility of substituting for imports it is sustainable. Such assumptions fail to consider the global impacts in terms of relative energy consumption or relative contribution to the deterioration of the global environment.

It is apparent that energy costs in terms of depletion of non-renewable resources and contribution to global pollution are central to an evaluation of sustainability. Such considerations elevate the role of Native harvesting as a relatively conservationist approach to obtaining food, although demographic pressures may mean that in the long run it is not sustainable. They also point to the paradox between optimizing at the local level and optimizing for the globe that exists in regarding energy consumptive northern agriculture and tourism as sustainable activities. Modification of Weeden's (1989) basis for identification of sustainable activities, adding consideration of relative energy costs to the criteria of environmental compatibility, economic feasibility, and social acceptability, would reduce the range of sustainable possibilities.

The relationship between the North and distant urban regions is, as Mulvihill and Jacobs (1991) argue, a key component of any rational approach to sustainable development. Urban regions, with their demands for industrial minerals and energy, have had a deleterious effect on the quality of the northern environment. At the local level this is manifest in visual degradation of landscapes, acidic leeching from tailing ponds (Duncan 1975), modification of tundra landscapes in the quest for oil and natural gas, removal of vegetation cover (with subsequent ecosystem modifications), and disturbance and destruction of fish and wildlife habitats. Some pollutants found in the North, such as mercury (associated with activities as diverse as mining, pulp and paper processing, and hydro-power generation), emanate from urban-oriented industries located in the North, but others, especially airborne toxins, have their origin in industrial heartlands.

Many opponents of northern pipeline construction who made submissions to the Berger and Lysyk inquiries in the mid-1970s argued that the obvious approach to maintaining the quality of the northern environment lies in changing attitudes towards consumption and conservation in southern urban areas, that any amelioration of demands from urban areas will have a beneficial impact on the environment of the periphery. The traditional refutation of this approach has been that the heartland and the periphery are so interrelated that prosperity, and the material quality of life in peripheral regions depend on urban output and that a portion resource consumption is related to peripheral demands.

When the question of whether sustainable development in the North should be viewed as a relative (i.e., ameliorate the adverse impacts of development) or absolute (i.e., prohibit any type of development that has a negative impact on the environment) is considered, it becomes clear that the North and urban regions are bound by a Gordian knot. Which approach best reflects the axiom "think globally, act locally" is debatable. From a crude utilitarian standpoint, "the greatest happiness for the greatest number," the global perspective would be that some degradation of the northern environment in the course of energy extraction is acceptable if the result is reduction of global pollution. The local perspective that local landscapes, economies, and social well-being would be adversely affected by such a course can best be cast in other than selfish terms. Using community empowerment to prohibit activities that are not sustainable, that (in the case of unresolved Native claims) may be illegal, or that detract from the viability of local activities would promote global sustainability by forcing urban regions that demand energy and minerals from the North to conserve.

Problems can be anticipated with both approaches. Increased northern energy extraction would be futile if not accompanied by stringent long-term increases in efficiency and conservation in urban areas. On the other hand, prohibiting energy extraction in the North could merely exacerbate the global situation if the response of industrial societies was to turn to more highly polluting energy sources from elsewhere.

Conclusions

Although no single framework exists for the identification of potential sustainable activities, it is clear that there is some dissonance between types of activities viewed as sustainable and the true requirements of sustainability. It is suggested that relative energy consumption should be considered alongside social, conventional economic, and environmental factors in identifying sustainable activities. At the local level the appropriateness of tourism and agriculture is questionable, while the value of Native harvesting as a relatively sustainable activity is cautiously endorsed, although the long-term prospects for this activity are debatable. Community-based agendas may not truly reflect the adage "think globally, act locally" and greater consideration has to be given to the appropriate and reciprocal relationship between the North and urban regions in global sustainable strategies.

References

Alaskan Agricultural Action Group. 1983. An economic assessment of Alaskan agriculture. [Juneau?]: Division of Finance and Economics, Alaska.

Asch, M.J. 1982. Capital and economic development: A critical appraisal of the Mackenzie Valley Pipeline Commission. Culture 2 (3): 3–9.

Berkes, F. 1987. Common-property resource management and Cree Indian fisheries in subarctic Canada. In McCay, B.M., and Acheson, J.M., eds. The question of the commons. Tucson: University of Arizona Press. 66–91.

Brightman, R.A. 1987. Conservation and resource depletion: The case of the boreal forest Algonquians. In McCay, B.M., and Acheson, J.M., eds. The question of the commons. Tucson: University of Arizona Press. 121–42.

Brody, H. 1981. Maps and dreams. Vancouver: Douglas and McIntyre.

Canada Gas Research Institute. 1990. Natural gas and the environment. Toronto: Canadian Gas Association.

Cruikshank, J. 1974. Through the eyes of strangers. Whitehorse: Yukon Territorial Government.

Daly, H.E. 1980. Economics, ecology and ethics: Essays towards a steady state economy. San Francisco: Freeman.

Dimitrov, P., and Weinstein, M. 1984. So that the future will be ours. Unpubl. report prepared for Ross River Indian Band. Available at Council for Yukon Indians, 18 Nisutlin Drive, Whitehorse, Yukon.

Duerden, F. 1983. Migration patterns and Yukon settlements 1968–78. Musk Ox 32: 10–21.

– 1986. Teslin: Community and village economy. Unpubl. report prepared for Teslin Indian Band. Available at the Economic Development Department, Council for Yukon Indians, Whitehorse, Yukon.

Duncan, D.W. 1975. Leachability of anvil ore, waste rock and tailings. Ottawa: Department of Indian Affairs and Northern Development, Arctic Land Use Research Program.

Farb, P. 1968. Man's rise to civilization: As shown by the Indians of North America from primeval times to the coming of the industrial state. New York: Dutton

Firey, W. 1966. Man, mind and the land; A theory of resource use. Glencoe, Illinois: Free Press.

Fuller, S., and McTiernan, T. 1987. Old Crow and the northern Yukon: Achieving sustainable renewable resource utilization. Alternatives 15 (4): 18–25.

Fuller, W.A., and Hubert, B.A. 1981. Fish, fur and game in the NWT; Some problems of, and prospects for increased harvests, in renewable resources and the economy of the North. Ottawa: Association of Canadian Universities for Northern Studies. 12–29.

Government of Yukon. 1986. Yukon 2000: Building the future. Whitehorse: Department of Economic Development.

Hardin, G. 1968. The tragedy of the commons. Science 162: 1243–8.

Heartwell, C. 1988. The forest industry in the economy of the Yukon. Whitehorse: Department of Renewable Resources, Government of Yukon.

Hodge, T. 1982. Energy in the Yukon: A soft path for a northern territory. Alternatives 10 (4): 3–12.

International Union for the Conservation of Nature and Natural Resources (IUCNNR). 1980. World conservation strategy: Living resource conservation for sustainable development. Morges, Switzerland: United Nations Environment Program, World Wildlife Fund.

Kassi, N. 1987. This land has sustained us. Alternatives 14 (1): 20–1.

Lovins, A.B., Lovins, H.L., and Bender, M. 1984. Energy and agriculture. In Jackson, W., Berry, W., and Colman, B., eds. Meeting the expectations of the land. San Francisco: Northpoint Press. 68–86.

Martin, C. 1978. Keepers of the game. Berkeley: University of California Press.

McCandless, R.G. 1985. Yukon wildlife; A social history. Edmonton: University of Alberta Press.

Mulvihill, P.R., and Jacobs, P. 1991. Towards new south/north development strategies in Canada. Alternatives 10 (2): 34–9.

Naess, S. 1988. Sustainable development and the deep long range ecology movement. Trumpeter 5 (4): 138–52.

Pell, D., and Wismer, P. 1987. The role and limitations of community based economic development in Canada. Alternatives 14 (1): 31–4.

Quigley, N.C., and McBride, N.J. 1987. The structure of an arctic micro-economy; The traditional sector in community economic development. Arctic 40 (3): 204–10.

Ray, J.A. 1984. Periodic shortages, Native welfare and the Hudson's Bay Company 1670–1930. In Krech, S., III, ed. The sub-arctic fur trade. Vancouver: University of British Columbia Press. 1–20.

Redpath, D.K. 1979. Land use programs in Canada; Yukon Territory. Ottawa: Lands Directorate, Environment Canada.

Rees, W.E. 1988. Sustainable development: Economic myths and ecological realities. Trumpeter 5 (4): 133–88.

Robinson, J.L. 1945. Agriculture and forests of the Yukon Territory." Canadian Geographical Journal (August 1945): 54–73.

Robinson, M., and Pretes, M. 1988. Beyond boom and bust: A strategy for sustainable development in the North." Polar Record 25 (153): 115–20.

Ross, W.G. 1981. Whaling, Inuit and the Arctic Islands 1880–1980. Ottawa: Royal Geographic Society of Canada. 33–50.

Rushforth, S. 1977. Country food. In Watkins, M., ed. Dene Nation – The colony within. Toronto: University of Toronto Press. 32–47.

Stager, J.K. 1974. Old Crow Y.T. and the proposed northern gas pipeline. Ottawa: Northern pipeline Task Force on Northern Oil Development. Report No. 74–21.

Staples, L., and Usher, P. 1988. Subsistence in the Yukon. Whitehorse: Council for Yukon Indians.

Statistics Canada. 1986. Human activity and the environment; A statistical compendium. Ottawa: Ministry of Supply and Services.

Statistics Canada. 1987. Census Bulletin 94–123. Ottawa: Ministry of Supply and Services.

Task Force to Review Comprehensive Claims Policy. 1985. Living treaties lasting agreements. Ottawa: Department of Indian Affairs and Northern Development.

Tisdell, C.A. 1983. An economist's critique of the world conservation strategy, with examples from the Australian experience. Environmental Conservation 10 (1): 43–52.

Usher, P.J. 1971. The Bankslanders: Economy and ecology of a frontier trapping community. Vol. 2. Ottawa: Northern Science Research Group. Department of Indian Affairs and Northern Development.

– 1976. Evaluating country food in the northern Native economy. Arctic 29 (2): 105–120.

– 1987. Indigenous management systems and the conservation of wildlife in the Canadian North. Alternatives 14 (1): 3–9.

Weeden, R.B. 1985. Northern people, northern resources, and the dynamics of carrying capacity. Arctic 38 (2): 116–120.

– 1989. An exchange of sacred gifts. Thoughts toward sustainable development. Alternatives 16 (1): 42–9.

Wolfe, R.J., and Walker, R.J. 1987. Subsistence economies in Alaska: Productivity, geography, and development impacts. Arctic Anthropology 24 (2): 26–81.

World Commission on Environment and Development. 1987. Our common future: The report of the World Commission on Environment and Development. Oxford: Oxford University Press.

52 The Northern Native Labour Force: A Disadvantaged Work Force

ROBERT M. BONE AND MILFORD B. GREEN

Introduction

One of the challenges facing Canada in the 1980s is the development of its North. Population is a key element in industrial, political, and social development. An understanding of the characteristics and dynamics of the northern Native population and its labour force is thus critical for effective planning and decision-making by public and private sectors in Canada.

According to the federal government's strategy document, the prime economic priority "lies in the development of Canada's rich bounty of natural resources" (Canada 1981: 2). This development strategy ensures that northern megaprojects based on mineral resources will dominate the economic pattern of industrialization in the 1980s. By quickly expanding the employment base of the North, these large-scale enterprises can have profound implications for social change by increasing the participation of Natives in the wage economy. Yet the past record of Native employment in the mineral industry is poor – less than 1 per cent of the national labour force in the mining industry is Native and only 6 percent of the Northwest Territories' labour force in the mining industry is Native (EMR 1981: 88).

With this model of development in mind, the question of social development of Native peoples and the issue of their greater participation in the wage economy should be of concern. This paper examines

SOURCE: *The Operational Geographer/Géographie Appliqué*, no. 3 (1984): 12–14. Reprinted by permission of the publisher.

the characteristics and role in a mixed economy of the northern Native labour force. Three social demographic elements affecting the disadvantaged Native work force are examined through 1976 data on the Metis of northern Saskatchewan.

The Northern Native Labour Force

The Native labour force in the Canadian North is often ill-equipped to take full advantage of the economic opportunities arising from large-scale development projects. Similar problems arise when modern industry locates in an underdeveloped part of the world where the local population has a traditional rather than an industrial culture. Since the benefit of the project to the region or country is partly measured by the number of local people obtaining employment, companies are under pressure to hire local workers. In turn, efforts by companies are limited by the "employability" of those seeking wage employment. Although the concept of employability is complex, this paper examines two basic factors affecting job attractiveness of a person: (1) the level of fluency in the nationally spoken language(s) and (2) the level of education. In both cases, the Native labour force in northern Canada has large numbers of its members who have difficulties communicating in English and who have had little formal schooling. These two factors combine to limit the type of wage employment available and to restrict employment to the local area where employers familiar with individuals can accommodate educational and language handicaps.

Although the northern labour force is not formally segmented, a de facto division consists of two types of jobs – permanent and temporary. Permanent employment generally is associated with government positions and jobs in large companies. Such employment offers security, a chance for advancement, and opportunities for learning new skills. Temporary employment, on the other hand, is characterized by seasonal and casual work. Such employment rarely offers any prospect for the future and commonly is characterized by poor wages and working conditions. A disadvantaged worker is more likely to be hired for a seasonal or casual job than for permanent employment because job interviews and required job qualifications frequently eliminate the disadvantaged worker. Recognizing the need for human resource development in the Northwest Territories, the federal government recently announced a $21 million economic agreement with the Government of the Northwest Territories (Canada 1982: 1–8229). Approximately $5.5 million will be provided to increase the participation of northerners, particularly Natives, in the wage economy by giving them the opportunity to obtain the basic skills, experience, and knowledge required for entry into government and business. This pre-employment training of currently unemployable residents of the Northwest Territories is one more effort by federal government to correct the disparities in employment between Natives and non-Natives, and to ensure a higher participation of Natives in future megaprojects.

Another factor encouraging temporary employment is the desire by some Native peoples to trap, hunt, and fish. This desire to live off the land from time to time and to supplement their urban food supply with wild game, fish, and berries is well known. The availability of temporary work in the summer offers the serious trappers the best of both worlds because wage income provides capital for purchasing better trapping equipment and snowmobiles. If only interested in obtaining country food, the Native urban dweller can hunt or fish at the most favourable times by either taking time off or by terminating employment. Temporary wage employment, however, is characterized by seasonal and erratic job opportunities in most Native communities. Hence, with long periods of unemployment, individuals often apply for welfare.

The Mixed Economy of the Canadian North

The mixed economy has evolved historically from a heavy dependence upon a combination of trapping and hunting to a system including wage income and welfare payments. The original mixed economy saw Native families living on the land, and moving from place to place in search of game and fur. The modern version sees Native families residing in settlements, and partaking in various related activities, including working for an employer. Because Native peoples now live in communities, their accessibility to wage employment has greatly increased. Residence in a village, however, has altered the character of the hunting and trapping parties. In the past, the entire family lived on the land for most of the year. Now the common pattern is for small groups of men to spend short periods trapping and hunting. This change is partly due to the modernization of transportation which allows greater distances to be covered in much less time, and partly from social pressure to return to the settlement where family and friends reside.

As Canada extends its infrastructure into the frontier, urban centres in the Arctic and Subarctic are likely to grow in size, and their transportation links with cities in the south will improve. Those northern settlements which experience the greatest economic growth may become regional centres. Because most regional development is likely to be associated with large-scale mining projects such as Pine Point and Norman Wells, wage employment likely will grow in importance and become concentrated in major urban centres. On the other hand, smaller communities may remain more traditional, and as part of the mixed economy, its labour force may achieve a blending of land-based resource exploitation and urban wage economy.

The Metis Labour Force of Northern Saskatchewan

Population cannot be considered a static feature by planners. Plans must take into account not only the present size and character of the labour force but also the factors likely to influence it in the future. Three such elements affecting the Metis labour force are population growth, language, and education.

The Metis people residing in thirty-four communities in northern Saskatchewan, like their counterparts in the territories, constitute an extremely youthful population. In 1976, for example, nearly 49 per cent of the Metis were under the age of fifteen. The demographic boom which produced this youthful character began in the late 1950s. This "bulge" by 1976 was beginning to affect the size of the potential labour force. The number of young people seeking employment increased substantially as this demographic wave continued to age. The potential labour force is likely to form 60 per cent of the total Metis population by 1986, up 9 per cent from 1976. Although their annual rates of natural increase are diminishing, they still remain well above the national average. Consequently, the proportion of young Metis entering the work force will be greater than that for other comparable provincial populations. This demographic fact has considerable importance for regional and social planning.

The geographer is often concerned in studying a region about the characteristics of the local labour force. Language skills are often a key indicator of the "employability" of a frontier local work force. The use of English by Native peoples increased substantially in northern Canada with their settlement in urban centres. This urbanization took place for the majority of Native peoples in the 1950s and 1960s. Consequently, most adults over the age of thirty-five taught themselves English while those under thirty-five probably had some schooling. The implications for wage employment are severe and affect family income (Table 1). One group speaks English at home (group 1) and one speaks Cree or Chipewyan at home (group 2). Metis families who spoke English at home had nearly double the proportion of families in high income brackets (incomes over $10,000 per year) – 38 per cent of group 1 are in the high income bracket, but only 20 per cent of group 2. A t test of these two language groups by income distribution confirms that they are statistically different at the .01 level of significance. We thus conclude that variations in English language skills have a direct bearing upon family income.

TABLE I
PERCENTAGE OF METIS SPEAKING ENGLISH IN THE HOME
AND NOT SPEAKING ENGLISH IN THE HOME, BY FAMILY
INCOME FOR NORTHERN SASKATCHEWAN, 1976

Income groups	Group 1 English-Spoken	Group 2 English-Not Spoken
$1,000 and less	4.3	0.9
$1,001–2,500	8.7	9.8
$2,501–5,000	18.2	31.1
$5,001–$10,000	30.9	38.7
$10,001–$15,000	21.1	15.7
$15,001–$20,000	11.2	2.7
Over $20,000	3.6	1.1
Total	100.0	100.0

t test: 2.94 with a two tailed probability of .003

SOURCE: Compiled from the authors' Housing Needs Survey.

The level of education found in the potential labour force is a key indicator along with English-language skills of "employability"; these levels are generally low, particularly for those over thirty-five years of age. Given the fact that the public school system was first introduced into some of these centres as late as the 1960s, the absence of a well-educated labour force is not surprising (Table 2). The Metis population over fourteen years of age contains 23 per cent with no schooling, and only 11 percent with ten or more years of public schooling. This table emphasizes that a large proportion of this labour force is ill-prepared to join the industrial labour force except as casual or seasonal workers. Even more striking is the difference in schooling by age groups.

The most dramatic statistic is the number of people of working age with no formal education. As age increases (Table 2), the proportion with no schooling increases from 7 per cent for the youngest group (15 to 34) to 67 per cent for the oldest group (over 54). Similarly the percentage of those who have completed grade ten or more declines as age increases; the youngest group has 16 per cent and the oldest only 0.5 per cent.

Although access to education facilities has improved for northern Natives over the last part of the decade, their average educational level is still far below that of southern residents. Innovative ways are needed to keep teenagers in school and to attract high school graduates to vocational centres and universities. Because these changes will take time before paying off, the longer implementation of such a program is delayed, the more distant and difficult will be solutions to the issue of

TABLE 2
EDUCATION LEVELS FOR METIS, BY AGE GROUPS, IN NORTHERN
SASKATCHEWAN, 1976 (PER CENT)

Grade Level	15 to 34	34 to 54	over 54	Average	Total
None	7.1	32.9	67.0	23.2	602
1–3	4.3	13.9	12.1	8.0	207
4–6	28.3	26.8	13.7	25.5	662
7–9	36.7	15.5	5.5	26.4	684
10–12	16.3	4.4	0.5	10.7	279
Other	7.4	6.5	1.2	6.2	160
Total	100.0	100.0	100.0	100.0	2594

SOURCE: Calculated from the authors' Housing Needs Survey

labour participation. In the short run, we cannot wait for these educational gains. Some immediate economic results are necessary, and attention must be focused not on the welfare system but on the mixed economy to redistribute wealth which over the century has remained "central to Native life in the north" (User 1982: 187). Geographers who are particularly knowledgeable about renewable-resource development should focus their attention on ways to strengthen the commercial viability and to increase the productivity of resource harvesting.

Implications

Regional development in the Canadian North, besides achieving economic growth, should meet certain social goals. One of these goals comprises increased participation of the northern Native labour force in the wage economy. Evidence in this paper stresses that this work force is currently ill-equipped to engage in full-time wage employment which normally leads to economic stability for the individual and family, and to a sense of social well-being and purpose. Short-term employment which is more available to those with social handicaps has none of the advantages of permanent employment. Although this type of work may provide one element in the mixed economy, its uncertainty, poor working environment, and low pay are decided disadvantages. Clearly, the present northern Native work force is disadvantaged. A major challenge thus is to overcome disparity in participation rates between the northern and the Canadian work force. Planners should have a hard look at the education and training system required to achieve this goal. The educational problem, however, is so deep-rooted and perhaps even a generation or two away from meaningful results, that short-term means of improving the income level and economic stability of Native families needed now include a more commercial version of the mixed economy involving the majority of Native northerners. With such a dual approach, education and employment programs can be put into action which will lead to immediate results as well as to long-term solutions.

Acknowledgments

The authors acknowledge the support of the Northern Population Committee and the Donner Canadian Foundation.

References

Canada. 1981. *Economic Development for Canada in the 1980s*. Ottawa: Department of Supply and Services.

Canada, Department of Energy, Mines, and Resources (EMR). 1981. *Mineral Policy: A Discussion Paper*. Ottawa: Department of Supply and Services.

Canada, Department of Indian and Northern Affairs (DIAND). 1982. *Stimulus, Job Possibilities Provided by Signing of Economic Development Agreement with Northwest Territories Government – News Release 1–8229*. Ottawa: Department of Supply and Services.

Usher, P.J. 1982. Unfinished business on the frontier. *The Canadian Geographer* 26, no. 3: 187–90.

53 Regional Perspectives on Twentieth-Century Environmental Change: Introduction and Examples from Northern Canada

JOHN D. JACOBS AND TREVOR J. BELL

At the end of the twentieth century, we see widespread recognition that human activities are changing the global climate, with the potential for major environmental impacts and socioeconomic dislocation (Houghton et al. 1996). The composite instrumental record of average global temperatures since 1860 shows 1997 to be the warmest year to that point (Bell and Halpert 1998). Records continue to be broken in 1998, adding credence to projections of global warming. There is an expectation, not yet confirmed by climate science, that an anthropogenically modified climate resulting in global warming will mean changes in intensities, trajectories, and frequencies of weather systems throughout the world (Houghton et al. 1996; Francis and Hengeveld 1998). It follows from the scale-transgressive linkages that exist among atmosphere, ocean, cryosphere, and biosphere, that global climate change should be characterized by heterogeneous regional environmental change.

The theme of comparative regional studies is a familiar one in geography, and this approach has special relevance to emerging problems of global environmental change. 'Global change' and even 'global warming' are vague and imprecise concepts until they are brought down to regional and local levels, where the impacts are most likely to be felt. At those levels, events may even run counter to global trends.

It was recognition of the complexity of regional environmental change that led us to call a special session of the 1997 Annual Meeting

SOURCE: The Canadian Geographer/Le Géographe canadien. Vol 42, no. 4 (winter/hiver 1998): 314–18. Reprinted by permission of the publisher.

of the Canadian Association of Geographers in St John's on this topic. This introduction and the five papers which follow have grown out of the presentations and discussions in that session. By way of background, we present some observations pertaining to northern Canada that exemplify the regional perspective.

The idea of contrasting regional environmental change is illustrated by trends over the last several decades in parts of northern Canada and adjacent Alaska. According to the 1995 report of the Intergovernmental Panel on Climate Change (IPCC), the strongest regional warming between 1980 and 1994 was in northwestern Canada and Alaska, with a second warm pole in north-central Siberia (Nicholls et al. 1996). For the 1961 to 1990 period, Chapman and Walsh (1993) found an annual warming trend of 0.5 to 0.75°C per decade in northwestern Canada and central Alaska and a cooling of 0.25 to 0.5°C per decade over western Greenland and Baffin Island, with the winter trends dominating. The extent of winter sea ice, which is closely related to winter temperature, increased by about 8 per cent in the Baffin Bay–Labrador Sea sector and decreased by about 3 per cent in the Beaufort Sea sector over that same period (Chapman and Walsh 1993). Maxwell (1997a) reported summer warming on the order of 0.25°C per decade from 1961 to 1990 for northwestern Canada and Alaska but no trend in the eastern Canadian Arctic and southern Greenland.

The practice of fitting time series to a regression line and testing for significance is commonly used to detect trends; however, the choice of starting and ending points can strongly influence the results. Thus, an updated look at the Canadian Arctic temperature series from 1948 to 1996 showed the winter cooling trend in the east to have lessened as a result of near- to above-normal temperatures since 1994, while the overall warming trend in the west continued (Maxwell 1997b).

Precipitation generally shows more spatial variability than temperature, which makes the determination of trends in rainfall and snowfall problematic in the Canadian Arctic, where climate stations are sparsely distributed. From surface-station data for the period 1961 to 1990, annual and summer precipitation totals appear to have increased by about 10 per cent in the western Arctic, while experiencing no change in the east (Chapman and Walsh 1993; Maxwell 1997a). Groisman and Easterling (1994) reviewed precipitation and snowfall records for the United States and Canada with attention to corrections for instrumental bias and spatial non-homogeneity and concluded that annual snowfall and rainfall had increased in Canada north of 55°N by 20 per cent between 1950 and 1990. The areal extent of springtime snow cover over North America as a whole was determined from satellite data to have diminished by 16 per cent in the 1973–1994 period (Groisman

et al. 1994). From station data for 1961 to 1990, the date of disappearance of snow cover was found to have become earlier by about one
week in the western Canadian Arctic, with no change in the east
(Foster 1989; Maxwell 1997a), which is consistent with the spring and
summer temperature trends in the two sub-regions.

Warmer and possibly wetter summers in high northern latitudes
should mean increased vegetation growth. Remote sensing permits
monitoring of plant growth on a regional and global scale. National
Oceanic and Atmospheric Administration (NOAA) meteorological satellites fitted with advanced Very High Resolution radiometers (AVHRR)
can detect enhanced absorption of photosynthetically active radiation
(wavelength range 0.4 to 0.7 µm) through a measure known as the nor-
malized difference vegetation index (NDVI). Using NOAA NDVI data,
Myneni et al. (1977) analyzed northern hemisphere vegetation for the
period 1981 to 1991 at ten-day intervals with a spatial resolution of
eight kilometres. July-August NDVI values averaged over ten-degree
latitudinal zones revealed a generally increasing trend in plant growth
in the order of one per cent per year in all zones, including the area
north of 65°N. When mapped in polar projection, the average growing
season NDVI values and the percentage increase in NDVI showed distinct regional differences (Myneni et al. 1997, Figure 2). North of 60°N,
from about 90°W to central Alaska, there was an increase in NDVI of
about 25 per cent between 1982 and 1990, while eastward across Baffin
Island and Greenland (as well as in the Canadian High Arctic), there
was no significant increase.

Ground temperatures, though driven by the surface energy balance,
tend to dampen the surface temperature signal with depth and thus
provide a naturally filtered record of surface temperature trends over
time. A sustained increase in surface temperature will result in changes
to the permafrost regime (Woo et al. 1992). Lachenbruch and Marshal
(1986) reported that near-surface ground temperatures in northern
Alaska increased by 2°C to 4°C over the previous few decades, while
Kwong and Gan (1994) estimated that the southern limit of discontinuous permafrost in the Mackenzie Valley had moved northward some
120 km between 1962 and 1989. Although permafrost observations
are few in the eastern Arctic, between 1988 and 1993, Allard et al.
(1995) found a decrease in ground temperatures along the Hudson
Strait shore of Quebec that was consistent with the decrease in regional air temperatures there.

Finally, glaciers are well-known indicators of regional climate
change, and overall, glacial mass balance has implications for global
sea-level change. Many middle- and low-latitude glaciers around the
world have experienced mass loss in recent decades (e.g., Haeberlie et

INDICATIONS OF LATE TWENTIETH-CENTURY ENVIRONMENTAL CHANGE
IN NORTHERN CANADA

Element	West	East	Period of Observation
Winter temperature	Increase (+1.0°C)	Decrease (−1.0°C)	1961–1990
Summer temperature	Increase (+0.4°C)	No change	1961–1990
Annual precipitation	Increase (+10%)	No change	1961–1990
Summer precipitation	Increase (+10%)	No change	1961–1990
Date of disappearance of snow cover	Earlier (ca 1 week)	No change	1961–1992
Terrestrial photosynthetic activity	Increase (25%)	No change	1982–1990
Ground temperature in permafrost	Increase (+0.08C yr⁻¹)	Decrease (−0.05C yr⁻¹)	ca 1900–1984 (West) 1988–1993 (East)
Glacier extent and mass balance	Accelerated decrease	Decrease to no change	1961–1990
Winter sea-ice extent	Decrease (3%)	Increase (8%)	1961–1990

SOURCES: Lachenbruch et al. 1986; Foster 1989; Chapman and Walsh 1993; Allard and Wang 1995; Haeberlie et al. 1996; Dyurgerov and Meier 1997; Maxwell 1997a; Myneni et al. 1997.

al. 1996), largely as a consequence of increased regional summer temperatures. According to a recent review by Dyurgerov and Meier (1997), while glaciers in Alaska and the Canadian Arctic experienced net loss over the period 1961–1990, the relative loss of mass was greater in the west than in the east.

This brief review, summarized in the table above, shows that in recent decades the eastern and western parts of northern Canada have experienced some environmental change of differing degrees and in some respects of opposite sign. These east-west contrasts are a consequence of the general configuration of atmospheric circulation across North America and its links to oceanic circulation. Sectoral differences of climate in middle and high latitudes persisting over decades are a normal feature of the global climate system (Palmer 1998). Among associated environmental responses, the apparent contrast in vegetation growth changes between east and west is consistent with the contrasting growing season temperature trends (i.e., warming in the west and no change in the east). Given the complex relationships between climate and the physiology and ecology of plants, it is likely that the actual response of vegetation in the western Arctic to the recent warming is more than simply an increase in plant leaf area or biomass, and it is therefore difficult to predict what the long-term consequences of a sustained warming trend might be (Callaghan and Carlsson 1997).

Regional contrasts in environmental change are not unique to the twentieth-century record. Studies of the paleoenvironmental record show that variable regional responses to global and hemispheric trends in climate are a normal feature of the Earth system. Proxy records from the Canadian Arctic show strong evidence for regional variability in climate and environmental change during the Holocene (the last ten thousand years). Differences in local land-ice cover were primarily responsible for regional variability during the demise of late Wisconsinan ice sheets, while sea-surface temperatures have been a source of variability during the entire Holocene period (cf. Dyke et al. 1996). The persistence of remnant ice masses of the Laurentide Ice Sheet in the eastern Arctic until at least 6 ka (Dyke and Prest 1987) delayed the onset of terrestrial warming, which elsewhere in the Arctic had been experienced soon after the summer insolation maximum, around 10 ka (at 65°N; e.g., Bradley 1990). Thus, the thermal maximum in the eastern Arctic lagged about four thousand years behind the insolation maximum (Williams et al. 1995), causing marked regional asynchroneity in the timing of postglacial environmental change (e.g., Ritchie and Harrison 1993; Webb et al. 1993).

Ocean currents affect regional climate patterns in coastal environments. For example, Dyke et al. (1997) have proposed that the contrasting regional chronologies of the Holocene driftwood incursion in Arctic Canada are most easily explained by irregular variations in the path and strength of the Transpolar Drift Stream (TDS) during postglacial times. Like other major currents, the TDS partly dictates the characteristics of currents downstream. Therefore, these variations, in addition to controlling the pattern of driftwood dispersal from the Arctic Ocean, likely had contrasting impacts on sea-surface temperatures and ice conditions along Arctic coastlines. The source of variation in TDS output may be related to precipitation patterns affecting freshwater discharge into the Arctic Ocean (Dyke et al. 1997; Mysak and Power 1992) or possibly atmospheric circulation patterns and regional wind stresses (Tremblay et al. 1997).

The timing of Holocene tree line advance in northern Canada varied in response to regional climate characteristics. Investigations in northwestern Canada indicate a northern extension of boreal forest by 9 ka (roughly coincident with the maximum in high-latitude summer insolation) and forest decline at 5 ka BP (Ritchie and Hare 1971; Ritchie et al. 1983). In contrast, MacDonald et al. (1993) document tree line advance at ~5ka BP and subsequent retreat after 4ka BP in north-central Canada. They attribute this delayed advance to shifts in the summer position of the Arctic front, whereby a northward shift would produce warmer and moister summer conditions, suitable for forest expansion, and a southward shift would re-establish cooler and drier summers and promote forest decline. Small changes in frontal wave geometry therefore likely caused tree line advance in north-central Canada with opposite or negligible impact in northwestern Canada (MacDonald et al. 1993).

Overpeck et al. (1997) compiled a paleoclimate record of the entire Arctic for the last four hundred years using a variety of proxy indicators (e.g., tree-ring widths, lake sediment varve thicknesses). Although their primary goal was to assess the climate events of this century from the perspective of the last four, their data also reveal the regional nature of Arctic climate change. For instance, the timing and duration of Little Ice Age cooling varied from region to region across the Arctic, as did the subsequent period of warmer conditions that began in the mid-nineteenth century.

Recent attempts at global synthesis of variations in climate over the past several centuries (Mann et al. 1998) point to the need for better understanding of spatial variability of environmental trends, present and past. This may be seen as a prerequisite for the construction of models that can effectively simulate the regional patterns of Holocene climate as a basis for making informed projections about future environmental change (cf. Jetté 1995). From this review and the papers that follow, it should be evident that there is still much to be learned from local and regional-scale studies of past and present environmental change.

References

Allard, M., and Wang, B. 1995. "Recent cooling along the southern shore of Hudson Strait, Quebec, Canada, documented from permafrost temperature measurements." *Arctic and Alpine Research* 27, 157–66.

Bell, G.D., and Halpert, M.S. 1998. "Climate assessment for 1997." *Bulletin of the American Meteorological Society* 79, 51–550.

Bradley, R.S. 1990. "Holocene paleoclimatology of the Queen Elizabeth Islands, Canadian High Arctic." *Quaternary Science Reviews* 9, 365–84.

Callaghan, T.V., and Carlsson, B.A. 1997. "Impacts of climate change on demographic processes and population dynamics in arctic plants." In Oechel, W., Callaghan, T., Gilmanov, T., Holten, J., Maxwell, B., Molau, U., and Sveinbjörnsson, B., eds. *Global Change and Arctic Terrestrial Ecosystems*. New York: Springer–Verlag, 129–52.

Chapman, W.L., and Walsh, J.E. 1993. "Recent variations of sea ice and air temperature in high latitudes." *Bulletin of the American Meteorological Society* 74, 33–47.

Dyke, A.S., and Prest, V.K. 1987. "Late Wisconsinan and Holocene history of the Laurentide Ice Sheet." *Géographie physique et Quaternaire* 41, 237–63.

Dyke, A.S., Dale, J.E., and McNeely, R.N. 1996. "Marine molluscs as indicators of environmental changes in glaciated North America and Greenland during the last 18,000 years." *Géographie physique et Quaternaire* 50, 125–84.

Dyke, A.S., England, J., Reimnitz, and Jetté, H. 1997. "Changes in driftwood delivery to the Canadian Arctic Archipelago: The hypothesis of postglacial oscillations of the Transpolar Drift." *Arctic* 50, 1–16.

Dyurgerov, M.B., and Meier, M.R. 1997. "Year-to-year fluctuations of global mass balance of small glaciers and their contribution to sea-level changes." *Arctic and Alpine Research* 29, 392–402.

Foster, J.L. 1989. "The significance of the date of snow disappearance on the arctic tundra as a possible indicator of climate change." *Arctic and Alpine Research* 21, 60–70.

Francis, D., and Hengeveld, H. 1998. *Climate Change Digest: Extreme Weather and Climate Change.* Ottawa: Minister of Supply and Services Canada.

Groisman, P.Ya., and Easterling, D.R. 1994. "Variability and trends of total precipitation and snowfall over the United States and Canada." *Journal of Climate* 7, 184–205.

Groisman, P.Ya., Karl, T.R., Knight, R.W., and Stenchikov, G.L. 1994. "Changes of snow cover, temperature, and radiative heat balance over the Northern Hemisphere." *Journal of Climate* 7, 1633–56.

Haeberlie, W., Heolz, M., and Suter, S. 1996. *Glacier Mass Balance Bulletin no. 4 (1994–1995).* Zurich: IAHA (ICSI) – UNEP – UNESCO.

Houghton, J., Meira Filho, L., Callander, B., Harris, N., Kattenberg, A., and Maskell, K., eds. 1996. *Climate Change 1995: The Science of Climate Change.* Cambridge: Cambridge University Press.

Jetté, H. 1995. "A Canadian contribution to the Paleoclimate Model Intercomparison Project (PMIP)." *Géographie physique et Quaternaire* 49, 4–12.

Kwong, Y.T.J., and Gan, T.Y. 1994. "Northward migration of permafrost along the Mackenzie Highway and climatic warming." *Climatic Change* 26, 399–419.

Lachenbruch, A.J., and Marshal, B.V. 1986. "Changing climate: Geothermal evidence from permafrost in Alaskan Arctic." *Science* 234, 689–96.

MacDonald, G.M., Edwards, T.W.D., Moser, K.A., Peinitz, R., and Smol, J.P. 1993. "Rapid response of treeline vegetation and lakes to past climate warming." *Nature* 361, 243–6.

Mann, M.E., Bradley, R.S., and Hughes, M.K. 1998. "Global-scale temperature patterns and climate forcing over the past six centuries." *Nature* 392, 778–87.

Maxwell, B. 1997a. "Recent climate patterns in the Arctic." In W. Oechel, T. Callaghan, T. Gilmanov, J. Holten, B. Maxwell, U. Molau, and B. Sveinbjörnsson, eds. *Global Change and Arctic Terrestrial Ecosystems.* New York: Springer-Verlag, 21–46.

– 1997b. *Responding to Global Climate Change in Canada's Arctic, vol 2* of the *Canada Country Study: Climate Impacts and Adaptation.* Downsview, ON: Environment Canada.

Myneni, R.B., Keeling, C.D., Tucker, C.J., Asrar, G., and Nemani, R.R. 1997. "Increased plant growth in the northern high latitudes form 1981 to 1991." *Nature* 386, 698–702.

Mysak, L.A., and Power, S.B. 1992. "Sea-ice anomalies in the western Arctic and Greenland-Iceland Sea and their relation to an interdecadal climate cycle." *Climatological Bulletin* 26, 147–76.

Nicholls, N., Gruza, G.V., Jouzel, J., Karl, T.R., Ogallo, D.F., and Parker, D.E. 1996. "observed climatic variability and change." In Houghton, J., Meira Filho, L., Callander, B., Harris, N., Kattenberg, A., and Maskell, K., eds. *Climate Change 1995: The Science of Climate Change.* Cambridge: Cambridge University Press, 137–92.

Overpeck, J., Hughen, K., Hardy, D., Bradley, R., Case, R., Douglas, M., Finney, B., Gajewski, K., Jacoby, G., Jennings, A., Lamoureux, S., Lasca, A., MacDonald, G., Moore, J., Retelle, M., Smith, S., Wolfe, A., and Zielinski, G. 1997. "Arctic environmental change of the last four centuries." *Science* 278, 1251–6.

Palmer, T.N. 1998. "Nonlinear dynamics and climate change: Rossby's legacy." *Bulletin of the American Meteorological Society* 79, 1411–23.

Ritchie, J.C., Cwynar, L.C., and Spear, R.W. 1983. "Evidence from northwest Canada for an early Holocene Milankovitch thermal maximum." *Nature* 305, 126–8.

Ritchie, J.C., and Hare, F.K. 1971. "Late Quaternary vegetation and climate near the Arctic tree line of northwestern North America." *Quaternary Research* 1, 331–41.

Ritchie, J.C., and Harrison, S. 1993. "Vegetation, ale levels, and climate in western Canada during the Holocene." In Wright, Jr, H.E., Kutzbach, J.E., Webb III, T., Ruddiman, W.F., Street–Perrott, F.A., and Bartlein, P.J., eds. *Global Climates since the Last Glacial Maximum.* Minneapolis: University of Minnesota Press.

Tremblay, L.-B., Mysak, L.A., and Dyke, A.S. 1997 "Evidence from driftwood records for centuy-to-millennial scale variations of the high latitude atmospheric circulation during the Holocene." *Geophysical Research Letters* 24, 2027–30.

Webb III, T., Bartlein, P.J., Harrison, S.D., and Anderson, K.H. 1993. "Vegetation, lake levels, and climate in eastern North America for the past 18,000 years". In Wright, Jr, H.E., Kutzbach, J.E., Webb III, T., Ruddiman, W.F., Street–Perrott, F.A., and Bartlein, P.J., eds. *Global Climates since the Last Glacial Maximum*. Minneapolis: University of Minnesota Press.

Williams, K.M., Short, S.K., Andrews, J.T., Jennings, A.E., Mode, W.N., and Syvitski, J.P.M. 1995. "The eastern Canadian Arctic at 6 ka BP: A time of transition." *Géographie physique et Quaternaire* 49, 13–27.

Woo, M.-K., Lewcowicz, A.G., and Rouse, W.R. 1992. "Response of the Canadian permafrost environment to climatic change." *Physical Geography* 13, 287–317.

54 Insights of a Hunter on Recent Climatic Variations in Nunavut

PETER ERNERK

Distinguished delegation I come from Rankin Inlet of Nunavut where, last week, we were finally able to use our snow machines. This was already one month later than they were able to start using them in Repulse Bay, four hundred miles to the north. I wonder if there is a conflict between the scientific knowledge of the experts and the traditional knowledge of the Inuit elders or whether in fact they are both the same. One body of knowledge is based on decades of experience, the other on the collection of scientific data.

Last July 18, 1991, in the Keewatin [Kivalliq], the area west of Hudson Bay, we had temperatures in the thirties for more than one month. On July 10, 1991 a friend and myself, my son and daughter, went caribou hunting on our ATV's. Along the way, we were amazed to notice that every small lake had scores of local people swimming. It was so hot that all day I hunted without a shirt. We also stopped to swim. This was the first time in my life that such a thing had happened to me.

This summer, on July 10, 1992, I was wearing windpants and a down jacket to work on the outside of my house. Ice break-up was one month late in most of the communities and the freeze-up seems now to have come so early. The first barges for freight transportation, scheduled for the end of July, took until mid-August to get rolling due to the late break-up.

SOURCE: Rick Riewe and Jill Oakes, eds., *Biological Implications of Global Change: Northern Perspectives*. (Edmonton: Canadian Circumpolar Institute, University of Alberta, et al., 1994): 5–6. Reprinted by permission of the publisher.

The statistics provided to me by Environment Canada indicate that the mean temperatures of July and August for the years 1989 and 1991 were indeed the highest for the past twelve years. This past summer, 1992 the temperatures were the lowest for the same twelve years.

The Roman Catholic Church representatives in Chesterfield Inlet, 80 miles to the northeast of Rankin Inlet have been recording weather for the past fifty years. Their observations indicate that in the 1940s there were frequent readings of −55° and −60° Fahrenheit [−48° and −51°C] in the winter. That now in the 1980s and 1990s the lows are more like −45° and −50° Fahrenheit [−43° and −46°C] in the winter.

This summer for the first time large tracts of ice along the Meladine River System, just outside of Rankin Inlet, did not melt. All summer, residents have been able to collect ice in addition to the fresh water for drinking and making tea. With another cold summer next year will this be the beginning of permanent ice buildup?

The elders do tell us that once every so many years cool summers repeat themselves, that also some years have freak snowstorms that occur in mid-summer. My father told me about one such year in his memory where there was such a cold and extended snowstorm in the middle of an otherwise warm summer, that this storm caused many baby birds to die, that all the mosquitoes appeared to die off, and there was quite an accumulation of snow on the ground. He remembered that after the storm was over the snow melted and things did return to normal. In discussing this particular storm with my boss, Ollie Ittinuar, who is presently seventy years old, he indicated to me that he remembered that particular storm. My father was at least twenty-five years his senior and we place this storm in the early 1930s. It was the storm and summer to remember.

In our part of the world scientists had observed in 1980 that the caribou were decreasing, that the numbers were down to 37,000. They advised that hunting would have to be severely restricted to protect the herds from extinction. Inuit kept pointing out that the herds merely changed migration routes periodically and that they would reappear in years to come.

Suddenly in 1982 the caribou counts indicated a vast number of animals that had not been there previously, now the numbers were at 300,000 individuals. Now there was no longer a need for restriction of the hunting rights of the Inuit. Through there traditional knowledge the Inuit had known this. I assume that this is another example of the cyclical variations in the caribou population.

On a related matter, an observation that I made this spring, and it has been confirmed by other travellers, is that the sun seems to be "stronger". Traditionally Inuit used animal oil and fat to protect lips from the rays of the sun and the wind, today we have lypsol from the "Northern Store". Traditionally we had bone glasses that reduce the glare of the sun to protect the eyes from snow blindness. Today we have a variety of "sun" glasses with assorted side protectors for the spring traveller. On an extended trip which I made this spring I found that contrary to other years these forms of protection were no longer adequate to protect me from the rays of the sun. Prior to my departure I purchased "Top of the Line" sun glasses with protective leather on the side, and I wore them religiously. Yet I spent several weeks upon my return seeing double due to the eye strain (and the check-up that I had at the optometrist proved that it was not yet my age catching up with me). Various nurses from the health centres have reported more burns, more sun allergies and sun sensitivities, amongst the Inuit population.

In earlier years the dark skinned Inuk did not need to protect himself as much as the fair skinned, 'Qablunaaq' but this seems to be changing, even dark skinned Inuit are suffering from burns the way their southern friends do. Is this due to the global warming or to the thinning ozone layer that is also being observed and documented or to a combination of both.

In spite of several summers being warmer than average we seem to have observed that the winters have been colder, and longer.

The scientists are telling us that this year the cold spell is due to the volcanoes in the Philippines, to the accumulation of dust in the atmosphere. The elders are aware of the effects of such occurrences and that they repeat on some kind of regular pattern.

Is there really global warming overall or is the phenomena just the cyclical variations of the weather that has been observed by the elders and passed down for the knowledge of the next generation?[1]

I hope that the brief time that I have shared with you, and my knowledge can contribute to your work in the field of global warming.

---

[1] The extent and causes of climatic change in the North continue to involve scientists, politicians, and local residents. In the spring of 2202 Dr Peter Wadhams of the Scott Polar Research Institute reported that the polar ice is now only 2.7 metres thick, compared with 4.9 metres twenty years ago, and predicted its total disappearance during the summer by 2080. Dr Ian Stirling of the Canadian Wildlife Service reported that collapsing snow dens due to early spring melting was killing young polar bears by suffocation (*The Sunday Times*, 24 March 2002). Delegates at the 2002 Inuit Circumpolar Conference in Kuujjuaq, Quebec, confirmed that the ice was forming much later in the year and breaking up much earlier in the spring; they warned of the wider repercussions of climatic change in the Arctic for many other areas of the world. – ed.

55 Introduction to Mackenzie Basin Impact Study: Summary of Results

S.J. COHEN ET AL.

Objectives

This is the Final Report of the Mackenzie Basin Impact Study (MBIS), a six-year collaborative research project which began in 1990 and was supported by the Canadian government, Northwest Territories government, BC Hydro, the University of Victoria, Esso Resources Ltd. and others. The purpose of the study was to look at the effect which a change in climate might have on the Mackenzie Basin, its lands, waters, and the communities that depend on them.

The story of the study starts in 1988. At the Changing Atmosphere Conference in Toronto, scientists warned the world's governments about the potential for a change in climate because of the increasing concentrations of carbon dioxide and other greenhouse gases. At the time, the Canadian government had just completed its first environmental action plan (Green Plan). It included a three-pronged strategy to deal with climate change:

- To promote the limitation of greenhouse gas emissions;
- To support research on the processes and effects of climate change;
- To understand the repercussions of the research results on government policy.

SOURCE: S.J. Cohen, ed., *Mackenzie Basin Impact Study Final Report* (Downsview: Environment Canada, 1997), 1–2 (extract). Reprinted by permission of the publisher and editor.

The timeliness of the study was underlined at the Earth Summit in Rio de Janeiro in 1992 when Canada signed the Framework Convention on Climate Change. It committed the country to stabilizing greenhouse gas emissions at 1990 levels by the year 2000. Four years later, in the spring of 1996 the Intergovernmental Panel on Climate Change, which was established by the United Nations, concluded that recent variations in the climate could not be due to natural forces alone. The panel said that continued increases in concentrations of carbon dioxide and other greenhouse gases would lead to a warming of the world's climate.

The Mackenzie basin was one of three areas in Canada which the scientific community selected for a detailed study. The other two areas were the Prairies and the Great Lakes–St Lawrence River Basin.

The Mackenzie Basin was chosen for a case study in order to look at a high latitude or northern region which had sensitive ecosystems and a large number of Aboriginal people who were still following the traditional ways or lifestyle. The question was: how would an economy which was based on natural resources and a Northern culture cope with climate warming? How could they deal with the changes which were expected to be the most significant in the world?

Indeed, there are signs that the climate has warmed up in the Mackenzie Basin. This area, which includes parts of the Yukon and Northwest Territories as well as northern British Columbia, Alberta, and Saskatchewan, has experienced a warming trend of 1.5°C this century. Scenarios of climate change, based on experimental results from General Circulation Models of the atmosphere, suggest that this region could warm up by 4°C to 5°C between the thirty-year baseline period of 1951–1980 and the middle of the twenty-first century.

This report is directed to governments, communities, researchers, and the private sector and all those individuals or organizations with an interest in climate change in this region. The report outlines the potential damage identified during the course of the study and provides recommendations for future action.

Four Key Findings

Despite the enormity of the task, the years of research, and the hours of sometimes animated discussion, it is possible to distill four points or key findings from the study.

1 Effect on the land – Most of the regional effects of climate warming scenarios are not positive. They include lower minimum water levels in the region's waterways and increased erosion from thawing permafrost, as well as a rise in the number of forest fires and landslides and a reduction in the yields of forests. These will probably

offset any potential benefits from a longer growing season. Some of these changes have been observed during the recent 35-year warming trend.

2 Effect on communities – Most participating stakeholders said the region can adapt if the changes occur slowly. But, if the area warms up quickly, adapting will be considerably more difficult. If vegetation and wildlife patterns are modified by climate change, then traditional Aboriginal lifestyles could be at risk. The effect of long-term climate change on communities, however, will also be determined by other factors, including lifestyle choices made by the region's inhabitants. Stakeholders did not know what role climate change would play in the future of the region's two economies – the wage economy typical of southern Canada and the non-wage economy of the traditional lifestyle.

3 Role of regional stakeholders – Increased local and regional control of land and water resources will help to reduce the area's vulnerability and help local residents adapt to climate change. That, however, may not be enough to respond effectively to global warming. Similarly, reducing regional emissions will not be enough to prevent the climate from changing. If the governments which signed the convention on climate change fail to slow down the change in climate, then regional stakeholders may need to intervene at national and international levels to warn others about the consequences to the Mackenzie Basin.

4 Role of the integration process – The effect of a change of climate on the Mackenzie Basin is more than the sum of changes to the trees, wheat, water, and permafrost. Governments, communities, industries, and people will respond to the combined effects of climate change on water and land resources. These responses will be tempered and shaded by the choices government officials, community residents, and industry leaders make in response to other issues such as the demands of the global economy, traditional lifestyles, and political realities. Computer-based models are one way to bring together or integrate many parts of the whole, but these models are limited in their abilities to describe how regions and people relate to climate change and other stresses. The experience of MBIS suggests that an integrated assessment requires a partnership of stakeholders and scientists, in which visions are shared and respected, and information is freely exchanged.

56 The Ecosystem Approach: Implications for the North

ROBERT F. KEITH

The circumpolar North, like much of the rest of the globe, has been subject to a succession of environmental crises. The "pristine" Arctic is not so pristine.[1] Ozone thinning and the long-range transport of industrial contaminants by both air and water systems are two of the most recent threats to northern peoples and the ecosystems on which they depend. The cumulative effects of hydroelectric development on riverine and marine ecosystems, the fallout of radionuclides on the tundra, the impacts of petroleum and mineral wastes, and the endangering of species also are important indicators that circumpolar ecosystems are threatened.

More than species and populations are at risk: entire ecosystems are affected.[2] The Arctic Environmental Protection Strategy,[3] initiated by Finland in 1991 and endorsed by the other seven Arctic nations including Canada, is recognition not only of the significance and urgency of these issues but also of their ecosystemic nature and, thus, of the need to take an ecosystem approach to their resolution.

---

[1] Twitchell, K. 1991. "The Not–So–Pristine Arctic." *Canadian Geographic*, Feb/Mar, 52–60.
[2] Canadian Arctic Resources Committee. *Northern Perspectives* 21 (4), 1994: 28.
[3] *Arctic Environmental Protection Strategy*. 1991. "Declaration on the Protection of the Arctic Environment" signed at Rovaniemi, Finland, 14 June 1991.

SOURCE: *Northern Perspectives* 22, no. 1 (spring 1994): 1–5. Reprinted by permission of the publisher.

What is an "ecosystem approach" and what is its significance to how society conducts its affairs and to the future of the North?[4] The ecosystem approach considers a whole system, rather than its individual parts and their connections. It applies systems perspectives to gain insights into the nature of ecosystems and the ways they respond to the stresses imposed by human activities. From an Aboriginal perspective, "traditional ecological knowledge" is a way of knowing and thinking about ecosystems. It is not just about wildlife, but about food chains and wildlife habitat, including land, waters, ice, and snow. And because people are a part of the environment, human activities are a part of traditional ecological knowledge. From a scientific perspective, researchers analyze ecosystems by studying population biology, food chains, trophic levels, landscape ecology, and stress-response models. It is becoming apparent that all ways of thinking about ecosystems are helpful in developing an ecosystem approach. No one way is sufficient.

Ecosystems are complex; they are systems within systems or hierarchies of systems. Contaminants may be seen as part of the metabolic systems of a beluga whale; the whale, as part of a population inhabiting a marine environment; the population of whales, as part of a community that is seen as a system of food chains in which the sun's incoming energy drives the whole system. All the subsystems are connected and any of them may affect the others. Some of these systems exist over large areas and for considerable periods of time.

Scientists think of ecosystems as being "driven" by the need to use (or, as scientists say, dissipate) large amounts of energy from the sun. The ecosystem's ability to use incoming energy increases with the complexity of its food webs. Biodiversity contributes to ecosystem complexity. In using incoming energy, an ecosystem may change. Such changes might include new life forms, thus enhancing biodiversity.

But the sun's energy is not the only force acting upon an ecosystem; external stresses such as pollution, declines in species populations, and natural calamities may challenge the integrity of an ecosystem. The ability of an ecosystem to manage incoming energy and other external forces that threaten its integrity is referred to as its capacity for self-

---

4 For further discussions on ecosystem approaches see: Allan, T.F.H., Bandurski, B.L., and King, A.W., 1993, "The Ecosystem Approach: Theory and Ecosystem Integrity," *Report to the Great Lakes Science Advisory Board*; Allen, T.F.H., and Hoekstra, T.W., 1992, *Toward a Unified Ecology* (New York: Columbia University Press); Edwards, C.J., and Regier, H.A., eds., "An Ecosystem Approach to the Integrity of the Great Lakes in Turbulent Times," *Proceedings of a workshop supported by the Great Lakes Fisheries Commission and the Science Advisory Board of the International Joint Commission*, Great Lakes Fishery Commission Special Publication; Hollings, C.S., 1986, "The Resilience of Terrestrial Ecosystems: Local Surprise and Global Change," in Clark, W.M., and Munn, R.E., eds., *Sustainable Development in the Biosphere* (Cambridge University Press).

organization. It is this feature of ecosystems that scientists now believe is particularly crucial for ecosystem sustainability.

An ecosystem may adapt to stress in several ways. It has been said that ecosystem adaptation consists of a temporary change followed by a return to the previous condition; however, recent thinking suggests a much more varied set of possibilities. Kay identifies five types of ecosystem response to stress:

1   The system can continue to operate as before, even though its operations may be initially and temporarily unsettled.
2   The system can operate at a different level using the same structure it originally had (for example, a reduction or increase in species numbers).
3   Some new structures can emerge in the system that replace or augment existing structures (for example, new species or paths in the food web).
4   A new ecosystem made up of quite different structures can emerge.
5   The ecosystem may collapse and no regeneration occurs.[5]

It is clear that a collapsed system is not "healthy," but it is equally clear that the original condition is not the only healthy state. Adaptations can be just as "healthy." Ecosystems possessing integrity should be able to

- Maintain normal operations under normal environmental conditions (this is often referred to as "ecosystem health");
- Withstand stresses, particularly those that threaten the system;
- Continue the process of self-organization.

Its ability to self-organize is the system's means of adapting to stresses to "preserve" itself, albeit at times in a new form. The complexity of ecosystems makes it impossible to predict precisely how the self-organization and regenerative processes will occur. The challenge of ecosystem management is to protect or restore the capacity to self-organize and therefore maintain or regain integrity. The objective of maintaining "ecosystem integrity" should be the overall purpose guiding human relations with the environment.

Biological and physical information are keys to the success of self-organizing processes in an ecosystem. Living systems that have evolved successfully have learned what constraints operate on them in the wider ecosystem and how to cope with those constraints. At the levels

5   Kay, J.J. 1993. "On the Nature of Ecological Integrity: Some Closing Comments." In Woodley, S., Kay, J.J., and Francis, G., eds. *Ecological Integrity and the Management of Ecosystems.* St Lucine Press.

of cells to species, the genes inform the self-organization process which pathways and changes are likely to be successful. At the level of ecosystems biodiversity supplies the information that guides the processes of adaptation and regeneration. Thus, biodiversity acts as the "library" for ecosystem regeneration.[6]

Implications for Managing Ecosystems

The continuing pressures of industrial and resource development inevitably mean more stress on Arctic ecosystems. Those with responsibility for ensuring the sustainability of the environment and people that depend on it must begin to plan and manage from an ecosystem perspective. What then are some of the implications of the ecosystem approach for planning and management?

Supporting Self-Organization Processes

Ecosystem dynamics suggest that human activities must protect and enhance the self-organizing capacity of ecosystems. The focus is not just on the ecosystem, but rather on the relationship between it and human uses of it. The challenge of ecosystem research is to understand better the processes of self-organization. The challenge of management is to attune human activities to these processes to protect the integrity of the ecosystems. Research and resource management should be directed toward managing change to support, rather than impair, self-organizing capacities. From a management perspective, it is important to protect the self-organizing capability of ecosystems, and one strategy to achieve this is to protect biodiversity.

The diversity of species is the "information bank" an ecosystem needs to function in normal conditions and to regenerate when stressed significantly. The reduction or loss of a species does not necessarily mean that an ecosystem will not continue to function in a healthy way. Some species "parallel" one another and when one diminishes others assume that part of the function previously carried by the diminished species.

Understanding Whole Systems

Much of contemporary science is based on selected elements within ecosystems. In the past, both scientists and resource managers have tended to be "locked in" to specialized pursuits of knowledge rather

---

[6] Kay, J.J., and Schneider, E., 1994, "The Challenge of the Ecosystem Approach." *Alternatives* 20 (3): 1–5.

than the study of whole ecosystems. The focus of science has been to reduce complexity to achieve observational and experimental control, an approach that seems counterintuitive to much of what we are beginning to understand about ecosystems. Although it is important, even necessary, that we gain more insight into the nature of cells, species, communities, and the like, these insights alone will not provide an understanding of complex ecosystems. There is a distinct need for us to think in terms of systems. Moreover, it is important to consider the broader context of whatever system we are observing and the influence of that context. Our analysis must be expansive enough to account both for the significant external forces acting on the system we examine and for the system itself.

No Single Expertise

For some time, ecologists have acknowledged that no single theory, discipline, or profession can describe the complexities of ecosystems. Yet, the institutional expression of both science and management too often demonstrates narrow focus and expertise. The ecosystem approach requires not only expertise in many categories but also a synthesis of the categories,. The sharing and exchange of many kinds of information is a basic necessity if we are to understand better the nature of ecosystems. As well, it is necessary that we integrate our shared knowledge into policy and practice.

Making Use of All Available Knowledge

The scientific knowledge we have accumulated so far is not nearly adequate for the challenges of the ecosystem approach. Hypotheses – more than solid information – characterize current ecosystem science. We are now beginning to understand, particularly in the North, that there are various ways of knowing and various modes of reasoning. The growing literature on Aboriginal knowledge of ecosystems indicates a potentially significant contribution to our overall understanding of ecosystem dynamics, societal impacts, and limits to human activity. Therefore, many observers and analysts can contribute to understanding and managing ecosystems. There is no "position of privilege" – no single point of view or methodology that prevails.

Precautionary Principle

Some still argue that limits to ecosystem uses are only another challenge in the long history of technological challenges to human ingenuity. More frequently now the view is that limits to ecosystem uses are real but are imprecisely known, and therefore our approach to resource use should be cautious. Faced with the reality that we cannot predict how ecosystems will evolve, especially when seriously stressed, it is argued that we should err on the side of caution; i.e., not act in a way that we suspect might impair their integrity and self-organizing capabilities. We should avoid actions that could endanger the regenerative processes of ecosystems.

Reverse Onus of Proof

Present-day logic requires opponents of development to prove that there will be unacceptable impacts from such development. Some now argue that the principle should be one of "reverse onus of proof," requiring proponents of certain actions to prove that their activities will not impair the health or the capacity of ecosystems to regenerate.

Ecosystem Approaches in the North: Cumulative Effects Assessment

A current study of the cumulative effects of human activity in the Hudson Bay and James Bay regions is assessing ecosystem changes from both scientific and traditional ecological knowledge perspectives.[7] The review of the scientific information provides a basis for suggesting a number of hypotheses about long-term, cumulative, and synergistic effects. The absence of predictable conclusions is recognition of the considerable complexity and limited scientific knowledge of the region's ecosystems. Most of the conclusions are guesses or hunches about how the systems are responding to a variety of stresses. Signs of both stress and change have been found, yet there is little evidence that actions to relieve the stresses of hydroclectric developments and the long-range transport of contaminants into the region are being pursued. Although the principles of caution and reverse onus of proof do not underpin present policy or practice, there has been some progress in the conduct of environmental impact assessments. The review guidelines for formal assessment of hydroelectric projects in both Quebec and Manitoba made significant requests of the proponents for assessment of cumulative effects and justification of the claim that the proposed develop-

---

7 The Canadian Arctic Resources Committee, the Environmental Committee of the Community of Sanikiluaq, NWT, and the Rawson Academy of Aquatic Science are collaborating on a study of cumulative effects and sustainable development in the Hudson Bay and James Bay region, using both traditional ecological knowledge and scientific information.

ments were consistent with the principles of sustainable development. Such requests contribute, in a small way, to ecosystem perspectives.

Traditional Ecological Knowledge: The Hudson Bay Program

The study of traditional ecological knowledge (TEK) also is a contribution to ecosystem thinking. Whereas much of the scientific data for the region has been gathered in selected sites, in relatively brief periods of time, and seldom over an extended period, the TEK study relies on the observation and knowledge of Inuit and Cree hunters, trappers, and elders whose experiences is added to knowledge accumulated from preceding generations. Thus, their data are based on multiple observers, over extended periods of time, and on areas and territories they know intimately.

Participants in the study came from more than twenty communities around Hudson and James bays. Their knowledge has contributed to a detailed picture of ecological changes in the bays and some of the major rivers flowing into them. Preliminary findings from the TEK study appear to both confirm and refute what some scientists have found. Particularly interesting is the Native view that a third population of beluga whales resides year-round in southern Hudson Bay and James Bay. The scientific evidence suggests that there are only two populations and that they migrate out of the bays in winter. If validated, the TEK evidence will have important regulatory and management implications. At present, federal fishery officials are basing their calls for restricted harvests on declining populations. A new population could change their policy. Moreover, Inuit hunters who have observed belugas in places where they have seldom gathered believe they are seeking sanctuary in areas away from noise and disturbance. Thus populations may not be declining; they may be only altering their habitat.

These findings demonstrate the value of acknowledging and accepting different ways of knowing. Through shared management, or "co-management," it is possible to enrich and expand the ecosystem knowledge base. Each perspective has valid information to bring to bear on the issues; neither alone is capable of fully accounting for ecosystem dynamics.

The Ecosystem Approach and Environmental Monitoring

Canada does not have a comprehensive environmental monitoring program for Arctic ecosystems. The Arctic Environmental Strategy outlined monitoring initiatives for airborne contaminants and water quality, but these are a limited response to urgent issues, rather than a system designed to provide information on attributes of ecosystems as a whole. The persuasive argument for ecosystems suggests that we must begin to devise ways of observing and of recording and exchanging information on critical indicators of ecosystem integrity. More specifically, sets of indices on ecosystem health, the capacity of systems to cope with stresses, and the self-organizing properties of ecosystems should be the principle focus of research and of monitoring methods. Monitoring initiatives should combine both scientific evidence and traditional ecological knowledge.

The ecosystem approach also calls for explicit recognition of the need to manage human activities within the context of ecosystem sustainability; human activities, as much as ecosystems, require monitoring. To this end it is essential that public policy, planning, management, and evaluation processes become linked with participatory processes so that intent is known before action and is subject to assessment against criteria for ecosystem integrity and societal sustainability.

57 The Polar Continental Shelf Project

E. F. ROOTS

The Polar Continental Shelf Project was authorized by the Canadian government, by Cabinet Directive, on April 5, 1958. It was designed to conduct general mapping, oceanographic, hydrographic, geological, geophysical, geographical, and related studies, to be undertaken on the Arctic Ocean Continental Shelf, on the islands of the Archipelago, and in the channels. It was to act as a logistic instrument for all divisions of the Department of Mines and Technical Surveys (Energy, Mines, and Resources) in conducting research in the Arctic regions of Canada. It was to supply logistic support and facilities to other agencies conducting research in the area. No fixed date was contemplated for the completion of the project. Yearly financial expenditures have varied between $1,500,000 and $1,800,000.

The initial headquarters was established in 1959, at Isachsen on Ellef Ringnes Island. At this location is maintained a United States–Canadian permanent weather station. To provide a means of accurately positioning the stations where data was collected, and to allow the precise navigation needed by airborne and surface transport engaged in the survey, experiments were carried out on the electromagnetic propagation characteristics of the area, with a view to designing and installing an electronic survey and positioning system to cover the major area of investigation. The initial program was supported by one Beaver and one Otter ski-equipped aircraft. In the first year, these aircraft made nearly five hundred landings on unprepared landing strips on

SOURCE: Canada, Department of Energy, Mines, and Resources; Polar Continental Shelf Project, "P.C.S.P. Paper No. 133," (unpublished typescript, 1967).

both land and ice. The first year pointed up the problems and difficulties to be experienced in the Arctic and led to refinements of procedures and techniques for the following years. In spite of environmental difficulties, the first year produced much of scientific interest.

Scientists were flown to the Arctic in early March, and some personnel remained for the entire summer. Some thirty scientists and technicians were employed on the project the first year. The main object of the field work in 1959 was to supply a surveyed base and to test equipment and methods.

The full-scale research and survey program got under way in 1960. In the first year, work was concentrated on the continental shelf to the northwest of Meighen, Ellef Ringnes, and Borden Islands comprising a block some 200 miles [320 km] long and extending some 250 miles [400 km] to sea. A low-frequency Decca hyperbolic position-fixing system was erected, with the master station on Ellef Ringnes Island and the slave stations on Meighen Island and Borden Island respectively, to give immediate determination of positions to an accuracy of 200 yards [180 m] or less within an area 200 by 300 miles [320 by 480 km]. A base line was run by means of a tellurometer and theodolite across the Prince Gustaf Adolf Sea from Isachsen to Borden Island and another line from Isachsen to Meighen Island by way of Amund Ringnes Island and Fay Islands. A third line was surveyed from Isachsen and out over the Arctic Ocean. These survey lines were required to determine the position of the Decca stations.

At intervals along the main base lines, hydrographic soundings were made; regular oceanographic measurements of temperature and salinity, and samples of water were taken at all standard depths, together with bathythermograph casts and current measurements. Tidal records were made and a study made of complications introduced by temperature, wind, and possibly other factors on the tides in ice-covered waters.

More than eighty gravity measurements were taken on land and sea ice, including a traverse across a typical gypsum piercement dome. A gravity and magnetic traverse of the Meighen ice cap was made to delineate the rock floor beneath the ice. Magnetic surveys were made on various parts of the Ellef Ringnes and Meighen Islands, which have been useful in tracing geological contacts and interpreting geological structures.

Geological studies included the taking of grab samples and cores from the floor of Prince Gustaf Adolf Sea and from the Arctic Ocean near Cape Isachsen, studies of the stratigraphy of northern Ellef Ringnes Island, an investigation of unusual igneous rock structures, and a re-examination of the gypsum structures. Oriented rock specimens were taken to study remnant magnetism.

The physiographic studies included a survey of the land forms of the Ringnes Islands, a study of sediment transport in seasonal Arctic rivers, examination of patterned ground, observations on the formation and distribution of ground ice, and the preparation of air photo interpretation keys for selected areas. A glaciological program was undertaken on the Meighen Island ice cap, and a geological and botanical reconnaissance was undertaken around the exposed margins of the ice cap.

The results of the initial year indicated that more air support and more suitable ground equipment was required to increase efficiency of data collecting. This was improved in later years by the addition of helicopters, both small and large, additional ski-wheel Otter aircraft in place of the Beaver, a twin-engined Beechcraft aeroplane and other multi-engine specialized survey aircraft, and the acquisition of more suitable vehicles for travelling over snow, ice, and frozen ground. The first year's work demonstrated that with a well-planned program and proper equipment it was feasible to conduct research under Arctic conditions with a reasonable amount of comfort and with considerable success.

It should be observed that the Canadian Department of Transport operates several icebreakers in the Arctic Archipelago, to assist ships taking part in the summer re-supply of villages and stations. These ships are made available for oceanographic and other scientific observations once the re-supply mission is completed. The scientific and technical personnel for this work have been provided by the Polar Continental Shelf Project and other units of the Department. The icebreakers have been used in this manner since the inception of the Project.

The basic plan of study of the Polar Continental Shelf Project has been maintained, although the scope of investigation has expanded year by year since 1959. At first, the Project depended on the Isachsen Weather Station for accommodation and some facilities. It soon became apparent that self-sufficiency was required and portable buildings were designed and constructed. The main headquarters of the Project was moved in 1962, southwest to another United States–Canadian Joint Weather Station at Mould Bay on Prince Patrick Island. The present Mould Bay camp can accommodate eighty men, and is self-sufficient, with its own power supply, radio station, and workshops.[1] Temporary camps are established anywhere in the Arctic Archipelago or on the ocean, as required.

[1] The Mould Bay camp is now closed. The Polar Continental Shelf Project now maintains two bases in the Canadian Arctic: one at Resolute on Cornwallis Island in the High Arctic and one at Tuktoyaktuk at the mouth of the Mackenzie River. – ed.

It has been the purpose of the program to investigate, as thoroughly as possible, the area covered by the Decca Lambda electronic positioning equipment before moving the Decca transmitter stations to new positions. The plan was to leapfrog the stations progressively to the southwest. This policy has been followed and at present, the most westerly slave-station is located on Banks Island.

One of the larger programs undertaken was the bathymetric mapping of the main channels between the islands. To facilitate this work, which often was some distance from the coastal Decca system, portable Decca Hi-Fix equipment of shorter range but greater accuracy than the "Lambda" equipment used on the ocean, and for reconnaissance surveys, was employed. This equipment could be transported to the transmission site by helicopter, and could be readily positioned at a new location.

The bathymetric survey was carried out initially by using a wire line and by regular marine echo sounders mounted on motor toboggans travelling on the ice or carried in aircraft and operated through holes drilled through the ice sea at landing places. Later, equipment and techniques were developed to obtain echo soundings through open leads and cracks from a helicopter which tows echo-sounding equipment while in flight; other equipment allows the water depth to be measured through fifteen feet or more of unbroken sea ice. Bathymetric mapping to a scale of 1:500,000, and regular hydrographic surveys at scales of 1: 100,000 and 1:50,000 are being undertaken. The cost of such surveys is now comparable with that of ship-borne surveys, on a square mile basis.

The seismic work was expanded as the program continued. The geological structures lying beneath the surface of the islands and under the sea have been investigated by a series of reflection and refraction seismic traverses. Information on the thickness of the sedimentary sequence of the so-called "Sverdrup Sedimentary Basin," of the underlying granite material, and of the depth to the still deeper denser layer of the "Upper Mantle" beneath the Earth's crust has been obtained. Special techniques of seismic sounding in ice-covered waters and on permafrost were developed. Motor toboggans, tracked trucks, and helicopters were used for transportation in these experiments. The seismic studies are continuing.

An intensive gravity program is being carried out. The force of gravity is being measured at twelve kilometre intervals as part of the regional Canadian Gravity Survey. A network of control "base loops" was established throughout the area, and these have been linked with the absolute gravity stations at Resolute and Mould Bay. Special gravity studies have been undertaken in the course of glaciological investigations, or over interesting anomalies.

The glaciological studies have been continued on the Meighen ice cap. Records were made of accumulation and ablation, mass wastage, ice movement, temperatures at depths, plasticity of the ice at different depths, electrical resistivity in three directions, and other parameters which serve to determine the response of the glacier to climatic environment, meteorological, and micro-meteorological conditions. A bore-hole has been drilled completely through the thickest part of the ice-cap, and has been instrumental to give a continuing measurement of the physical nature and behaviour of the ice mass; the cores from the hole have been preserved and give a complete section, for laboratory study, of this cold, "high-polar" ice cap that apparently is not a remnant from the Ice Age but which appears to have grown in comparatively recent times. The results are being studied in relation to those of the seismic, gravity, and botanical research in the area to help give a picture of the climatic history. Farther to the southwest, in a still more arid part of the Archipelago, four thin, apparently nearly stagnant ice-caps have been surveyed for movement, accumulation, and wastage each year since 1963. It might be noted that considerable glaciological research is being conducted in Canada, by universities and other governmental agencies. A comprehensive program is in progress on the Barnes ice cap on Baffin Island by the Geographical Branch of the Department of Energy, Mines, and Resources, and other studies are under way in western and northwestern Canada, the investigations of the Polar Continental Shelf Project are planned to contribute to a continent-wide study.

The program of making oceanographic observations over the Continental Shelf was not pursued vigorously after the first two years. A good reconnaissance picture of the previously unknown oceanographic characteristics of the waters of the outer Arctic Archipelago and over the continental shelf was obtained as a result of this work; further oceanographic mapping was not warranted in view of the pressing need to employ the country's limited oceanographic research resources in other areas of more urgent and complex problems. Since 1963, the oceanographic work of the Polar Continental Shelf Project has been carried out as part of special studies of sea ice, acoustics, biological productivity, sedimentation, electronic or sonic energy transmission, or geothermal heat flow.

The studies in submarine geology have been continued. The program is designed to provide information on the character and stratigraphy of the sediments on the floor of the channels of the Arctic Archipelago and of the continental shelf. This consisted of a study of the offshore sediments by grab samples and short cores, and a detailed study of the sediments carried by the Arctic rivers, and of their distribution in the channels between the islands. These investigations have required the taking of hundreds of samples which are later carefully studied for their mineralogical, geochemical, organic, and fossil content.

A study of the geology of the Arctic Archipelago has been in progress for many years. Systematic geological mapping started in 1949 and has continued since. At present, the reconnaissance mapping of the area by the Geological Survey of Canada, is practically completed. The Polar Continental Shelf Project has provided logistic support and assistance where needed and in some cases carried out special geological assignments. Detailed geological studies in areas of special interest are in progress. Since the Sverdrup Basin is potentially an oil-producing area, much geological assessment of the region is being undertaken by industry under permit.

Studies on the magnetism of the area have been continued as part of the work of the Project. Temporary magnetic stations have been established at various points to supplement the permanent magnetic observations at Mould Bay, Resolute, and Alert, to aid in determining regional geomagnetic gradients and to investigate areas of special interest. Magnetic traverses have been run across selected geological features to increase knowledge of certain geological structures or formations. Areas of non-isotropic magnetic behaviour, and of locally enhanced electromagnetic conductivity, are receiving special study. Aeromagnetic surveys, mapping the total magnetic intensity on a scale of 1:250,000, have been undertaken systematically since 1961, and to date have covered approximately 145,000 square miles [375,000 km²].

Studies of the flow of geothermal heat from the ocean floor have been made in various parts of the continental shelf area, both in places where conditions are apparently "normal" and where from geomagnetic or seismic or other evidence there is reason to suspect an anomalous energy pattern in the Earth's crust. It is hoped that these studies will lead to a better understanding of what is proving to be a tectonically active part of the North American continent.

Botanical, entomological, and marine biological studies have been conducted on the Polar Continental Shelf Project by scientists from the Department of Agriculture, the Fisheries Research Board, and other agencies, in various regions of the Arctic Archipelago. The material secured has augmented the national collections in their fields.

The mapping of the formation, nature, and movement of sea ice of the Arctic Ocean and the waters of the Archipelago has been an important part of the program of the Project since 1960. Aerial surveys are supplemented with ground observations and physical measurements, and related to meteorological and oceanographic conditions in an attempt to understand the behaviour of sea ice and the causes of its remarkable fluctuation in character and movement from year to year. The work is leading to better forecasts of ice conditions in the shipping routes of Subarctic Canada. The movement of the large tabular ice-bergs, or "ice islands," which break off from the Ward Hunt Ice Shelf on Ellesmere Island and drift both to the east and west around and through the Archipelago, is recorded; these bodies can be followed from year to year and give a positive record of the effects of wind and ocean current.[2]

Topographical mapping of the Arctic Archipelago is now completed to a scale of 1:250,000, and in areas of special interest, to 1:50,000. The mapping is based on vertical air photographs taken at thirty thousand feet. Ground control was provided by the Army Survey Establishment of the Canadian Armed Forces in the Western Islands, and by the Surveys and Mapping Branch of the Department of Energy, Mines, and Resources, in the Eastern Islands. Surveyors working with the Polar Continental Shelf Project provided ground control on many of the Northern Islands. It can now be said that the mapping is adequate, except when special requirements arise.

The safety record of the Project has been good. One Otter aircraft and one helicopter were lost when landing on ice floes. One transport plane, containing a helicopter, was lost on a trip from Montreal to the Arctic Islands. The only persons killed on the Project have been the pilot of the helicopter and the crew of the transport plane.

Since its inception, the Polar Continental Shelf Project has contributed greatly to the scientific knowledge of Arctic Canada. In recent years, more than one hundred persons of various disciplines have, each year, taken part in the operation. It can be said that the project has met the expectations and hopes held for its progress. It has proven definitely, that scientists and technicians can work in the unfavourable climate with a minimum of discomfort and danger. With the passage of time, more and more of the studies are being integrated into the normal programs of the Branches of the Department of Energy, Mines, and Resources. The great value of the Project continues to be its logistic support and coordination of planning and transport to the advantage of all scientific groups working in the area.[3]

2 In 1985 the Polar Continental Shelf Project was involved in 235 different projects throughout the Arctic. One of the most dramatic was the establishment of a major camp on ice island T3 in the Arctic Ocean. – ed.

3 The Polar Continental Shelf Project is now included in National Resources Canada. Its mandate is to provide cost-effective coordinating logistics, advice, and support to Canadian-government, university, and independent groups, and on a cost-recovery basis, to private sector and non-Canadian groups that conduct research programs in the Arctic. The Polar Continental Shelf Project provides information about scientific operations in the Arctic to clients, northern inhabitants, and the general public. – ed.

58 Requiem for a Fossil Forest

ED STRUZIK

On a Sunday evening in early July, James Basinger flew in to the tiny settlement of Resolute on the south shore of Cornwallis Island, some 1,500 kilometres northeast of Yellowknife. The Saskatoon scientist and his six-member team were to bunk for the night in the single-storey headquarters of Polar Continental Shelf Project. The following afternoon, they would board a twin-engine plane for a two-and-a-half-hour flight to the Eureka weather station on Ellesmere Island, and then a helicopter for the final thirty-minute leg of their journey. Their destination: a 45-million-year-old fossil forest on Axel Heiberg Island.

Basinger was familiar with the routine. Head of the department of geological sciences at the University of Saskatchewan, he had made the same trip nearly every summer for the past thirteen years. As he and his colleagues sat at a kitchen table in the small cafeteria drinking coffee and chatting, he says he was relaxed and "looking forward to visiting the old homestead." A university student and a post-doctoral fellow with him, who were about to view the prehistoric oasis for the first time, were nearly giddy with excitement. Then, the centre's cook stopped by for a chat. So, she said, you must be joining the American scientists who stopped over last week on their way to the same campsite. "Our jaws hit the floor," says Basinger. "We were completely stunned."

It wasn't that the veteran researcher objects to foreign scientists working in the Arctic. But the only outside project involving the fossil forest that he had heard about involved large-scale excavation. Indeed,

SOURCE: *Canadian Geographic* 119, no. 7 (November/December 1999): 42–8. Reprinted by permission of the author.

nearly a year earlier, when two researchers at the University of Pennsylvania sent him an outline of their plan for studying the environment and climate of the ancient site, Basinger says he was "horrified" and conveyed his concerns to Polar Continental Shelf Project and the Nunavut Research Institute, the lead agency for science, research, and technology development in the new territory. When he heard nothing more, he assumed the proposal had been dropped. He was incredulous at the cook's comment.

Sure enough, twenty-four hours later, as the chopper crossed the frozen fiord at Eureka Sound in the yellow glow of a brilliant Arctic evening and began its descent over the rugged terrain, Basinger could see at least a dozen people hovered over trenches and holes that had been dug in a hilltop. At the bottom of the slope between two picturesque streams was a profusion of blue and yellow nylon tents. And outside one of two much larger, bright-red shelters, which turned out to be a portable kitchen and laboratory, a skull-and-crossbones flag flew at half-mast.

It was a sign of the times for polar research in Canada. Despite an additional $1 million in funding this spring, Polar Continental Shelf Project – a Natural Resources Canada agency that provides ground and air support services for scientists in the Arctic – is operating at half the strength it was five years ago. After a bold start, it has become, in the words of polar scientists John England, Art Dyke, and Greg Henry, "a rusted-out VW beetle with four flat tires, foregoing its historic role of research facilitator for that of travel agent."

Polar Shelf, as it is more commonly known, provides free support to approved Canadian government and university researchers, but charges foreign scientists for its services – an obvious incentive to court big-money enterprises from outside the country. And, recently, the three largest Arctic expeditions have been predominantly foreign: American scientists, mostly from NASA, have been on Devon Island for the past two summers studying Haughton crater for clues about life on Mars; last year, an international group of oceanographers, ice scientists, and atmospheric researchers in a Canadian Coast Guard ship that had been deliberately frozen into the drifting ice pack in the Beaufort Sea; and, this summer, a tundra ecology project was conducted by Swedish scientists.

In hindsight, Basinger says, perhaps he should have expected outside researchers at the fossil forest. Still, he cannot help feeling let down that he was not consulted by Polar Shelf or the Nunavut government or even informed that the study had been approved. When he heard the news in Resolute, his first instinct was to go home. There simply was not room for two scientific crews on such a small, fragile site, he de-

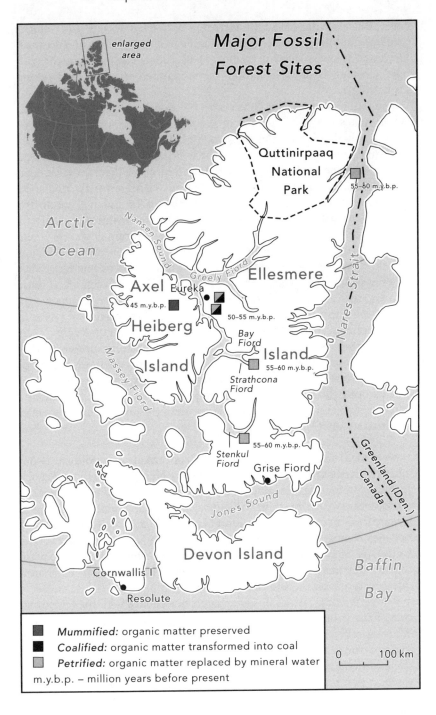

Major Fossil Forest Sites

enlarged area

Quttinirpaaq National Park

55–60 m.y.b.p.

Arctic Ocean

Nansen Sound

Greely Fiord

Ellesmere

Axel Eureka

45 m.y.b.p.

50–55 m.y.b.p.

Heiberg

Bay Fiord

Island

Island

55–60 m.y.b.p.

Massey Fiord

Strathcona Fiord

Nares Strait

55–60 m.y.b.p.

Stenkul Fiord

Grise Fiord

Jones Sound

Greenland (Den.) / Canada

Devon Island

Baffin Bay

Cornwallis I

Resolute

Mummified: organic matter preserved
Coalified: organic matter transformed into coal
Petrified: organic matter replaced by mineral water

m.y.b.p. – million years before present

0   100 km

cided. But his wife, Elisabeth McIver, a research associate and adjunct professor at the University of Saskatchewan, persuaded him to push forward. He owed it to the scientists and students he had brought with him, she said; besides, he should see for himself what was happening to the forest.

The remains of ancient forests can be found throughout the Arctic. What makes the fossils on the northeast coast of Axel Heiberg particularly valuable is that the site is one of the largest, oldest, and most exquisitely preserved of its kind in the world. Analyses by Basinger and others have shown that some 45 million years ago the area resembled the Carolinian forests found in the state of Georgia today. Redwood swamps, deciduous flood plains, and boreal forest uplands supported soft-shelled turtles, alligators, and rhinoceros-like creatures, as well as a host of small mammals. The mean annual temperature was somewhere between 12° and 15°C and it got as high as 25°C during the warmest month. Exactly what caused the warm weather remains unclear.

What is apparent from Basinger's studies is that the remains of these ancient trees and other flora were buried in freshwater sediment in a broad, flat basin over the course of hundreds of thousands of years. In an environment where little or no mineralization could undermine their organic integrity, some of the specimens are so perfectly preserved that they are almost indistinguishable from the litter on the floor of a modern coniferous forest. Their extraordinary quality has led some scientists to hope the fossils will reveal important clues about the climatic and environmental history of the planet – and the possible consequences of today's global warming.

Not surprisingly, Basinger was more concerned with political issues when he arrived at the site this summer. He spoke first with Dick Jagels, a University of Maine plant anatomist, who had spent a year at the University of Alberta in the 1960s; Jagels was amiable and down-to-earth but never ventured beyond chit-chat. A short time after, Ben LePage, Basinger's former PhD student and now a researcher at the University of Pennsylvania, sought out his one-time mentor to explain that he was unaware that Basinger and his students would be resuming their work on the site this summer.

It was two days before Basinger came face-to-face with the real architect of the project, University of Pennsylvania ecologist Art Johnson. Wearing a hard hat, smoking cigars, at least one of which he ground into the forest floor with his boot – and wielding a chainsaw to cut some of the largest trees out of the permafrost, Johnson did not create the best first impression; still, he proved to be forthright, even charming in his own way. Before meeting with Basinger, he explained in an interview that his team's goal was not to compete with, or displace, a

Canadian scientist; it was to reconstruct the forest environment in a manner that might shed light on global climate change during a time when humans had no impact.

His work was supported by a three-year, $1.2-million (US) grant from the Andrew Mellon Foundation in New York, Johnson said, adding that he expected, given the exquisite nature of the site, to get even more money for a larger study with greenhouses, growth chambers, and additional scientists. "What we have here is unique. Not only is it spatially explicit – the wood, the bark, the leaves, the imprints in the rock are so well-preserved and remarkable – there is really nothing quite like it. I can tell you that there are a lot of scientists like me in the United States who would give anything to have a piece of this."

The fact that no one saw fit to tell him or Basinger about each other's work this summer was, in Johnson's opinion, another story. "Someone obviously dropped the ball big time there." Still, he was prepared to make a deal. "Tell Jim that we have money for him, in the amount of six figures over three years. That's money that all goes to science, not overhead, if he is prepared to join us."

Later that evening in the shadow of a large Canadian flag, which Basinger's students had raised in a symbolic show of protest, Johnson made the offer directly to Basinger. The meeting was initially awkward, Basinger later recalled. Once he and Johnson dispensed with the formalities though, the atmosphere warmed considerably. "Art is basically a nice guy," says Basinger, "but I'd be a hypocrite if I accepted this, even if it does represent far more money than the Canadian government would ever give me."

Basinger's concern is not with the science, but with the methods, which he fears were destructive to the fossil forest. His suspicions are not unfounded. In 1986 he excavated one of the large mummified stumps and sent it to the Canadian Museum of Nature in Ottawa. Although it was wrapped in damp newspaper and plastic film, it quickly began to crack and shatter. David Grattan, a wood conservationist with the Canadian Conservation Institute (CCI), called upon in hopes of saving the specimen, found that when the saturated fossil was expose to air, its interior and exterior shrank at different rates. Unless an effective treatment could be developed, Grattan concluded, other unearthed fossils would also disintegrate.

When he visited the fossil forest itself, Grattan delivered more bad news. Erosion was already a problem due to the region's extreme weather conditions and the absence of any significant vegetative cover. And exacerbating the natural wear and tear were military helicopters landing on the hill to disgorge visitors from a base in Eureka, tourists from cruise boats exploring the site and collecting souvenirs, and, incredibly, evidence that someone had used the mummified wood for campfires. The CCI subsequently issued a warning to territorial and federal government departments about the fragility of the site, along with a list of dos and don'ts for anyone travelling there. A second alert was issued after Art Johnson invited CCI scientists to get involved in this summer's project – an invitation they declined.

Yet, in spite of the cautionary flags, Johnson's project was approved. In August, an editorial in the *Nunatsiaq News* of Iqaluit accused the Nunavut Impact Review Board (NIRB) of sleeping while the application was considered. Its executive director, Joe Ahmad, demanded an apology, saying the board acted on the recommendation of the territorial agency responsible for culture and heritage, which stated in a letter to NIRB: "The proponent's proposed activities do not constitute a threat to known archeological resources."

"Of course not," says Basinger, "there are no archeological sites" in the fossil forest. "Anything prehistoric would fall outside their direct interests." And that, he says, is the crux of the problem: "At some point, we need to have a national interest in things. It should not be the role of Nunavut [to assess the impact of scientific work in the fossil forest.]"

By summer's end, Bruce Rigby, executive director of the Nunavut Research Institute, who says he was on leave when the permit was granted to the American scientists, was involved in "ongoing discussions with the Nunavut government to decide what to do about the fossil forest ... But from what I've reviewed, the file was handled quite well by the institute." Rigby added that his agency is meeting with a representative of the US research team in late fall to discuss plans for the upcoming season.

At the same time, many of the country's top polar scientists and two of the country's largest conservation groups – the Canadian Nature Federation and the Canadian Parks and Wilderness Society – were calling for measures to protect the fossil forest. One option, says Bill Peters, director general of the CCI, would be to extend the boundaries of Quttinirpaaq National Park over to the northeast corner of Axel Heiberg. That plan was dismissed by Parks Canada, however, on the basis that park status would only encourage unwanted traffic in the area. Moreover, both Heritage Minister Sheila Copps and then Environment Minister Christine Stewart maintained it is up to the Nunavut government, which would not resume sitting until the fall, to decide how to protect the fossil forest.

The inaction came as no surprise to such scientists as University of Alberta geographer John England, who has watched in frustration as Polar Shelf's annual budget has fallen from a peak of $7.2 million in the early 1990s to just $4.2 million this year. In England's opinion, the buck passing was just one more sign that Canada is allowing scientific sovereignty in the Arctic to slip from its grasp. What he wants to know is, "What are the long-term costs to Canada of eliminating the Arctic scientific capability it paid to create over the past fifty years? When the next national imperative – an oil crisis, for example – pushes us back into the Arctic on short notice, what price will we pay for the lost corporate and scientific memories?"

Ian Stirling, another prominent polar scientist, says he believes collaboration with scientists outside Canada is both necessary and important, but he worries that current policies will relegate Canadian researchers to the status of scouts for big-money projects from other countries. "I simply don't understand how a permit could be given to conduct research that might be destructive on an Arctic site of such enormous scientific importance without a full assessment of the methods and their potential environmental impact," says Stirling. "And I am shocked that a senior and respected Canadian scientist who has been actively and carefully researching this site for the past fifteen years could apparently just be pushed aside without due consideration or notification. Does that mean anyone's science projects in the Arctic are vulnerable if well-funded outsiders are interested?"

Back home in Philadelphia, Art Johnson was feeling unjustly demonized in the debate that his research sparked. Still, he plans to resume his research on Axel Heiberg next year. He hopes by then Canada will have its act together. "I think the Canadian government has got to decide what it wants with the fossil forest. Once that's done, I'm confident that everyone's interests, including ours, can be accommodated."

59 Sovereignty, Security, and Surveillance in the Arctic

CANADIAN ARCTIC RESOURCES COMMITTEE

In 1994 two joint committees of the House of Commons and Senate held extensive public hearings on Canada's foreign and defence policies. Both committees were charged with advising the federal government on priorities and issues into the next century, and both were doing so in the uncertain political context of the collapse of the Soviet Union and the demise of the Cold War.

There is much to applaud in the committee reports released in autumn 1994. For example, sustainable development is suggested as an "overarching foreign policy theme," and environmental security is acknowledged as a policy objective of the first order. Both themes resonate strongly in northern Canada, where pollutants generated through non-sustainable development in the south and in other countries are contaminating the food web and endangering the health of northerners, particularly Aboriginal northerners, who eat large quantities of "country food."

Yet neither parliamentary committee said much about the North. Nor did either suggest how its general recommendations should be implemented in this vast area. This is disappointing and illustrates a lack of foresight, for wide-ranging political and economic changes are under way in the circumpolar Arctic that will likely propel the region into the political spotlight in coming years.

The Arctic is the only portion of the globe bordered by the United States, Russia, and Canada. Past certainties – superpower confronta-

SOURCE: *Northern Perspectives* 22, no. 4 (winter 1994–95): 1–2. Reprinted by permission of the publisher.

tion over the pole – are giving way to new possibilities, including production of the region's extensive proven and prospective energy and
mineral resources. In a few short years the Arctic Ocean may become
a route for general cargo vessels linking Japan, east Asia, and the western seaboard of North America with western Europe and the eastern
seaboard of North America. Such a development would have profound
new policy implications for Canada.

In the past the circumpolar Arctic was dominated by defence needs
– a policy arena in which Canada played only a minor role. Today this
region is emerging as a venue in which all facets of foreign policy can
be exercised, if we have sufficient imagination. Sustainable development and environmental security promise to be the policy touchstones
in the circumpolar Arctic well into the next century.

Canada has persevered through the eight-nation Arctic Environ-
mental Protection Strategy (AEPS) to promote circumpolar solutions to
shared problems. Since 1989 Canada has sought to establish an Arctic
Council to bring together senior ministers of the Arctic nations.[1] The
Hon. André Ouellet, Minister of Foreign Affairs, recently appointed an
Arctic Ambassador and gave her a mandate to pursue the Arctic
Council initiative. In the press conference introducing the ambassador,
Mr Ouellet endorsed the concept of a legally binding arrangement –
presumably a treaty – to better equip the Arctic nations to deal with
environmental and sustainable development issues.

All this is to the good. We cannot achieve environmental security
through unilateral action. Moreover, multilateral approaches that serve
our national self-interest sit well with the Canadian body politic and
reflect our sometimes diffident attitude. Certainly Mr Ouellet's treaty
idea deserves very serious consideration.

But a third objective should guide our foreign and defence policies in
the Arctic: Canadian sovereignty over the Northwest Passage. While
Canadians believe the Northwest Passage to be "internal waters" subject to our control, this is neither the perception nor the position of the
United States, which sees the passage as an international strait. In 1985
the American icebreaker *Polar Sea* transited the Northwest Passage
uninvited and in so doing created a political storm that echoes still. In
response, the federal government took both political and legal measures to assert Canada's sovereignty over the passage, including effecting an agreement with the United States regarding Canadian consent
for future transits. But that agreement applies only to icebreakers and
does not prejudice the legal position of the United States.

These arrangements did not fully quiet the controversy. Sovereignty
on paper is one thing: Being there is another. To exercise our sover-

---

[1] The Arctic Council was established in Ottawa in September 1996. See chapter 62. – ed.

eignty and jurisdiction over the passage effectively and to ensure that our environmental standards and laws are enforced, we have to control how, why, and by whom it is used. To provide for this, the federal government proposed to construct a polar class 8 icebreaker for year-round operations in Arctic waters and to acquire a fleet of nuclear-powered submarines to allow our navy to patrol under the Arctic ice. An Arctic Subsurface Surveillance System (ARCSSS) – acoustic listening devices similar to those used in the north Atlantic – was proposed to monitor underwater use of the passage.

None of these technological components to Canada's Arctic sovereignty has been implemented. Nuclear submarines and the powerful icebreakers were abandoned as too expensive. The Department of National Defence (DND) has, apparently, been negotiating for some time with potential contractors to install an ARCSSS, but it is unclear whether or not Canada is actually committed to installing such a system.

The defence and foreign policy committees of Parliament discussed Canada's territorial integrity and sovereignty, but little of this discussion was directed to the North, the very area in which our sovereignty is most directly and obviously questioned. Neither committee recommended technologies the federal government should use to affirm Canada's sovereignty in the Arctic, although both looked forward to further bilateral and multilateral arrangements in the North and beyond to promote sustainable development and environmental security. The proposed acoustic surveillance system was not mentioned in a recently released defence white paper – an oversight of extraordinary proportions, if an oversight it was.

The sovereignty component of our national agenda in the North must not be lost even as we welcome past adversaries as new friends and collectively come to grips with common economic, environmental, and social problems. Donald McRae, a respected law professor at the University of Ottawa, when appearing before the parliamentary committee on national defence, said that "subsurface transits undertaken without Canada's consent are a serious encroachment on Canada's sovereignty over Arctic waters. He noted that Canada's sovereignty over Arctic waters can be lost through "dereliction," or neglect of duty. That's just what could happen if Ottawa fails to act.

60   Arctic Sovereignty: Loss by Dereliction?

DONALD M. McRAE

"Sovereignty" in the Arctic

"Arctic sovereignty" is a symbol of Canadian identity. The "North" is integral to Canada and to how Canadians perceive themselves. Canadian sovereignty over the lands and waters of the Canadian Arctic Archipelago[1] is of the essence of Canada as a nation. The defence of Arctic sovereignty is therefore crucial to Canada's defence policy.

The term "sovereignty" evokes many images and, while the claim to Arctic sovereignty partakes of many of those images, there is at the core a question of law and a question of fact. Is it possible for a state to claim sovereignty over such an area, and has Canada in fact established such a claim? In the context of this submission, there is a third question: If Canada has established its sovereignty over the land and waters of the Canadian Arctic Archipelago, is that sovereignty liable to be undermined by future events?

In law the term "sovereignty" is more readily applied to the authority, or "jurisdiction," of a state over land territory. It signifies the full and complete authority of an independent "sovereign," or in more

---

[1] The term "waters of the Canadian Arctic Archipelago" refers to the water *between* the islands of the archipelago and not to the waters in the open seas of the Beaufort Seas and the Arctic Ocean to the west or to the waters of Davis Strait and Baffin Bay to the east.

SOURCE: *Northern Perspectives* 22, no. 4 (winter 1994–95): 4–9 (extract). Reprinted by permission of the publisher.

modern terms "state," over the lands within its territorial limits. The test in law for determining whether a state has obtained that authority, or sovereignty, over land is one of effective occupation and control manifested through continuing acts of authority. As essential is the acquiescence of other states to the claim of sovereignty or their formal recognition of the claimant state's authority.

In respect of the lands of the Canadian Arctic Archipelago, Canada's title and "sovereignty" are not in doubt. No state disputes Canada's claim over this territory, and thus no legal issues arise. Sovereignty over the waters between the islands of the archipelago, by contrast, is more complex, since historically the principle of freedom of the seas has meant that the jurisdiction of a state ends at its coast. The seas have been free and open to all.

The doctrine of the freedom of the seas runs contrary to any claim to Canadian sovereignty over Arctic waters. It would deny Canada the right to control access to those waters, to preserve the unique and fragile Arctic environment, or to protect the way of life of the indigenous inhabitants. For these and other reasons, successive Canadian governments have framed Canada's claim to the waters as a claim to sovereignty – a claim to full and complete authority and jurisdiction over the waters.

An enquiry into Canadian sovereignty over Arctic waters involves the questions of what jurisdiction a state may claim over waters off its coasts and whether Canada has done what is necessary to "perfect" a claim to these waters.

The Jurisdiction of a State over Waters off its Coasts

Traditional legal doctrine granted states authority over areas of sea off their coasts known as the "territorial sea." After years of controversy, it is now generally accepted that the breadth of the territorial sea is twelve nautical miles. The authority of a state within that territorial sea is akin to sovereignty, with one important exception: A state must grant foreign vessels a right of "innocent passage" through the territorial sea.[2] In other words, although some limitations may be placed upon them, foreign vessels have a right of access through the territorial sea of any state.

Since the starting point for the territorial sea (the "baseline") was the low-water line on the coast, full "sovereignty" over waters – that is, sovereignty not subject to any right of passage – was limited to areas of the sea essentially enclosed by the land, bays having a narrow entranceway into the sea. In such circumstances, states drew a straight "base-

---

[2] This right of "innocent passage" also includes a right of overflight.

line" for the territorial sea across the mouth of the bay.[3] The waters seaward of the line were the territorial seas (through which there would be a right of innocent passage), and the waters landward of the line were the "internal waters" of the state. "Internal waters" are treated in law as land territory. They are subject to the full sovereignty of the state, and no right of innocent passage exists through them.

In some instances, states have claimed that waters that would otherwise not be enclosed as internal waters be treated as internal by virtue of an "historic title." Such a title is established through a long-standing claim to the waters acquiesced to or recognized by other states. Canada's sovereignty over the waters of Hudson Bay is generally recognized as falling into the category of "historic title."

In 1951 the International Court of Justice accepted that states could draw "straight baselines" for the measurement of the territorial sea across areas of coast heavily indented with many off-lying islands. The example before the Court was the west coast of Norway, where the coastline is cut into by fjords and there are many small islands off the coast (known as the "skjaergaard").[4] Instead of following the low-water line, these "straight baselines" would be drawn from the most seaward points on the mainland and island coasts, linking mainland and islands and enclosing significant areas of water. The effect of these "straight baselines" was that the waters behind them would be "internal waters"; in other words, waters over which the state would have full sovereignty.

A potential qualification to the sovereignty of a state over its "internal waters" lies in those cases where the waters include a "strait used for international navigation." Such a "strait" is generally defined as a body of water joining two areas of high seas or joining the high seas and a territorial sea that is used for international navigation.[5] In such straits, vessels have a right of passage equivalent to the right of innocent passage in the territorial sea or, where the regime of "transit passage" applies, a right even greater than that of innocent passage.[6] Although the extent of use necessary to constitute a strait as "international" is a matter of some controversy, there must be some evidence that foreign shipping does in fact use the route for navigation.

3 According to Article 7 of the 1958 Convention on the Territorial Sea and Contiguous Zone, such a baseline could be no more than 24 miles in length. Article 10 of the 1982 Convention on the Law of the Sea contains the same provision.
4 *Fisheries Case (United Kingdom v. Norway)* (1951) I.C.J. Rep. 116.
5 Article 37 of the 1982 Convention on the Law of the Sea applies the international straits regime also to straits between the high seas and the exclusive economic zone of a state.
6 The concept of "transit" passage embodied in Part III, Section 2 of the 1982 Law of the Sea Convention prevents the coastal state from impeding "continuous and expeditious" transit by foreign vessels.

The implications of the foregoing are that for Canada's claim to sovereignty over the waters of the Canadian Arctic Archipelago to be justified in law, it must be demonstrated that the waters are the internal waters of Canada *and* that the waters of the Northwest Passage do not constitute an international strait.

The Legal Basis of Canada's Claim to Sovereignty over the Waters of the Canadian Arctic Archipelago

Canada's claim to sovereignty over the waters of the Arctic Archipelago has been expressed in a variety of ways, not always consistently but always with the objective of ensuring Canadian control over the waters and over passage through them. An early expression of this claim is the "sector theory" associated with the famous resolution asserting Canadian sovereignty up to the North Pole introduced into the Senate in 1907 by Senator Poirier. According to the sector theory, polar states are entitled to exercise sovereignty between their mainland territory and the North Pole in an area of longitude running from their east and west coasts to the Pole.

The rationale for the sector theory has never been clear. To the extent that it is based on contiguity – that is, the claimant state happens to be next to the territory claimed – it does not provide a sound basis for founding a territorial claim. Nor can a solid foundation for the theory be found in the practice of states. Canadian officials have made statements from time to time indicating that Canada's claim to sovereignty over Arctic islands and waters is based on the sector theory, but such statements have been neither uniform nor consistent. Certainly Canada has never disavowed the sector theory as a basis for its claim to sovereignty over the waters of the Arctic Archipelago, but neither has it made the theory a principal plank of its position.[7]

Similarly, Canada has not claimed that its sovereignty over Arctic waters is based upon an historic title.[8] Such a claim in respect of the waters has not been made consistently; official statements refer to historic title over the land only. Moreover, it would be difficult to argue that other states have recognized or acquiesced in any claim to historic title by Canada to all of the waters of the Canadian Arctic Archipelago.

The principal foundation in law for Canada's sovereignty claim over the waters of the Arctic Archipelago is that the waters lie behind the proper baselines for the measurement of the territorial sea (and hence are internal waters of Canada) and that the Northwest Passage does

---

7   For a detailed discussion of the sector theory, see Donat Pharand, *Canada's Arctic Waters in International Law* 3–87 (1988).

8   Ibid., 89–130.

not constitute an international strait (and hence there is no right of passage through it for foreign vessels). Canada's approach to asserting its claim to sovereignty over these waters has been largely a reactive one: that is, rather than seeking the express approval of other states for its position, Canada has reacted to events that might be interpreted as challenges to its sovereignty. In the face of such events, the Canadian government has taken actions designed to reinforce Canada's authority and to make clear where sovereign authority lies. In other words, it has exercised authority where that has appeared necessary.

In the early 1970s, such an approach was known as the "functional" approach to the assertion of Canadian sovereignty. Following the voyage of the American oil tanker Manhattan through the Northwest Passage, the Canadian government adopted the Arctic Waters Pollution Prevention Act (AWPPA),[9] under which Canada asserted the jurisdiction necessary to control future tanker traffic through the Northwest Passage.

Canadian officials followed this assertion of jurisdiction with a strategy to secure its international acceptance. Not prepared to have its authority challenged directly, Canada made a reservation to its acceptance of the jurisdiction of the International Court of Justice to prevent other states from challenging the Arctic Waters legislation before the Court. Ultimately, the Canadian strategy was successful, and Article 234 of the 1982 Law of the Sea Convention contains what is known as the "Arctic exception," a provision that recognizes the jurisdiction of states in ice-covered areas to take measures affecting shipping – for the purpose of preventing, reducing, and controlling marine pollution – that go far beyond those they could take in other ocean areas off their coasts.[10]

The Arctic Waters Pollution Prevention Act was not an assertion of full sovereignty; indeed, critics argued that by asserting a jurisdiction of less than sovereignty, Canada had diminished its sovereignty claim. But the AWPPA was a manifestation of sovereignty, and its ultimate international acceptance helped to consolidate Canada's authority over the waters of the Canadian Arctic Archipelago.

In 1985 the voyage of the US icebreaker Polar Sea through the Northwest Passage raised again the question of whether Canada really had sovereignty over the Passage. This clear indication that the United States did not accept Canada's claim to sovereignty over the waters of the archipelago reinforced some of the concerns about the Manhattan

---

9 Statutes of Canada, 1969–70, c.47.
10 The Law of the Sea Convention will enter into force later this year, although it is unlikely that either Canada or the United States will be parties to it at that time. This does raise some questions for Canada about the exact status of the provisions of Article 234.

voyage some fifteen years earlier. An important factor in ensuring the international validity of Canada's claim is acceptance by other states, particularly by the only state having an overt interest in using the Northwest Passage for transit purposes.[11]

The question for Canada at that time was whether everything possible had been done both to assert its claim to sovereignty over Arctic waters and to ensure that it was in a position to demonstrate a de facto capability to exercise the sovereign authority it claimed. Two types of action were taken by the Canadian government. First, "straight baselines" were drawn around the outermost islands of the Arctic Archipelago to indicate that these were internal waters of Canada. Second, measures were announced to reinforce Canada's presence in the area and to enhance its ability to detect the actions of others: building a Class 8 icebreaker and increasing surveillance overflights.[12]

At the same time, Canada began exploring with the United States mechanisms to ensure that Canada could consent to the US interest in the transit of the Northwest Passage. The result was the 1988 Arctic Cooperation Agreement, under which the United States pledged that all navigation by US icebreakers in waters claimed by Canada to be internal would "be undertaken with the consent of the Government of Canada." However, the agreement also provides that nothing in it or any practice under it affects the position of either government in respect of "the Law of the Sea in this or other maritime areas" – in other words, both sides are preserving intact their respective positions on the status of the waters of the Northwest Passage. Thus, while the agreement does not advance Canada's claims, it does negate the impact of US actions that would otherwise be detrimental to the claim.

Canada's claim to sovereignty over the waters of the Arctic Archipelago stands or falls on whether the drawing of straight baselines enclosing the waters as internal waters can be justifiable in law and on whether the waters of the Northwest Passage constitute an international strait. The argument supporting the use of "straight baselines" in the context of the Arctic Archipelago derives from the decision of the International Court of Justice in the *Fisheries Case*. The geographic relationship between the Canadian mainland and the islands of the archipelago and among the islands themselves; the unique nature of waters frozen and used as land for much of the year; the particular economic dependency of the indigenous peoples of the area on

---

11 The United States did attempt to downplay the significance of the voyage for the sovereignty issue. See Griffiths, "Beyond the Arctic Sublime" in Franklyn Griffiths, ed., *Politics of the Northwest Passage*, 248 (1987). Nevertheless, the voyage was undertaken without seeking the consent of Canadian authorities.

12 Statement of the Secretary of State for External Affairs to the House of Commons, 10 September 1985. *House of Commons Debates*, 6462–4.

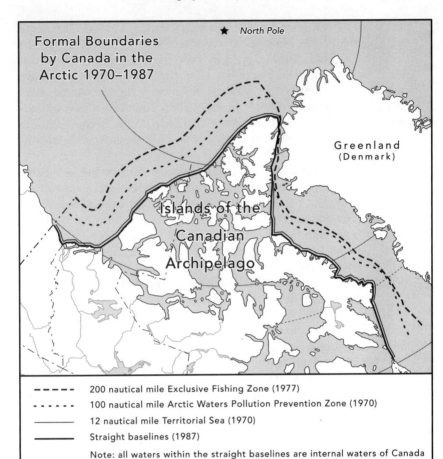

Formal Boundaries
by Canada in the
Arctic 1970–1987

★ North Pole

Greenland
(Denmark)

Islands of the
Canadian
Archipelago

- - - - -   200 nautical mile Exclusive Fishing Zone (1977)
· · · · · ·   100 nautical mile Arctic Waters Pollution Prevention Zone (1970)
————————   12 nautical mile Territorial Sea (1970)
————————   Straight baselines (1987)

Note: all waters within the straight baselines are internal waters of Canada

SOURCE: *Looking North*, DIAND, 1989

the waters; and the highly irregular and indented nature of the coast-line and islands lead to the conclusion that this is almost a classic case for departure from the low-water line rule. As this writer said in an article some eleven years ago, "The Canadian Arctic is nothing more than the Norwegian *skjaergaard* writ large."[13]

Canada's drawing of straight baselines was protested by some states, although none took the opportunity to take the matter to the International Court of Justice, an option that Canada had expressly invited by removing its 1970 reservation to the Court's jurisdiction. Thus, the matter has not been resolved definitively. However, after an exhaustive

---

13 McRae, "Arctic Waters and Canadian Sovereignty," *International Journal* 38 (1983): 483–4.

analysis of the law on this matter, Canada's leading legal scholar on the waters of the Arctic, Professor Donat Pharand, has concluded that the "straight baselines" promulgated in 1986 are justified in law. Moreover, Professor Pharand has pointed out that most writers who have considered the matter of "straight baselines" for the Arctic have reached a similar conclusion.[14] In short, the preponderant view of legal authorities is that the waters of the Canadian Arctic Archipelago are properly enclosed by straight baselines and are the internal waters of Canada.

Nevertheless, could the waters of Northwest Passage still be regarded as constituting an international strait, through which a right of innocent or "transit" passage exists? The test for determining whether a body of water amounts in law to an international strait consists of two elements: a geographic test and a functional, or "use," test. There is no doubt that the Northwest Passage meets the geographic test; it is a body of water joining two oceans or two areas of high seas.[15] The application of the functional test, however, suggests that these waters do not constitute an international strait.

The functional, or "use," test requires that the waters be "used for international navigation." The key question is whether a certain volume of shipping is necessary. Clearly, the fact that a body of water *could* be used for navigation does not constitute it an international strait. And, it is generally accepted that the use must be more than token or isolated; there must be evidence of actual use and some kind of widespread interest in continuing use. In the *Corfu Channel Case*,[16] where the test was laid down, there were something in the order of 3,000 transits of the North Corfu Channel over a 21-month period. In 1984 Professor Pharand pointed out that in an 80-year period there had been only 11 foreign transits of the Northwest Passage, all "with Canada's consent or acquiescence, either expressed or implied."[17] On that basis he was prepared to conclude that, "The Northwest Passage is not an international strait, because it has never been used for international navigation."[18]

Since that time, the voyage of the *Polar Sea* is the only known transit of the Passage undertaken without Canadian consent. This single isolated incident could hardly alter the validity of Professor Pharand's conclusion, particularly in light of the United States having made it clear at the time that it did not regard the voyage as establishing a

---

[14] Pharand, *Canada's Arctic Waters*, 159–79. See note 7.
[15] Law of the Sea Convention, 1982. Articles 37 and 45.
[16] [1949] I.C.J. Rep. 1.
[17] *Northwest Passage: Arctic Straits*, 102 (1984).
[18] Ibid., 120.

precedent that would challenge the Canadian position on the waters of the Northwest Passage. Moreover, the subsequent Arctic Cooperation Agreement suggests that there will be no more *Polar Sea* voyages – that is, no more American icebreakers transiting the Northwest Passage without Canadian consent. So, even if the *Polar Sea* was a precedent, it is no more than an isolated, single instance. Thus, the conclusion remains: The Northwest Passage is not a strait that is "used for international navigation" and hence cannot constitute in law an international strait.

The Future Vulnerability of Canadian Sovereignty over Arctic Waters

Can Canada rest satisfied that its claim to sovereignty over the waters of the Arctic Archipelago is secure and not subject to future challenge? The answer is no, for three reasons. First, the arguments to support the Canadian position have not been tested in international litigation – and may never be – but until such a test occurs some uncertainty will remain. Second, it is not possible to say that the Canadian position on sovereignty over Arctic waters has received universal acceptance by other states. In particular, the position of the United States continues to be troubling for Canada. Third, the discussion of transit has always been about surface transit. What implication does subsurface transit have for Canadian sovereignty over Arctic waters?

There is no Canadian interest in having the question of Arctic sovereignty litigated before an international tribunal. No state is currently challenging Canada on this matter, and thus there is no question to be placed before such a tribunal. Moreover, the longer states refrain from active challenge to Canada's position, the stronger that position grows. In this respect, the words of Ivan Head written some thirty years ago remain valid today: "the passage of time enures to the benefit of the Canadian claim."[19]

Equally, it does not appear that any action by the Canadian government designed to change the position of the United States would be fruitful. Clearly, the United States has neither endorsed the view that the Northwest Passage is part of the internal waters of Canada nor accepted that the Passage does not constitute an international strait. At the same time, the United States does not appear interested in actively challenging the Canadian position. Moreover, the 1988 Arctic Cooperation Agreement assures that future voyages by US icebreakers will be conducted only with Canada's consent. In effect, the major threat to

---

19 Head, "Canadian Claims to Territorial Sovereignty in Arctic Regions," *McGill L.J.* 9: 219 (1962–63).

Canadian sovereignty posed by the voyage of the *Polar Sea* has now been eliminated.[20] However, it is incumbent on Canada to ensure that any future voyages, whether by vessels of the United States or of other states, take place only with Canadian consent.

The principal issue that remains, therefore, is that of subsurface transit of the Northwest passage. That such transits occur appears to be widely accepted, although their extent is a matter of speculation.[21] The position taken by the Government of Canada has been that any submarine transit of Arctic water is undertaken pursuant to bilateral and multilateral defence arrangements and hence is, at least implicitly, with Canadian consent.[22] However, this not only leaves unanswered the question of transit by states with whom no such defence arrangements exists, but also assumes that the United States would not in future invoke these transits to the detriment of the Canadian claim. Furthermore, and perhaps most importantly, Canada's argument that these voyages have been consented to assumes that Canada knows of each transit.

What would the consequences be for Canada of submarine transit of the Northwest Passage without Canada's knowledge or consent? Real doubt would be cast on Canada's claim that it is exercising sovereign functions over Arctic waters. Incursions into the land territory of a state without that state's consent are regarded as serious encroachments on sovereignty, and the Arctic sovereignty claim treats the waters of the Northwest Passage as if they were land territory. At the very least, therefore, subsurface transits undertaken without Canada's consent are a serious encroachment on Canada's sovereignty over Arctic waters.

Would such transits weaken Canada's sovereignty claim? To a certain extent, the matter is complicated by the fact that these submarines would undoubtedly be military vessels. The ability of warships to partake of passage rights through the territorial sea and international straits has been a matter of controversy in international law.[23] For its part, the United States is a strong advocate of warships having such

---

[20] It should be noted that some have questioned the effectiveness of the Arctic Cooperation Agreement as a means for protecting Canadian sovereignty. In particular, it has been argued that the reference to the consent of Canada does not guarantee that consent will be requested before a voyage takes place. See Purver, "Aspects of Sovereignty and Security in the Arctic," in McRae, Donald, and Munro, Gordon, eds., *Canadian Oceans Policy: National Strategies and the New Law of the Sea*, 172–3 (1989).

[21] Ibid, pages 712 and 177. See also Critchley, "Defence and Policing in Arctic Canada," in *Politics of the Northwest Passage*, 209. See above, note 11.

[22] Purver, 172 and 176. See above, note 20.

[23] See Brownlie, *Principles of Public International Law*, 3rd ed., 206–7 (1979).

rights, and the tenor of the 1982 Convention on the Law of the Sea is to treat warships as having access to international straits and to recognize that submarines may transit such straits submerged.[24] As a result, any submarine transit of the Northwest Passage without Canada's consent could have an important impact on Canada's claim to sovereignty over those waters.

Such an impact would be twofold. First, if Canada does not know of these transits, or knows about them but does nothing, then the Canadian government is not exercising the functions of a state in that area.[25] Those functions imply having authority and control over the area, both of which would be lacking if such transits continued unimpeded. Second, a pattern of submarine transit of the Northwest Passage over a period of time could give credence to the argument that the Northwest Passage was being "used for international navigation," and hence was subject to the regime of international straits. Again, this would defeat the Canadian sovereignty claim.

What Must be Done to Preserve Canada's Sovereignty Claim?

The question arises, what must Canada do to ensure that its rights over Arctic waters remain secure? In respect of surface transit, Canada has acted to ensure that in practical terms its rights will not be challenged by the United States, although the question of enforcement has not been addressed clearly. Has Canada done the same in respect of subsurface transit? As a starting point, does Canada have an adequate surveillance capacity to ensure that it has knowledge of any subsurface transit of the Northwest Passage?

In 1971 the *Defence White Paper* indicated that Canada had "only very limited capability to detect submarine activity in the Arctic."[26] By 1983 the situation did not appear to have improved significantly: in that year the Sub-Committee on National Defence suggested that a bottom-based sonar system might be considered.[27] Finally, the provision of a fixed sensor system in Arctic sea routes was announced in the 1987 *Defence White Paper* along with plans for the acquisition of

---

24 The reference in Article 39 (1)(c) of the 1982 Law of the Sea Convention to transit by vessels in their "normal modes" is understood to mean submerged transit by submarines.

25 As Purver has written, "the occurrence of activities that Canada has little ability to detect, let alone prevent, cannot but derogate from its claim to full sovereignty." See above, note 20, 178.

26 *Defence in the 70's*, 18 (1971).

27 *Canada's Maritime Defence: Report of the Sub–Committee on National Defence*, 51 (1983).

nuclear-powered submarines.[28] Such a system still remains to be established,[29] and plans for the submarines have been abandoned.

Are there any limits to what Canada must do to preserve its sovereignty claim over Arctic waters? In other words, does Canada have to take all measures possible to assert and continue to assert its sovereignty over these waters? Does this imply that it must be able to detect all transits of the Northwest Passage by whatever means they occur and have the capacity to enforce its laws against all who contravene them? In 1985 the government announced that it would construct a Polar Class 8 icebreaker to "provide more extensive support services, to strengthen regulatory structures, and to reinforce the necessary means of control."[30] Such a vessel was designed to signify that Canada had the means to exercise control over the waters of the Northwest Passage. In 1990 the project was cancelled. This raises questions about how Canada plans to exercise the sovereignty functions for which the icebreaker was intended.

Failure by a state to exercise its sovereign authority can lead to an abandonment of its sovereignty claim. Failure to object or to take action in the case of prejudicial activities of another state can constitute acquiescence to the activities of that state. In the context of sovereignty over land territory, it has been said that, "absence of a reasonable level of state activity can lead to a loss of title."[31] This suggests that the standard is one of reasonableness. Thus, the question becomes, what is a reasonable level of activity in the context of the waters of the Canadian Arctic Archipelago?

Certainly, failure to take any steps at all to become aware of subsurface transit of the Northwest Passage would cast doubts on the seriousness of Canada's sovereignty claim and could, in the long term, contribute to a perception that it had been abandoned. By contrast, having the full capacity to become aware of any transit of the Northwest Passage and to prevent unauthorized entry into waters claimed by Canada to be internal would be a clear and unequivocal manifestation of sovereign authority. Obviously, this capacity is unrealistic in the Arctic environment, in the light of the technological, economic, and political considerations that have to be taken into account. Nevertheless, some capacity for enforcement of Canada's sovereignty claim is essential.

---

28 *Challenge and Commitment: A Defence Policy for Canada*, 51–2 (1987).

29 The development of an Arctic subsurface surveillance system was announced by the Minister of National Defence on 3 January 1991, but the matter has yet to be proceeded with.

30 See above, note 12.

31 Brownlie, 148. See above, note 23.

The starting point for any enforcement function is knowledge. Thus, adequate provisions for surveillance become an essential element in maintaining Canada's sovereignty claim over Arctic waters.[32] Knowledge of any subsurface transit gives Canada a variety of options, diplomatic and other, for dealing with potential challenges to its sovereignty. The objective is not necessarily to prevent such transits – indeed, past Canadian governments have indicated that they wish to ensure that properly controlled traffic does have access through the Northwest Passage. The objective is to ensure that such transits occur with Canadian consent and in accordance with regulations established by Canada – precisely what a sovereign state would expect of anyone entering its territory.

Conclusions

Canada's claim to sovereignty over the waters of the Canadian Arctic Archipelago is well-founded in law. It rests on the fact that the unique geography and environment of the Arctic Archipelago justifies the drawing of straight baselines and enclosing the waters as the internal waters of Canada. The relatively small number of transits of the Northwest Passage over history prevents it from being regarded as a strait "used for international navigation" to which the legal regime of international straits would apply.

Canada's sovereignty over Arctic waters cannot, however, be taken for granted. Sovereignty can be lost; it can be abandoned. And it can be abandoned by dereliction. Failure by Canada to exercise its sovereign authority over the waters will diminish the credibility of its claim of sovereignty, and continued and frequent transit of the Northwest Passage, whether by surface or subsurface vessels, could lead to the Passage becoming a strait "used for international navigation."[33] In such circumstances, Canada could no longer claim sovereignty over the waters.

Canada has taken measures to assure that surface transits are with its consent. In this regard the Arctic Cooperation Agreement diminishes the threat of unilateral transit by US government icebreakers. And Canada has the capacity through overflight and surface vessels to monitor foreign surface passage or overflight. Subsurface passage, by contrast, remains a matter over which Canada is not in a position to assert its sovereign authority.

---

32 As Professor Pharand has said: "Surely [Canada] has the right to know what goes on in its own waters; indeed, it has a duty to find out." See above, note 7, 243.

33 Recent reports of diminishing Arctic sea ice cover with the possibility of a lengthened navigation season may result in greater government concern about Arctic sovereignty. – ed.

A precondition for exercising enforcement jurisdiction – for taking measures against unauthorized subsurface traffic – is knowledge of occurrence. To exercise the sovereign authority it claims and to preserve its claim to sovereignty over Arctic waters, Canada must at least be in a position to monitor subsurface use of the waters of the Arctic Archipelago.

61 Commentary: The Canadian Polar Commission

WHIT FRASER

To properly understand the importance of the Canadian Polar Commission and its role in the polar regions of Canada and the world, it is best to look at the Arctic, its people, and their communities in the broadest sense.

Northern observers are well aware of the tremendous changes taking place in northern Canada and the circumpolar regions of the world. The political change is the most evident and visible. The recent vote in the Northwest Territories approving the boundary line that may lead to the creation of two new territories is indeed an historic development in Canada and a step towards changing confederation itself. There is every possibility that a new territory will be established, called Nunavut, where Inuit make up more than 80 per cent of the population and as a result may realize their dream of an Inuit homeland. The western territory, becoming more and more known by the name Denedeh, will have a much different make-up, an almost equal mixture of Dene, Metis, and whites. In each, this is a time of change, challenge, and excitement. It is not confined to the political development. Interwoven are the complex land claims negotiations and Aboriginal self-government.

By every measure and standard, these developments should be seen as important steps in the growth and development not only of northern Canada, but of our country, and in the end we should all be richer for them.

SOURCE: *Arctic* 45, no. 2 (June 1992): iii–iv. Reprinted by permission of the publisher.

However, what northerners and Canadians must now ask themselves is whether or not we are ready and capable to come to terms with the other pressing challenges facing the northern regions. Are we ready to address head-on the environmental questions? Are we ready to look at the social issues facing northern communities and northern people? And finally, can we overcome the tremendous economic obstacles that lie ahead? These are not simply the issues of self-government, they are the issues of the North and Canada and they are the concerns and challenges of the Canadian Polar Commission.

The Polar Commission was created by the federal government in September 1991. It was the result of two federal studies and much consultation with Canadian research institutions, universities, and northern residents. Both reports were clear on the need for a Polar Commission. The mandate of the twelve-member Board of Directors directs it to monitor and report on the state of knowledge in polar regions through a variety of means. The Commission also has the responsibility to help disseminate knowledge about polar regions, nationally and internationally, to work with northern and southern institutions fostering understanding about science and research in polar regions, and to advise the federal government and others on northern issues.

To the Commission itself that means ensuring that the scientific research carried out in polar regions is of the highest possible standards and at all times takes into account the concerns and interests of the northern peoples.

Seven months into that mandate, it is clear that the need for research of the highest quality has never been greater and that the Polar Commission and indeed polar science cannot ignore the environmental, economic, or social realities of the Arctic.

The scientific community, northerners, and all Canadians are daily becoming more aware of the possible consequences of climate change and global warning. The need for knowledge and understanding of climate change is essential in the polar regions. It will not happen overnight, but the need for study, knowledge, and understanding is apparent and it appears that, to its credit, Canada is responding.

There is a concentrated national initiative under way to study and understand the effects of climate change, but there is also the need for extra initiatives to ensure that northern concerns and interests are addressed in the research. There is also the need to involve northerners in the science itself, to work closely with northern people to understand their particular concerns and to include in the science the traditional knowledge of the Aboriginal people as that relates to climate and the environment.

The other major environmental concern is known in the Arctic by the simple term "the contaminants issue." It refers to the very high levels of toxins, heavy metals, chemicals, and other compounds that have been detected in the northern food chain. There is again a growing national concern about these contaminants. There is also a growing body of scientific research developing to better understand the source of the contaminants, the extent of the problem, and the effects on the northern environment and people. In the North today, understanding of this issue is most critical. What may be at stake in future years is the very substance of the Arctic peoples. They fear that in years to come they may be faced with the difficult choices of substituting highly nutritional northern foods such as seal, fish, and caribou with less nutritional and much more costly imported southern foods such as chicken, beef, and pork. The issues are being addressed, but again, there is the need for northern understanding and input.

Much has been said and written in recent years about the social conditions of the Arctic and Arctic peoples. Many have tried to explain why northern peoples suffer painfully from excessively high rates of suicide, alcoholism, and family violence. The Polar Commission has taken the position that the social sciences in the Arctic should be put on an equal footing with the physical and natural sciences in terms of commitment and importance.

The Commission is clear that social studies in the North must begin to address the social conditions that now exist and the importance of the direct involvement of the northern peoples cannot be overstated. Southern-based social scientists working alone cannot and certainly have not been able to solve the "social ills" of the North. Indeed, for reasons few can explain, much of the social disintegration of the North happened over a period when there was a vast amount of study.

Economically, the northern regions are as delicate as the environment. There has been an unprecedented explosion in the birthrate in northern Canada in the past thirty years. Close to half the population is under twenty-five years old. The question northerners are asking themselves is who will provide jobs for that growing population? The problem is not made easier by the fact that in many of these northern communities there are already staggering levels of unemployment.

For all of the apparent problems of the North, there is also growing optimism. Northern leaders have been telling members of the Board of Directors of the Polar Commission that appropriate community-based research is the best way to begin addressing many of the concerns. They add that there are a number of important initiatives in these areas now underway through some government departments, the Science Institute of the NWT, the Science Institute of the Yukon, and Yukon College, as well as a number of southern-based universities.

A priority of the Polar Commission will be to expand on those initiatives, provide some coordination where appropriate, and develop a plan to encourage northern involvement and participation in research activities in the northern regions.

Hand in hand with that, the Commission would like to see the use of traditional knowledge expanded. We believe that it is important that all Canadians and all northerners recognize the tremendous benefit indigenous knowledge can play in northern research in many areas, including health, social issues, and justice. We also think that there should be initiatives with northern communities to develop methods for recording, disseminating, and using traditional knowledge.

The Board of Directors also wants to enhance academic study on northern regions. We would like to see scholarly study on important northern issues in law, politics, economics, and the natural and applied sciences increased. In addition, we see the need for more interdisciplinary study. We hope to either encourage institutions to develop journals to record and distribute this material or identify journals that are already in existence and encourage them to expand into broader areas of scholarly study. These two initiatives are given to show the breadth of what constitutes science in northern regions.

We have taken the advice of Dr Tom Symons in the *Shield of Achilles* that recommended to the federal government the establishment of the Polar Commission. Symons said, "the Commission must be concerned with the full range of knowledge in and about the polar regions and not limited to the contemporary restrictive meaning of the word science." Symons encountered no member of the research community who disagreed with that approach.

Certainly no member of the Polar Commission disagrees with it, and northerners simply never looked at science or knowledge any other way. What the Commission strives for are ways to bring high-quality scientific research in all its forms and all its levels directly to bear on the critical issues now facing the northern regions and our country.

62 The Arctic Council: Will It Be Relevant?

CHESTER REIMER

The Arctic Council was established with great fanfare on September 19, 1996 in Ottawa, Canada. Ambassadors, politicians, academics, and other dignitaries were on hand as the Foreign Ministers and other senior officials of the eight Arctic countries signed the *Declaration on the Establishment of the Arctic Council*.

Years of negotiations preceded this historical occasion. In the early 1990s, the USA refused to take part in talks but the other seven countries drafted a declaration anyway. In 1993 they shied away from actually signing it and when, in 1995, the USA finally agreed that there was merit in a multilateral forum such as an Arctic Council, it seemed other governments were less sure they wanted it. Furthermore, the indigenous peoples of the Arctic who had been enthusiastic and welcomed participants in the negotiations were ready to walk away from the Council because major clauses affecting them had been altered or removed in the later stages of negotiation. At the end of the day, however, the Arctic's indigenous leaders decided to give support to the Council and promised to vigorously work from within to make it relevant to the peoples they represent.

The various degrees to which each Arctic government, respectively, gave its support are due to diverse factors within each nation and will not be addressed in depth here. National budgets, political will, other regional fora, and the ways in which each government views the importance of bilateral versus multilateral cooperation are some of the

SOURCE: *NSN Newsletter*, no. 20 (December 1996): 16–18. Reprinted by permission of the publisher.

factors affecting the different levels of governmental support. It may also be these factors that will determine the future impetus that the eight governments give to the work of the Council.

Notwithstanding the need for financial and political commitment from governments, the quintessential variable that will determine the success of the Arctic Council is what the indigenous peoples' leaders spoke about at the inauguration in September: relevance. In other words, will the Arctic Council be relevant? Will it address the needs of the people living in the Arctic? Or will the Arctic Council be a place where minister meet, ambassadors host fine dinners, and government officials congratulate themselves for having created yet another glorious institution? The indigenous leaders agreed that it is worth the try to make the Arctic Council relevant.

As currently planned, the Arctic Council's members (the eight Arctic governments) and its so-called permanent participants (three indigenous peoples' organizations and possibly more in the near future) will oversee and coordinate four of the five working groups of the previous *Arctic Environmental Protection Strategy* (AEPS) and will develop terms of reference for the implementation of a sustainable development initiative. As in the case of the AEPS, Senior Arctic Officials (SAOs) will continue in their roles of supervising the work of the Council between ministerial meetings.

It is clear from the *Declaration* and from what indigenous leaders have been saying in this embryonic stage that the Arctic Council's success will depend on its relevance to northern peoples. Its relevance, in turn, will depend on (1) how well the work of the AEPS is incorporated and expanded upon; and (2) how meaningful and effective the new sustainable development initiative created through the *Declaration* will be.

Since its establishment in 1991, the AEPS has a fair record of monitoring pollution as it enters the Arctic environment and has assessed its possible impact on the Arctic food chain. It has a fair record of analyzing the various existing legal instruments and how they may be best applied to the Arctic in providing protection to the marine environment. The AEPS also has a good record in collecting data on the state of the Arctic's flora and fauna and is on its way to developing a biodiversity strategy for the Arctic. It has also increasingly, with some notable exceptions, incorporated the views and knowledge of the Arctic's indigenous peoples. More work in all these so-called *environmental protection areas* needs to be done, however, and the existing working groups of the AEPS are excellent and natural bases from which the work should continue. In addition, those responsible for the *environmental protection* side of the Arctic Council must take seriously a new mandate introduced in the Arctic Council Declaration, that of disseminating information, encouraging education, and promoting interest in Arctic-related issues. If this is done specifically with the Arctic's indigenous peoples in mind, the environmental work will become even more relevant at the community level. Let us assume that governments continue to recognize the importance of this work and that at the upcoming Arctic Environment Ministers' meeting in June 1997, a clear message from the Ministers will be made to this effect. If it is, then a major pillar of the Arctic Council, that of environmental protection, will be relevant and contribute to the Council's success.

What then is to be made of this new creature, the Arctic Council's *sustainable development initiative*? Its terms of reference have yet to be developed and, in fact, the rules of procedure for the Arctic Council as a whole have still not been agreed upon. At a November 1996 Senior Arctic Officials meeting held in Oslo, there was fear among the three indigenous peoples' organizations that the discussion on terms of reference was getting bogged down in procedural matters and that in the end, nothing might ever get done.

At the meeting, it also became clear what the indigenous leaders of the Arctic wanted the new *sustainable development initiative* to be. They wanted it to get on with concrete projects that are relevant to their peoples at the community level. They wanted the *initiative* to have a human focus. Sheila Watt-Cloutier of the Inuit Circumpolar Conference and Sven-Roald Nystø of the Saami Council spoke of some of the social problems their people are experiencing at this time. In giving direction to what the terms of reference for the *initiative* should look like, Nystø reminded the government officials what, in part, should be considered:

Increasing unemployment, alcohol abuse, and identity problems, especially among youngsters, are serious matters we have to take into consideration when we are talking about sustainable development in the Arctic.

As such, Nystø said that policies which encourage the transfer and development of indigenous peoples' "identity, language, and cultural heritage from one generation to the next" and which foster "strong and living communities with a well-developed industrial and social life" are paramount for the Saami people of Norway, Sweden, Finland, and the Kola Peninsula in Russia.

Watt-Cloutier spoke about how the *initiative* should combine both indigenous and scientific knowledge and, in this context, explained how once Inuit in Greenland, Canada, Alaska, and Russia were a highly successful nation. She said:

Historically, we were a highly independent people with our own health, education, and justice systems, conducting our lives with incredible wisdom. We were, and still are, scientists in our own right. If you have trouble believing this, try in minus 40C, not only to build a house of snow but a house of snow warm enough for your naked children to sleep comfortably. Then tell me scientific knowledge was not at work here ... Researchers: work with us – work with our elders who know their land, the sea, and animals like the palm of their hand. Involve us every step of the way in partnership through the research and analysis of data.

Watt-Cloutier also reminded that Senior Officials that in today's world, many Inuit have adapted well and have become successful in modern business and have exchanged hunting equipment for briefcases, while others continue to live successfully off the land and the sea. In response to rapid changes in the last fifty years, many Inuit have survived, grown, and thrived, she said. But she also added the following:

In other areas we have not managed so well and in fact we have become highly dependent upon substances such as alcohol and drugs, institutions such as welfare and unemployment, and processes such as gambling and religion. These unhealthy choices are ways that our people are trying to change the quality of their experience, only to be going in a downward spiral of further self-destruction ... Inuit ... have the highest suicide rate in North America.

Watt-Cloutier does not want the Arctic Council's sustainable development initiative to attempt to cure all of these ills. She says, however, if the Arctic Council does not at least aim its programs and projects at addressing some of these issues, it will immediately lose relevance at the local level and support from the indigenous leaders will vanish.

The eleven minority indigenous peoples of the Russian Arctic represented by the Association of Minority Indigenous Peoples of the North, Siberia, and the Far East of the Russian Federation (AIPON) are also facing difficult economic and social hardships. At the Arctic Council's inauguration in Ottawa in September 1997, AIPON's president, Yeremey Aipin, spoke of how many of the peoples he represents are on "the edge of an abyss and above their heads is the threat of physical disappearance." Aipin wants the Arctic Council to recognize that there is a debt owed to the Arctic's indigenous peoples who have for too long been exploited. He said in Ottawa:

The oil and gas companies, the forest logging operations, the gold and silver industry on the lands of the indigenous peoples have given nothing to the indigenous peoples themselves.

Aipin also does not think that the Arctic Council can fix all of the problems of Russia's northern indigenous peoples, but that it must try and that it must be relevant. He said "we hope that the Arctic Council will find a mechanism for resolving some of our deepest problems of survival."

Also at the Ottawa inauguration, Leif Halonen representing the Saami Council, warned:

Let us not forget that environmental problems cannot be treated as something separate from the society in which they exist ... Saami life was throughout history, based on a comprehensive approach to existence in general; the human is an integral part of his surroundings. Exploitation of natural resources is necessary, but it must fit into a sustainable context.

It is clear from the statements of the Arctic's indigenous leaders, expectations are high. They have been part of several years of negotiations and have given their support throughout. They have worked hand in hand with government officials and have sent a message of cooperation to them. They fear, however, that if the Arctic Council gets bogged down in procedural matters and does not attempt to be relevant to the real questions at the community level, including those of land rights, mineral exploitation, social issues, cultural matters, unemployment, economic development and human dignity, then it will all have been a waste of time.

The indigenous peoples are optimistic that the Arctic Council can make a difference, that it can be relevant. In Yeremey Aipin's opening remarks at the inauguration he stated: "in creating an Arctic Council, there is no doubt that we are perfecting the world."

Note on the Editor

William C. Wonders is University Professor and Professor Emeritus of Geography at the University of Alberta. He received a BA (Honours) and a PhD in geography at the University of Toronto and an MA in geography at Syracuse University. He taught at the University of Toronto until 1953, when he received the first appointment in geography at the University of Alberta. On establishment of a separate Department of Geography, he served as department head from 1957 to 1967. Dr Wonders was chairman of the organizing committee for the establishment of a northern research institute at the University of Alberta and was also the first chairman of the directorate of the resultant Boreal Institute (now the Canadian Circumpolar Institute). He was a guest professor at Uppsala University in Sweden in 1962–63 and was awarded an honorary doctorate from that university in 1981. He is a fellow of the Royal Society of Canada, the Royal Canadian Geographical Society, and the Arctic Institute of North America. His research interests and publications have focused on settlement geography in the Canadian Northwest, Alberta, and Scandinavia. He is now retired and lives in Victoria, BC.